Sensory Ecology

Sensory Ecology

■ ■ ■

HOW ORGANISMS ACQUIRE AND RESPOND TO INFORMATION

David B. Dusenbery

W. H. Freeman and Company

NEW YORK

Library of Congress Cataloging-in-Publication Data

Dusenbery, David B.
Sensory ecology : how organisms acquire and respond to information
/ David B. Dusenbery.
p.   cm.
Includes bibliographical references and index.
ISBN 0-7167-2333-6
1. Senses and sensation. 2. Ecology. 3. Physiology, Comparative.
I. Title.
QP431.D87   1992
574.1'88—dc20   92-2278
                                                            CIP

Copyright © 1992 by W. H. Freeman and Company

Printed in the United States of America

1 2 3 4 5 6 7 8 9 0   VB   9 9 8 7 6 5 4 3 2

To my parents, who taught me to seek the truth

To my teachers, who showed me how

To Sharon, who showed me a new side to life

Contents

■ ■

Symbols

■ ■

Capital letters are used for stimuli and responses (I, S, N, R, J, C, H, Q), lowercase letters for distances and velocity (x, y, z, d, r, v), but capitals for dimensions (L, W, H, A, V).

A	Area (m^2)
a	Absorbance; the distance over which $1/e$ is absorbed (m^{-1})
a	Atto; the unit prefix for 10^{-18}
B	Bandwidth of a channel
C	Concentration
C	Contrast
C_0	Average or true concentration
c	Centi; the unit prefix for 10^{-2}
c_0	Cost with no signal present
c_s	Cost when signal is present

D	Diffusivity; the diffusion coefficient
d	Depth; distance (m)
d	Dimensionality
d	Distance between light receptors
e	Base of natural logarithms
$\mathrm{erfc}(\xi)$	A function called the complement of the error function; the area under the normal curve from ξ to infinity
$\exp(\xi)$	A function meaning e is raised to power ξ
F	Tension
f	Femto; unit prefix for 10^{-15}
f	Frequency, in Hz
f_0	Fundamental or transmitted frequency
G	Gibbs free energy
G	Relative gradient: $(1/C)(dC/dx)$
G''	Second derivative of Gaussian distribution
g	Gram; unit of mass
H	Potential information content per symbol
H	Height (m)
H_p	Height of plume
Hz	Hertz; unit of frequency equal to one cycle per second
I	Intensity of stimulus
I_0	Intensity with no signal
I_s	Intensity with signal
I_{SD}	Standard deviation of intensity noise
I_t	Average rate at which stimulus events are received
I_{Th}	Threshold intensity
J	Flux; total quantity of stimulus per second leaving a source in all directions or through a specific area

J	Joule; unit of energy
K	Kelvin; unit of absolute temperature
k	Kilo; unit prefix for 10^3
k_B	Boltzmann constant
L	Length (m)
L_A	Attenuation length; the distance in which intensity is reduced by $1-1/e$
$L_{0.5A}$	Half-attenuation length; the distance in which intensity is reduced by $1/2$
ln	Natural logarithm (to base e)
L_p	Length of plume
log	Logarithm to base 10
$\log_2$	Logarithm to base 2
M	Mega; unit prefix for 10^6
m	Meter; basic unit of length
m	Milli; unit prefix for 10^{-3}
N	Newton; unit of force
N	Number of different symbols
N	Noise power
N_t	Number of stimulus events in time t
n	Nano; unit prefix for 10^{-9}
P	Probability
P_s	Prior probability of a signal being present
P	Power (watts); energy per unit time
p_s	Sound pressure
p	Pico; unit prefix for 10^{-12}
ppb	Part per billion; 1×10^{-9}
ppm	Part per million; 1×10^{-6}

Q	Quantity of stimulus
R	Real gas constant
RH	Relative humidity
R_M	Maximum rate of information transmission
$R(t)$	Response at time t
$R_i(t)$	Response to an impulse stimulus at time t
r	Radius: range:distance from source (m)
r_D	Range of detection
r_d	Damping distance of periodic stimulus
r_R	Effective radius of receiver in sounding
r_{ST}	Distance between source and target in sounding
r_T	Effective radius of target in sounding
rms	Root mean square: $X_{rms} \equiv <X^2>^{1/2}$
S	Thermodynamic entropy
S	Salinity
S	Signal power
S_i	Impulse signal
s	Second; unit of time
T	Temperature, in degrees C or K
t	Time (s)
V	Volt; a measure of electrical potential difference
V	Volume (cm^3)
V_T	Transmission volume (cm^3) = Q / I_{Tb}
$\dot{V}_T$	Transmission volume rate of increase (cm^3/t) = $\dot{J}/I_{Tb}$
v	Velocity or speed (m/s)
v_a	Velocity of animal
v_c	Velocity of current

v_s	Velocity of sound
W	Watt; unit of power (J/s)
W_S	Effective sweep width of a search
W_p	Width of plume
x	Distance along one dimension (parallel to flow, if any)
y	Distance along a second dimension
z	Distance along a third dimension (perpendicular to surface)

GREEK LETTERS

α	Attenuation coefficient (m^{-1}) [alpha]
α_m	Angle of heading from magnetic north
δ	Thickness (of unstirred layer) [delta]
γ	Dip angle of earth's magnetic field (from horizontal to north pointing) [gamma]
Δ	Change or uncertainty in following parameter [Delta]
ΔC_{rms}	Root mean square variation in concentration
ε	Rate of energy transfer between vortices of different size [epsilon]
θ	Angle [theta]
κ	Compressibility [kappa]
λ	Wavelength (m) [lambda]
μ	Micro; unit prefix for 10^{-6} [mu]
μ	Mass per unit length
ν	Kinematic viscosity [nu]
ξ	Dummy variable [xi]
π	A number equal to 3.14. . .[pi]
ρ	Density; number per unit area or volume [rho]

σ   Cross section for interaction [sigma]

Σ   Sum of terms like [sigma]

τ   Time constant (s) [tau]

φ   Angle between directions observed by adjacent light receptors [phi]

ω   Angular frequency ($2\,\pi f$) [omega]

MATHEMATICAL SYMBOLS

≪   Much less than

≫   Much greater than

≠   Not equal to

≤   Less than or equal to

≥   Greater than or equal to

±   Plus or minus

≈   Approximately equal to

∞   Infinity; $1/0$

×   Multiplication

∝   Proportional to

°C   Degree Celsius; unit of temperature

⟨ξ⟩   Average of ξ

$F(\xi)$ F is a function of ξ; i.e., the value of $F(\xi)$ depends on the value of ξ, where F stands for any of various functions

Preface

■ ■

In setting out to write this book, my original purpose was to call attention to the fascinating questions of how organisms acquire information about their environment. I particularly wanted to emphasize that simple organisms, which are often considered not to exhibit interesting behavior, face their challenges with abilities that we are still discovering. Consequently, a major objective for this book was the comprehensive treatment of all kinds of organisms, which would shed light on higher animals by contrasting their behavior with that of "simple" organisms. Such an all-inclusive approach led to problems in terminology. Because biologists studying different types of organisms or stimuli rarely interact with one another, they have developed different sets of terminologies. I have therefore tried to introduce a system of terminology that is based on distinctions of universal importance and consequently useful to all.

I have also sought to produce a work that would be a useful reference, providing access to the relevant physical principles for all biologists interested in the study of animal behavior in the natural environment. As the project progressed and received much enthusiastic support, I moved toward making the book suitable also for use as a textbook. I believe it will be found appropriate for all three purposes.

The book contains a significant amount of physics, because physical constraints are universal and it is necessary to understand them to recognize why organisms behave in the ways they do. Because many biologists are not at ease with physics, I have sought to present these relations in ways that do not require familiarity with a great deal of math or physics. Equations are thus provided primarily as a reference so that readers can use the relations to make predictions in situations of interest to them, but the primary emphasis is on presenting the relations verbally and graphically.

Citations are provided more as entry into the literature rather than as credit for discoveries. Thus there is a bias toward recent publications (which in turn cite older literature) and review articles rather than primary reports.

CHAPTER 1

Introduction

■ ■

Life may be defined operationally as an
information processing system—a
structural hierarchy of functioning
units—that has acquired through
evolution the ability to store and process
the **information** necessary for its own
accurate reproduction.
—Lila Gatlin, 1972

Life may be defined in many ways, but the most fundamental focus on information. In her book *Information Theory and the Living System* (1972, 1), Lila Gatlin emphasizes the internal, genetic mechanics of information storage and processing. Similarly, most studies of animal behavior today focus on the evolution of behavior, which results from changes in internal information processing. However, in addition to genetic information, all organisms obtain information externally through their senses in order to respond more rapidly to environmental conditions. The external mechanics of how sensory information is obtained from the environment is the subject of this book.

Ecology is the study of the interaction of organisms with their environment, including other organisms. Much of the work in this field has focused on the flow of energy and nutrients between organisms. Another important interaction between organisms and their environment occurs in the transmission of information. Tracing the flow of information, as opposed to flows of materials and energy within organisms, was the basic innovation that gave rise to molecular biology out of traditional biochemistry. This new point of view led to an enormous breakthrough in understanding the molecular basis of genetic processes. Although it would be optimistic to predict that tracing the flow of information might have the revolutionary impact on ecology that it did on biochemistry, this point of view deserves to be examined in detail.

In addition to being motivated by their basic curiosity about how the world works, humans have practical reasons for maintaining an interest in how organisms obtain and use information about their environment. Hunting and fishing are ancient examples. Our ancestors could hunt and fish more successfully if they understood the sensory abilities of their prey so as to avoid alerting them while stalking or trying to attract them with lures. The exploitation of the sensory limitations of fish has been developed to a fine art in fly-fishing, which employs thread, feathers, and fur to imitate insects sufficiently well to catch many kinds of fish. Of more importance today, explosive advances in the technologies of processing information are generating numerous opportunities to use environmental information to develop new machines such as robots. In exploiting these opportunities a common problem for the developer is finding strategies that are efficient. Living organisms throughout the course of evolution have tested many such strategies, and it may be useful to examine which have proven effective.

Early studies of animal behavior often focused on identifying raw abilities, often in the laboratory. This approach reached an extreme in the behaviorist attitude that running white rats through a maze could reveal useful information about general mechanisms of learning that would apply even to humans. In contrast, the approach called ethology insisted that animal behavior could be understood only by considering the natural context in which the animal in question lived; it was important to study behavior in the natural environment.

In addition to its focus on environmental interaction, ethology grew to have a strong emphasis on the evolution of behavior, the focus that dominates research on animal behavior today. A consequence of the emphasis on evolution is that comparisons are usually made between closely related species, and general factors affecting a wide variety of organisms are overlooked. The emphasis in this book is on the broad scope of problems that all organisms—not just animals—must solve, the physical constraints on possible solutions, and general strategies for solution.

Behavior differs widely among different kinds of organisms and is often very complex. The approaches to studying behavior are similarly diverse. In order to clarify the various questions that arise and put in context the view taken in this book, it is useful to present first a few basic ideas about scientific investigation.

1-1 SCIENTIFIC INVESTIGATION

In general, scientific questions are answered by formulating and testing **hypotheses.** Ideally, one considers several alternative hypotheses (Platt 1964; Chamberlin 1965). Good hypotheses must make predictions that are testable; if they do not, they have no value to scientific investigation. Strictly speaking, a hypothesis can be disproved by observation but not proved (Popper 1959, 33); a positive observation is at best consistent with a hypothesis. Consequently, the most useful observations are those that exclude one or more hypotheses. For example, the question of how owls locate prey at night might lead to the hypotheses that (1) they have eyes that allow them to see in dim light and locate prey by vision; (2) they have a keen sense of hearing and can locate prey by the sounds the prey make; (3) they use echolocation like bats, which make high-frequency sounds and hear the echoes from their prey; or (4) they use "X-ray vision."

If several hypotheses are consistent with the available evidence, the one that postulates the fewest unknown relationships is preferred. This principle is frequently referred to as **Occam's razor,** after the fourteenth-century philosopher William of Occam. A related criterion is that the more observations a hypothesis is consistent with, the more valid the hypothesis. Thus, a single hypothesis that explains a number of facts is superior to several hypotheses that each explain only some of the facts. This principle makes the hypothesis of X-ray vision for prey finding by owls less attractive than the alternative hypotheses, because they are simple extensions of the known abilities of other animals, whereas the hypothesis of X-ray vision postulates a novel mechanism. However, if the other hypotheses were to become excluded by further evidence, the X-ray hypothesis might then be reconsidered.

In practice, the reliability of observations or **evidence** varies, and the art of scientific investigation often hinges on deciding how to weigh different pieces of evidence. A common type of evidence is the **correlation** of one parameter (A) with another one (B). For instance, one could ask if the hunting success of owls correlated with how dark the night was or if the structure of owls' eyes has features that correlate with the eyes of other kinds of animals that see in dim light (they do). This type of evidence suffers from the fact that the direction of cause and effect is ambiguous: The correlation could result from A causes B, B causes A, or C causes A and B. Perhaps bright light causes changes in eye structure. Thus, correlation alone is considered weak evidence, although it is often the best available.

An improvement over simple correlation is to create an **experiment** in which one parameter, A, is altered by an external agent—usually the investigator. The effects of the alteration are assessed by correlating it with another parameter, B, but now one knows that A caused B. For example, it has been shown that, when all light (A) is excluded from a room, barn owls can still accurately locate mice (B). This contradicts the hypothesis that vision is the only means of finding prey. Further experiments have shown that in a dark room owls find prey (B) accurately only when the prey make noise (A), which therefore excludes echolocation as a mechanism of finding prey. It is clear that experimental evidence is usually much more dependable than mere correlation.

Experimental observations are often evaluated by statistical tests, which in many cases simply test whether an experimental manipulation has an effect of any kind. This is much weaker support for a hypothesis than a demonstration by statistical testing that a change of a specific nature predicted by the hypothesis has in fact occurred. This important distinction is often glossed over in biological research.

The ultimate evidence that one understands a particular mechanism is to be able to build working **models** of it. The construction of models is useful because it forces one to think about the details of how a mechanism works (Leggett 1984). A dramatic example is that few, if any, scientists appreciated how sophisticated our visual system is until they tried to build machines to use vision (Marr 1982, 16). In studies of animal behavior, common examples of the use of physical models include dummies of prey or mates (Wehner 1981) and the broadcasting of synthesized songs.

It is usually not practical to build physical models of behavior, but mathematical modeling or computer simulation is often a useful substitute. For example, a model based on measurements of the responses of tethered flies to artificial visual stimuli (Figure 1-1) can predict the path of a male fly pursuing a female in a high-speed chase (Wehner 1981, 411). This model provides strong evidence that the important information used by the pursuing male is limited to the visual information incorporated in the model. In the case of prey location by owls, one might use a model of the visual process to calculate, from one's knowledge of optics and physiology, the visual information that can be obtained at various light intensities with simulated eyes, then determine if the visual hypothesis is consistent with this knowledge. Mathematical modeling and computer simulation are usually less rigorous than using physical models, because more assumptions must be made to implement them. The next step along the road to more facile but less rigorous models is the use of logical or verbal models,

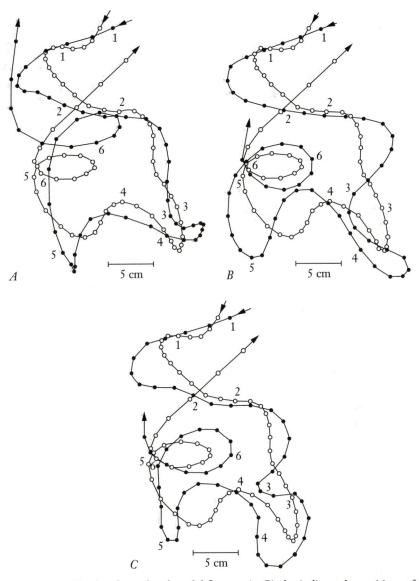

Figure 1-1 Tracks of actual and model fly pursuit. Circles indicate the positions of two flies every 20 ms; the open circles indicate the positions of the leading fly, the closed circles, the chasing fly. *A.* The actual chase. *B, C.* The movement of the fly being chased is assumed to be identical, but it is followed by one or another computer model. The similarity of the modeled chasing tracks with those of the actual track indicates that the information used by the model is all the fly actually needs. The flies were the lesser housefly, *Fannia canicularis.* (From Land and Collett 1974.)

which is common practice in studies of behavior (Hinde 1966, 7). In the case of the X-ray vision hypothesis, an important question is whether one can construct a plausible verbal model of how it would work.

1-2   BEHAVIORAL QUESTIONS

Let us now turn more specifically to the investigation of behavior. Because it is such an interesting and complex phenomenon, there are many different kinds of questions about behavior and approaches to its study. Confusion and conflict often arise in the study of behavior because researchers have different questions in mind, which frequently are not stated clearly. In addition, descriptions of behavior often mix in ideas about its functions or mechanisms as if they were objective fact. To a certain extent this is necessary, because it is impossible to describe the behavior of an organism precisely and completely, and reference to mechanisms or functions provides a convenient shorthand. Nonetheless, it is desirable to strive to minimize assumptions about mechanism and function in descriptions (Baylis 1982).

To place possible questions about behavior in perspective, consider the organization in Table 1-1. The synthetic approach asks what organisms do, how they do it, and why they do it. It is most frequently employed in studies of animal behavior and is often carefully described in textbooks (Hinde 1966, 4–7; Alcock 1975, 4–6; Drickamer and Vessey 1982, 7; Krebs and Davies 1987, 5–9). On the other hand, the analytical approach asks what problems organisms must solve, what strategies they use, and what mechanisms they use to implement these strategies. This approach may be used in textbooks but is rarely described. In his influential book on vision, David Marr (1982) emphasizes the importance of the latter approach. He argues that what has been missing in studies of vision is what he calls the computational theory, which includes identifying the problems to be solved (the goals) and devising a logical, rigorous strategy for solving these problems. The strategies are implemented by particular mechanisms not specified by the computational theory.

As with other areas of biology, sensory ecology benefits from both the synthetic and the analytic approaches; both are included in this book. However, relatively little attention is paid here to development and evolution. Rather the focus is primarily on what organisms can do, how they do it, and what functions are served in terms of the strategies implemented and the problems solved.

Table 1-1 Types of Behavioral Questions

Synthetic Approach: from behavior to causes

What do organisms do? (Description)

How do they do it? (Proximate causes)

Developmental mechanisms

Physiological mechanisms

Behavioral mechanisms

Why do they do it? (Ultimate causes)

Functions (Adaptive values)

Evolution (History of the species)

Analytic Approach: from problems to behavior

What problems must organisms solve?

What strategies do they use to solve these problems?

What mechanisms are used to implement the strategies?

There are many specific questions about behavior within each of the different types of questions. It is helpful to organize them into a **hierarchical pyramid** (Table 1-2) depending on the scope of the question. The apex of the pyramid has questions pertaining to the organism's reproductive fitness and overall life history strategy, which is the most general set of questions concerning an animal's behavior: What is its reproductive fitness? How does it manage to obtain such fitness? What are the strategies it uses? At the next level down are questions about survival and reproduction, at the third level questions about resource finding, then questions about locomotion, and finally, at the bottom of the behavioral hierarchy, questions about limb movement. Below this level would be questions in the realm of physiology, followed by the concerns of cell and molecular

Table 1-2 Scope of Behavioral Questions: A Hierarchical Pyramid

Reproductive fitness

Survival, Reproduction ...

Resource finding, Consummation, Defense ...

Locomotion, Eating, Mating, Listening, Looking ...

Limb movement, Jaw movement, Head movement, Growling, Defecation ...

biology, and so on through chemistry to physics. At any given level, questions about mechanism have answers that refer to the next lower level, and questions about function have answers referring to the next higher level. For example, leg movement is a mechanism of locomotion and locomotion is a function of leg movement.

1-3   SENSORY ECOLOGY

The body of knowledge concentrating on the top three levels of the hierarchical pyramid is often called **behavioral ecology** (Krebs and Davies 1987). This field focuses on the interactions between organisms and their environment but, like ethology, is currently more concerned with determining the adaptive advantages and evolution of behavior than in identifying the mechanisms of interactions. Behavioral ecologists have generally adopted a "top downward" synthetic approach, for example, estimating search areas from the proportion of targets located and the density of the targets and searchers (Varley et al. 1974, 61). In contrast, this book emphasizes a "bottom upward" analytic approach, for example estimating search areas from the speed and pattern of the movement of a searcher and its range of detection of the target. Behavioral ecologists have demonstrated that animals know many things about their environment without our having any understanding of the mechanisms by which they acquire the information (Kamil 1988). This work emphasizes the mechanisms.

A good example of an attempt to determine the information mechanisms underlying a given behavior is Tinbergen's study of the stimuli that release the begging response of the herring gull chick (Tinbergen 1967, 221–244). In this classic study an extensive series of experiments was carried out to determine exactly which features of the visual stimulus of the parent's bill were used to elicit the response of the chick. A sample of this work is shown in Figure 1-2. In short, **sensory ecology** concentrates on the third and fourth levels from the top of the behavioral pyramid, addressing questions such as what strategies are used to locate resources, what information is used, and how it is obtained.

Below the behavioral pyramid is the realm of physiology. Data on **sensory physiology** will be important to our considerations, but physiologists who measure sensory abilities are usually more interested in mechanisms of sensation than in the stimuli that are actually present in the natural environment. In contrast, here we will primarily treat the or-

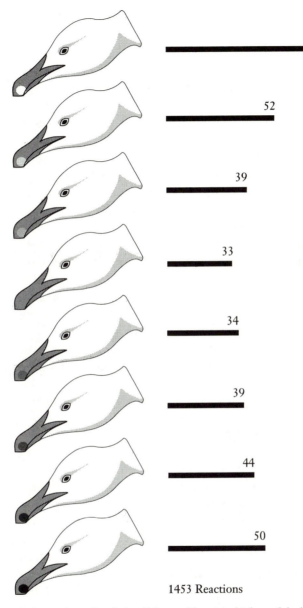

1453 Reactions

Figure 1-2 Tinbergen's study of stimuli for pecking, in which models differing in contrast between the bill and the colored spot near the tip were tested for their ability to elicit pecking by herring gull chicks. The bars represent the number of pecks elicited as a percentage of those elicited by a standard model. High contrast is more effective, whether the spot is darker or lighter than the bill. (From Tinbergen and Perdeck 1951.)

ganism as though it were a "black box" and leave questions about the mechanisms of sensation to physiology (Halliday and Slater 1983). Nonetheless, it is often useful to consider sensory mechanisms in order to estimate what information an organism can obtain. This approach is often expedient because anatomical studies are simpler than behavioral investigations.

Much of what is presented here is an attempt to lay out the limits of what is physically possible and then to compare these limits with our knowledge of what organisms actually do. It is now widely assumed that living systems always obey the laws of physics, even though this has not always been believed. There have been many examples of biological phenomena that seemed to violate the known laws of physics, but increased understanding eventually showed that physical laws were obeyed (Perutz 1987). We can thus assume that the physical world sets limits on what organisms are able to do, and we can ask to what extent organisms make use of their available opportunities.

To define sensory ecology more specifically, let us consider ecology to be the study of interactions of all kinds between an organism and its environment. The **interactions** themselves can be divided into three categories: physical, trophic, and informational. Each of these categories includes both beneficial and detrimental interactions. The simplest category, which is often ignored, consists of **physical** interactions such as mechanical forces or heat flow. A chair supports you by exerting forces on your body. You apply forces to it and to the ground in order to move. The chair is certainly a resource apart from any information it provides to you or any indirect effect it has on your metabolism. You have a physical interaction with the chair. Similarly, hydrodynamic forces are important to fish and other aquatic organisms, as are aerodynamic forces to birds and trees. A lizard basking in the sunlight to raise its body temperature has an interaction with its environment that is not trophic, because the energy gained is not used for doing metabolic work. (It may speed up chemical reactions, but it will not drive them.) In short, physical interactions are those that can be predicted by the laws of physics without recourse to biological knowledge, are not mediated by biochemical processes, and involve forces or energies that are large enough to move or heat the organism.

Even though all the above examples of physical interactions between an organism and its environment are clear-cut, it is not always easy to distinguish a physical interaction from other types. Microorganisms that swim upward or downward are often labeled as geotactic without determining that their behavior is based on a response to a gravitational stimulus rather than on a physical torque arising from differing buoyant densities at opposite ends of the organism. Similarly, in naming "magnetotactic" bacteria (Blakemore 1982) microbiologists have confused a physical interaction with an informational one (see Section 10-4).

In addition to these physical types, there are two other groups of interactions: **trophic,** which involve material or energy and have specific effects on the metabolism of the recipient, and **informational,** which provide information to an organism or transmit it to another one. For example, plants usually have a trophic interaction with light (photosynthesis), but light is usually an agent of an informational interaction for animals (vision). These two classes of interactions are distinguished by their functional importance to the recipient. In flowers, nectar mediates a trophic interaction with pollinating animals, while color patterns and scent function as advertisements and mediate informational interactions (Waser 1983). Although logically there is no need for an interaction to be exclusively either trophic or informational, in practice this division of function is usually clear. In any case, it is important to understand the nature of such interactions. For one thing, they have profound implications for evolution—organisms can easily come to ignore informational but not trophic or physical interactions.

The distinction between trophic and informational interactions is important to understanding two problems currently of widespread interest: defensive chemicals and mate selection. It has become clear that many plant tissues contain **defensive chemicals** that deter herbivores. A fundamental question about these chemicals is whether they are toxic (creating a trophic interaction) or are avoided because they so to speak "taste bad" (an informational interaction). Because of the flexibility inherent in dealing with information, the physiological mechanisms that produce bad taste should not develop during evolution unless the chemicals are toxic or at least resemble toxic chemicals. Thus, it is expected that defensive chemicals are either toxic or mimic toxic chemicals.

In sexual reproduction, by definition the female invests more in producing an egg than the male invests in the one sperm that fertilizes the egg. This unequal situation makes it advantageous to the female to be selective about which male is to fertilize her eggs. A basic difference among the vast variety of **mate selection** systems is whether the choice between males is based on trophic or informational interactions. In some species the male literally offers food to the female, although most courtship behaviors appear to provide the female with information about the male's ability to provide food in the future or about the quality of his genes (Krebs and Davies 1987, 172–86).

A case where the distinction between trophic and informational inter-actions is not obvious is in the phototropism of plants. When leaves bend toward light it is natural to assume that they are responding to the increased photosynthesis occurring in the lighted parts; the response might thus be considered a consequence of a trophic interaction with sunlight. However, experiments have shown that red light is ineffective in producing phototropism, while it is most effective for photosynthesis, indicating that the two processes are mediated by different receptor pig-ments (Salisbury and Ross 1978, 278). Since the blue-light receptor has no known trophic function, it is appropriate to conclude that phototropism is produced by an informational interaction with sunlight. Consequently, plants have both informational and trophic interactions with sunlight.

A special kind of informational interaction between organisms in-cludes the transfer of **genetic information.** The most fundamental conse-quence of sexual reproduction is that genetic information is transmitted from the male and female to their offspring. Although it is highly special-ized, this type of informational interaction is extremely important. Its study has become a field in itself, that of genetics. Genetic information provides directions for making a new organism from its parent(s) and functioning in a particular environment. However, genetic information about the environment is information about past environments that has accumulated in the organism's ancestors and has been obtained through chromosomes. Information about the environment to which the individual has itself been exposed is obtained through sense organs and is called **sensory information.**

Types of interactions that are important to distinguish are summarized in Table 1-3. In principle, ecology includes the study of all these interac-tions between organisms and their environment. Logically, there ought to be a field of information ecology to focus on informational interactions and would include both genetic and sensory information. However, gene-tics is such a distinct and well-developed field that it is more practical to break down information ecology into the two separate fields of genetics and sensory ecology (Ali 1978).

As indicated previously, sensory ecology overlaps the field of animal **behavior,** and the relationship between the two fields is important to consider. Most animal behavior textbooks (Hinde 1966; Marler and Hamilton 1966; Alcock 1975; Gould 1982) do not attempt to define behavior but often assume that behavioral responses involve movement

Table 1-3 Classification of Interactions Between an Organism and Its Environment

Physical Interactions
    Mechanical forces
    Gravitational forces
    Thermal energy
    Electromagnetic energy

Trophic Interactions
    Energy for metabolic work
    Nutrients
    Toxins

Informational Interactions
    Genetic information
    Sensory information

and muscular contraction, which is certainly the usual case. One dictionary (McKechnie 1978) defines behavior as "an organism's muscular or glandular response or responses to stimulation, especially those that can be observed." The problem is that there are responses that are observable and convey information yet do not involve muscles or glands. The question is therefore precisely what kinds of responses to include in behavior.

Is the flash of a firefly or the pulse of an electric fish (which transmit information to others but involve neither muscle contraction nor secretion) a behavioral act? And is the release of pheromones by a female moth or of adrenaline into the blood behavioral? The same dictionary states that behavior "expresses external appearance or action." The field of behavior focuses on responses, and therefore it seems reasonable to include within behavioral biology any response that is readily observable to an external observer of any species. Responses that are internal should be in the realm of physiology. Consequently, behavioral acts are particularly relevant to sensory ecology because they have the potential to convey information. However, sensory ecology is broader than behavior in that it includes all responses to sensory input—physiological and developmental as well as behavioral. The relationships between these fields are summarized in Figure 1-3.

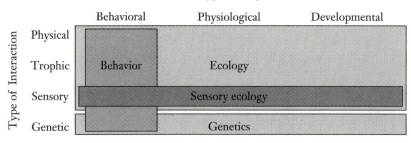

Figure 1-3 Some of the relationships between sensory ecology and related fields.

This book examines the basic principles influencing the flow of sensory information and discusses illustrative examples. Throughout, the emphasis is on fundamental relations. A wide range of possibilities is considered and an attempt made to understand why specific types of interactions occur. For instance, why is it advantageous for an organism to obtain information in the ways it does? And how is information used to solve the general problem of moving to the location of some resource?

The advantages and limitations of various sensory modalities are described herein and optimal strategies are considered. The physical limitations of time, distance, and the accuracy with which information can be obtained are discussed. The analysis is as quantitative as is feasible. However, the science of information is in a much more primitive state than those of energy, materials, and physical forces; there is no such simple relation between information income and outgo as there is with energy or materials. The known laws of physics prevent organisms from giving to other organisms more energy or materials than they take in, which leads to the pyramid of trophic levels seen in Figure 1-4. No such general laws regarding information flows have yet emerged.

The increasing specialization of scientists hinders communication between those studying different types of organisms. As a result, terminology often develops that is specific to one group of organisms. Thus one of the aims of this book is to present a system of terminology that is based on fundamental principles and consequently applies equally well to all kinds of organisms. This goal is facilitated by basing terms on categories of strategies rather than on specific behaviors. Behavior is complex and is species specific, but many of the problems facing organisms and the possible strategies for solving them are universal.

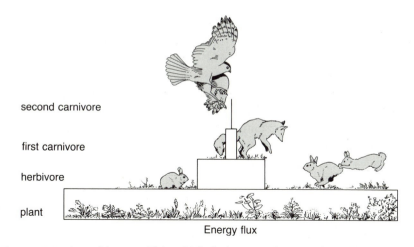

second carnivore

first carnivore

herbivore

plant

Energy flux

Figure 1-4   A trophic pyramid, in which the horizontal extent of each bar represents the net productivity of each trophic level in the ecosystem. In this particular system the efficiencies of transfer between levels are 20, 15, and 10 percent, but these values vary widely between ecosystems. (From Ricklefs 1990:199.)

A wide variety of organisms, environments, and mechanisms is considered here: even bacteria obtain information about their environment and swim toward more favorable conditions. Studying bacteria has the advantage that their abilities are relatively limited and one can therefore expect to gain a more complete understanding of their behavior. Plants obtain precise information about day length in order to predict the seasons. They can probably do this more accurately than can humans—without the aid of tools. Flowers are to a great degree signboards designed to send messages to insects or other pollinators. The abilities of humans without tools are often discussed here, as are the vocalizations of whales. The full range of living organisms is considered, within a wide variety of environments, from the darkness of the ocean abyss to the aerial domain of birds, and from the scale of a bacterial cell to that of the earth.

Obviously, it is impossible to review all that is known about every type of organism, and only selected examples can be presented. The selection is generally based on describing the simplest organism that exhibits a given use of sensory information. One advantage of this approach is that it demonstrates the surprising sophistication of organisms often considered to be too simple to have interesting behavior.

In summary, this book covers a wide variety of topics related to how and why organisms obtain information about their environment, and attempts to combine them in a manner that will provide an improved perspective. What is information used for? What type of information is obtained? What are the limitations on various types of sensory systems and information channels? What constraints limit the sensory system? Would it be adaptive to have a more-sensitive system, or is the noise level in the environment high enough that such a system would not be advantageous? How fast is information obtained? Would it be adaptive to obtain it faster? What processes carry information to organisms? What are the relative advantages of the different information channels?

In the following chapters, Part 1 deals with the basic characteristics of information, Part 2 with the properties of specific types of stimuli, Part 3 with behaviors in which stimuli are generated, and Part 4 with mechanisms for using information to move to spatial goals.

Information
Basics

■ ■ ■

In the following four chapters, some of the more general characteristics of information are covered, including a synopsis of the biological functions served by sensory information, the definition and properties of information, general mechanisms of its transmission, and strategies for signal detection.

CHAPTER 2

Functions of Information

■ ■

In full, fair tide let information flow;
That evil is half-cured whose cause we know.
—Charles Churchill, *Gotham*, 1774

Nearly all organisms obtain information about their environment to help solve a wide variety of problems—maintaining an appropriate environment, timing activities, locating resources or threats—and many organisms transmit information to other individuals in order to persuade them to do something. Such functions of sensory information are the aspect of sensory ecology that has been most studied. The main objective of this chapter is to review this material briefly as a background for the more detailed discussions of mechanisms that follow.

2-1   HOMEOSTASIS

The most fundamental problem organisms face is to maintain conditions suitable to their continued existence and functioning. Homeostasis refers to the processes by which feedback information is used to maintain a constant state of certain basic conditions such as temperature, water content, or position. In many cases, the feedback information is sensory in nature, informing the organism about its external environment. A few examples of the uses of sensory information to control conditions of prime importance to the organism are presented in this section.

Temperature is significant to the functioning of all organisms, many of which employ environmental information to help maintain an appropriate internal temperature. Small organisms cannot maintain a temperature different from that of their environment; many simply sense a temperature gradient and move to an appropriate temperature. This process, frequently called thermotaxis, occurs in bacteria (Maeda et al. 1976), protozoa (Jennings 1904; Nakaoka and Oosawa 1977), nematodes (Hedgecock and Russell 1975; Dusenbery 1988; Pline et al. 1988; Diez and Dusenbery 1989), and terrestrial arthropods (Madge 1961; Osuji 1975). Larger animals may employ additional information and complex responses to regulate their temperature. Turtles sometimes bask in the sun with their webbed feet held out perpendicular to the sun's rays, for maximum heat gain, and other animals burrow underground to avoid temperature extremes. Thermal stimuli may also serve purposes other than regulating internal temperature by conveying information about heat sources or heat sinks that might represent food or other resources (see Chapter 6).

Water is also significant to all organisms; most must maintain a certain level to survive. In air the availability of water is usually measured by the humidity, in aquatic environments by salinity. Most insects and other terrestrial arthropods must avoid prolonged exposure to air that is too dry. Some respond directly to gradients of humidity by moving to more humid regions (Bursell and Ewer 1950). Most aquatic organisms must remain in a given range of salinity. The simpler organisms such as nematodes merely move in salinity gradients toward an optimal one (Culotti and Russell 1978; Dusenbery 1983).

Maintaining **position** in a flowing medium is often important and difficult for motile organisms. Animals that swim or fly often use visual information from the substrate to provide feedback concerning their movement. This type of information may be used in an optimotor reflex to help maintain position (Dijkgraaf 1962; Möhl 1989).

Many organisms that move in a three-dimensional medium tend to maintain a fixed **distance** from a specific surface such as the bottom or top of aquatic bodies or the ground in the case of flying animals. This behavior has been called strato-orientation by Jander (1977). These animals probably make use of visual information: binocular cues or the flow of visual patterns (David 1986). A more difficult problem is to maintain a fixed intermediate depth in the ocean or in soil. In the former case, light's intensity is a likely cue (Blaxter 1988). As an example of the latter, moles seem to burrow at a constant depth below the soil's surface; they probably acquire some kind of information about their depth, perhaps from temperature.

Another problem related to maintaining a constant environment is the appropriate **spacing** of individuals. Minimal territories are usually established by organisms sending out species-specific signals such as bird songs or scent markings. The spacing within herds, flocks, and schools is probably maintained mostly by visual cues. Proper spacing may be important to provide improved efficiency in locomotion (as in the "drafting" of racing cars). An important question is whether animals can sense the improved efficiency directly, from the effort required to maintain their speed, or whether they employ their powers of vision to maintain a distance that is determined by learning or inheritance. Similarly, plants typically space out their leaves to minimize shading, using the intensity of blue light as a cue to their exposure to photosynthetically active wavelengths (Salisbury and Ross 1978, 278), and insects often avoid laying eggs near previously laid eggs of conspecifics (Haynes and Birch 1985).

The **direction** of orientation is important to a wide variety of organisms. As humans use the sense of gravity to remain oriented, which is crucial for bipedal locomotion, most plants similarly employ a sense of gravity to grow upright. Fish and crustacea that live in a three-dimensional world would seem to have much less need to maintain a fixed orientation. Nonetheless, many have a strong "dorsal light reaction" (Marler and Hamilton 1966, 526–31) that maintains their orientation by keeping their dorsal surface aligned toward the prevailing light source (Figure 2-1). In addition to the organism's direction of orientation, its direction of locomotion is significant. This becomes obvious when the direction is clearly toward a goal, but some organisms use environmental cues simply to maintain a constant direction of locomotion, to avoid searching an area already visited, so that the specific direction may be of no importance. The stimuli that serve such a function have been termed collimating stimuli (Pline and Dusenbery 1987).

2-2 TIMING

For many organisms it is advantageous to coordinate the timing of their developmental, physiological, or behavioral activities with their environment. This is especially significant for reproductive activities. For example, the flowering or germination of seeds from plants growing in temperate climates often requires a period of exposure to cold, which

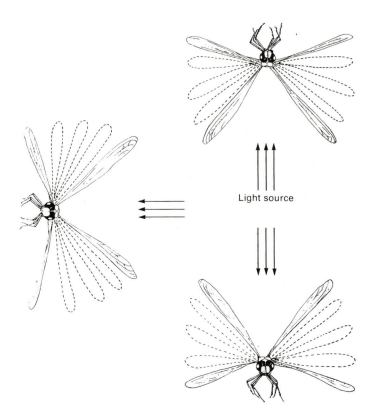

Light source

Figure 2-1   An example of the dorsal light reaction, showing that a flying dragon-fly keeps its dorsal side to the light. (From Bullock, Orkand, and Grinnell 1977.)

probably functions to ensure that winter is really over. And the seeds of certain pine trees require exposure to the heat of fire, which is likely to produce favorable growing conditions in coming years.

The problem of timing is particularly interesting when a significant interval is necessary for preparation before a response can be made, making it useful to be able to anticipate future events. There are two types of situations in which such anticipation is possible. One includes the anticipation of regular events in the physical environment, such as the time of day, the ocean's tides, and the seasons. The other type includes coordinating activities with other individuals of the same species, cooperating to synchronize behavior.

The adaptive advantage of the **synchronization** of activities within a species is easy to see in the case of reproduction, where it ensures that both sexes are active at the same time. For example, the sexual readiness of male and female goldfish is synchronized by the release of a pheromone (Dulka et al. 1987). Likewise, the reproductive state of female mice is influenced by the odor of male mice (Wilson 1970).

Apart from its importance in reproduction, the most common advantage of synchronization may be that it provides a mechanism for saturating predators. A good example of this may be the periodical cicadas such as the "seventeen-year locust," which lives seventeen years in the ground before emerging to reproduce. This strategy makes them a major resource for potential predators, but with long intervals of unavailability. It would thus be difficult for a potential predator to adapt to such an extreme and arbitrary schedule of prey availability. A basic informational question is to ask how the cicades control their emergence. Do they have a timing mechanism with a seventeen-year duration, or do they count seventeen seasonal cycles? Does each individual have its own accurate mechanism, or is there some form of communication that aids their coordination?

There are many cases where the timing of a regular event in the environment is important. Appropriate stimuli permit the **anticipation** of such future events as high tide (for laying eggs on a beach), nightfall (for seeking shelter), and winter (for bird migration or the cold hardening of plants). In other cases, regular events provide a mechanism for individuals to synchronize their activities. For instance, the plants of a given species may all flower within a week of one another year-after-year. Similarly, male and female moths and fireflies often seek mates during relatively narrow time intervals at night, but different species use widely different times of the day.

Part of an organism's mechanism for anticipating temporal cycles may be an internal clock such as that exhibited by circadian (daily) rhythms, but for accuracy such a clock must be reset frequently using sensory information. Plants generally determine the season by the current day length, a strategy probably much more reliable than simply using temperature changes. At the mid-latitudes the day length changes by one to two hours per month during the spring and fall. Determining the seasonal time to an accuracy of one week would therefore require measuring the day length to an accuracy of 2 to 4 percent (15 to 30 min). Plants are probably much better at this than are humans without the aid of tools.

2-3 SPATIAL ORIENTATION

A common problem for motile organisms is to identify the position or direction of certain resources (spatial goals), which of course requires environmental information. The resources might be food, prey, mates, new nest sites, leks, or certain habitats. Even bacteria swim toward higher concentrations of food (Adler 1969). Plants use information about the distribution of sunlight to grow toward areas with more light. Organisms that release spores find it advantageous to do so in the open, where the spores can be dispersed by air currents. Consequently, many microorganisms release their spores from the tops of stalks that have been positioned with the help of environmental stimuli. For example, the sporangiophores of the fungus *Phycomyces*, which have been particularly well studied, grow against gravity, toward light, into air currents, and away from objects (Galland and Lipson 1987; Shropshire and Lafay 1987). Pseudoplasmodia of the cellular slime mold *Dictyostelium* have a sensitive response to temperature (Bonner et al. 1950; Whitaker and Poff 1980) that may lead them toward the surface (Dusenbery 1988), and they give off and avoid ammonia, which causes them to grow into air, away from surfaces (Bonner et al. 1986; Kosugi and Inouye 1989). In addition, plants can sense the color of light in their surroundings and grow taller if green light, indicating competing vegetation, comes from below (Kasperbauer 1987; Bradburne et al. 1989; Ballaré et al. 1990).

One of the most interesting uses of sensory information is in guiding locomotion toward distant resources. This function occurs in all motile organisms from bacteria to mammals; its scale extends from travel over only a few centimeters to thousands of kilometers across the face of the earth. This use of sensory information is discussed in detail in Part 4.

2-4 DEFENSE

Most organisms have features that provide protection from predators or other threatening organisms. In many cases these defensive features are responses to environmental information. The most commonly observed defensive response is avoidance of a threat that has been identified and localized by sight or sound, as in a cat running away from a dog. In other cases the response will not be directed toward or away from the threat; the cat may simply run up a tree knowing innately that it is safer there.

Similarly, a spider threatened in its web will simply drop out of the web, and a mollusk contacted by a starfish will swim away in a random direction (Willows 1967). An even simpler response is for the organism to withdraw into a protective location; thus bivalves close their shells, turtles withdraw into their shells, and earthworms contract into their holes.

Although it is a less recognized response, probably the oldest form of defense is chemical. Many aquatic plants and animals contain chemical feeding deterrents (Carr 1988). Many terrestrial arthropods, including insects, ooze or spray noxious chemicals when they are disturbed (Eisner 1970). Some can spray with great accuracy, and some seem to identify likely predators with a high degree of specificity; in any case both abilities require environmental information. Plants often contain chemicals that interfere with herbivores. Some recent experiments have indicated that these chemicals are sometimes induced in response to environmental signals from injured neighbors (Baldwin and Schultz 1983; Rhoades 1985).

Often there is confusion as to whether a chemical that causes avoidance is primarily carrying information or is in fact damaging (toxic). Many researchers ignore this distinction, but it is a basic question of sensory ecology. If a certain chemical carries only information, there is no reason for the receiver to continue avoiding the source, unless the chemical is a warning of another hazard, in which case communication is occurring and the chemical is aposematic, or carrying a warning (see next section). Otherwise the chemical itself is probably damaging and is not primarily a signal. In the examples of chemical defense discussed in this section, what is of interest is the stimuli that trigger or direct the synthesis or release of protective chemicals. The study of the toxic chemicals themselves is very interesting, but the subject of chemical ecology rather than sensory ecology.

Animals that are preyed upon by vertebrates that hunt visually experience strong selective pressure to reduce their visibility by some sort of camouflage. The body color patterns of animals that have visual predators often blend with their background environment and disguise their true body shape. These defense mechanisms can be surprisingly successful and can reveal much about the natural environment of the camouflaged organism and the perceptual abilities of its predators (Saidel 1988). Some of the commonly used strategies include (1) creating color and textural resemblance to the background; (2) countershading to reduce the perception of roundness; (3) patterning to disrupt the perception of the true outline; and (4) the elimination of shadows by altering the organism's shape or posture. In some species individuals can modify their appearance to better

Figure 2-2   A bird with a camouflaged eye. The eye is surrounded by black surface coloration so that it blends with the necessarily black pupil. In addition, it is near a high-contrast border, which makes it difficult for an observer to pick out the features that do not match the background. (Carolina chickadee (*Parus carolinensis*) photographed by author.)

match different backgrounds, but this ability seems limited and does not seem to involve detailed perception of the environment (Saidel 1988). For example, most fish are dark on their upper surface and light colored below to blend with the background whether viewed from above or below. Certain fish that live at the edge of the lighted zone in the ocean even carry

Figure 2-3   The eyespot display of an automerid moth from Panama. *Top*: Normal resting attitude. *Bottom*: Reaction when touched. (From Ricklefs 1990.)

this strategy to the extreme of generating just the right amount of light on their bottom surface to match the light from above (Lythgoe 1988). Eyes are particularly hard to disguise, however, because of necessity they include the predictable features of a black hole and a circular shape, needed for rotation. Many color patterns of vertebrates (Figure 2-2) incorporate the eye into a dark stripe or spot to help disguise it (Cott 1941, 82–93).

Camouflage sometimes occurs in sensory modalities other than vision. For example, the warning calls of birds are often camouflaged by being soft sounds that are hard to localize (see pp. 329–30).

Many animals, particularly fish, also gain protection from predators by misleading them as to which direction they are faced. This misdirection is usually accomplished by placing conspicuous patterns resembling eyes near the rear of the animal. The frequency with which such eye patterns are found in body markings is testimony to the universality of the appearance of vertebrate eyes. Many butterflies and moths also seem to use this pattern to startle a potential predator by suddenly revealing huge eye patterns appropriate to an animal larger than the potential predator itself (Figure 2-3). At least one frog has adopted a similar eye-pattern strategy (Purves and Orians 1987, 1140).

2-5 PERSUASION

Sometimes an individual organism deliberately generates a signal in order to influence the behavior of another individual, a behavior called persuasion (Dawkins and Krebs 1978). When the receiver also benefits, the interaction is true communication; if not, it is deception (see Section 3-1). True communication usually occurs only between individuals of the same species. However, in a few cases it occurs between species that derive mutual benefits, such as flowering plants and their pollinators.

Probably the most common function of communication is **identification**. The striking color patterns and the songs of birds primarily transmit the message that "here is a male of species X." Other signals, such as the size of antlers, send messages about the fitness of males. Such information may be used by a female to select the most desirable mate or by a male to determine whether to challenge another male for access to females. Females often provide signals about their readiness for mating. Flowers provide both a particularly common and an unusual case of communication in that their primary functions are to attract an appropriate animal and to transfer pollen to and from it. Their conspicuousness and beauty to us are a consequence of their developing effective visual (Waser 1983) and chemical (Robacker et al. 1988) cues to identify them to potential pollinators.

Another common type of signal is that of a **warning** or **aposematic** signal. The alarm calls of birds and social mammals communicate the simple message that there is danger in the vicinity. Many organisms are defended by being unpalatable or dangerous to potential predators and by generating distinctive signals that a predator can learn to associate with their unpalatability, thus avoiding attack. We are most familiar with aposematic coloration in animals such as the venomous coral snake, but acoustic signals as in the rattlesnake and certain moths and probably chemical signals also occur (Eisner 1970; Miller 1983; Bailey 1991, 165).

The highest level of communication besides our own occurs in social insects, which must **coordinate** the activities of many individuals. A colony is a complex network of individuals exchanging information. Indeed, colonies of social insects have been likened to individual organisms, with the information exchange being analogous to the hormonal and neuronal communication in an individual animal (Seeley 1989). Most of this communication is carried out by chemical messengers (Blum 1985), but the incidental transmission of information also plays an important role in integrating the activities of the colony (Seeley 1989). Aside from human communication, the most sophisticated message we understand is the dance language of honey bees (see pp. 336–47), which is thought to communicate to the bees' nest mates the direction, distance, and quality of patches of flowers (von Frisch 1967).

Some organisms have been selected to generate signals that lead to responses that are not in fact beneficial to the receiver. Since the nature of information is such that it can easily be ignored, the receiver will continue making such a maladaptive response only if the response is one that is beneficial to it on the whole. This occurs when the receiver mistakes the signal in question for another signal. When the source of the signal has been selected to produce the signal, it is appropriate to speak of **deception**.

One well-known form of deception is **mimicry,** in which one species gains an advantage by resembling another species, called the model (Figure 2-4). For instance, a variety of insects can gain protection by adopting the yellow and black banding pattern of venomous yellowjackets. Similarly, a number of nonvenomous snakes gain protection by adopting color patterns resembling the red, yellow, and black banding of coral snakes.

Ground-nesting birds often attempt to divert predators from their nests by acting as **decoy.** Since birds are normally difficult to catch, the decoy is made more effective when it creates the impression that it is injured, as in the so-called broken-wing display. At the other end of the phylogenetic spectrum, certain algae are known to release attractants for the gametes of competing species, which presumably interferes with the reproduction of competitors (Carr 1988).

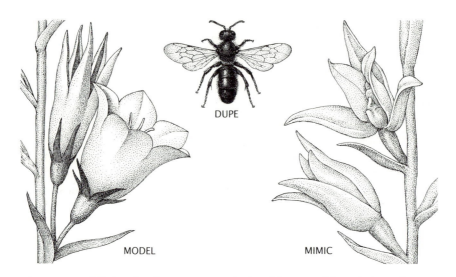

DUPE

MODEL

MIMIC

Figure 2-4 Mimicry is a three-part complex involving a model, a mimic, and a dupe. Here the model (*left*) is the bellflower *Campanula persicifolia*, which produces abundant nectar, a food for the leaf-cutting bee *Chelostoma fuliginosum* (*center*). The red helleborine orchid *Cephalanthera rubra* (*right*) produces no nectar but mimics the bellflower closely in color and thus deceives the duped bee into visiting the orchid and thus pollinating it. [From *Scientific American* 257 (1987): 78.]

A few organisms use a **lure** to attract other organisms to their vicinity. By definition, a lure provides no benefits to the organism attracted and it functions by resembling a model that does have benefits for that organism. For example, certain turtles can wiggle a tongue that looks like a worm, to attract the fish they prey upon. Ophrys orchids lure males of certain bee or wasp species for pollination by a series of chemical, visual, and tactile cues which suggest that the appropriate female insect is present (Robacker et al. 1988). Certain spiders mimic moth sex attractants to lure male moths within range of attack (Eberhard 1977; Stowe 1986).

Other deceptive strategies contribute to an organism's ability to escape. A variety of solitary but mobile insects produce defensive secretions that contain chemicals identical to the alarm pheromones of ants. Their strategy is apparently to momentarily disperse attacking ants while they make an escape (Blum 1985). Certain moths produce ultrasonic clicks when they hear an echolocating bat nearby and closing in on a target (see Section 13-A). These clicks, which resemble the echolocation clicks used by the bats, often cause the bats to turn away as if an obstacle were present (Simmons and Kick 1983). These strategies in which information is transmitted in order to interfere with the acquisition of accurate information by another individual might be called **jamming**, which is analogous to radio jamming.

This brief review demonstrates the wide variety of ways in which information is important to organisms of all kinds. The following chapters examine more closely what information is, the ways in which it can be obtained, and provide more detailed examples of how information is used by organisms to move themselves to where resources are located.

CHAPTER 3

Measuring Information

■ ■

Since information is the central component of sensory ecology, it is important to define it and consider how it is measured. This chapter gives an experimental definition of information transmission, describes the units for measuring information, discusses examples of transmission rates and processing capacities, energy costs of transmitting information, and difficulties of measuring the information content of stimuli.

3-1 DEFINITIONS

Although information may seem an obvious concept in this age of computers, this abstraction is actually a relatively recent intellectual development (Stonier 1990, 109–10). Just as the science of thermodynamics developed only after a century of experience with steam engines, a science of information is still emerging after a few decades' experience with telephone systems and computers. A quantitative definition of information is still not agreed upon. The best formed theory to date focuses on the maximum amount of information that can be transmitted in a given situation but says nothing about the actual information content.

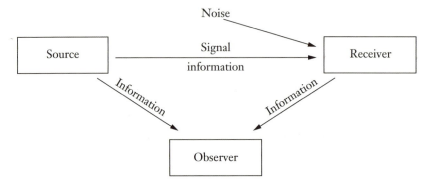

Figure 3-1 A schematic diagram of an informational interaction in which the observer concludes that a signal has passed between the source and the receiver because of correlations in information from the source and the receiver and of knowledge that neither a physical nor a trophic interaction occurred between them. This conclusion can be strengthened by performing an experiment in which the source is modified independently of the receiver and an appropriate change occurs in information about the receiver.

The transmission of information is an interaction between a **source** (or transmitter) and a **receiver** (Figure 3-1). Between the two is a transmitting **channel** involving the movement of energy or material from the source to the receiver. In sensory ecology, the receiver is an organism and the source is part of its environment (perhaps another organism), and the channel includes a stimulus, with the environmental factors that affect it. One experimental test for interaction between a source and a receiver is to remove or alter the source, then observe whether the change alters the behavior of the receiver sooner or later. Interactions, other than the transmission of information, can occur by providing either materials or energy (trophic interactions), or through physical forces. If trophic and physical interactions can be excluded, it may be concluded that information has been transmitted.

This exclusion is often accomplished by considering the quantities involved. An interaction mediated by physical forces must involve forces large enough to move the organism. An interaction mediated by trophic agents usually involves a quantity of material or amount of energy that is comparable to the total quantity of food or energy taken in, with the exception of such trophic factors as trace minerals or toxins, which may have trophic effects in very low quantity. In contrast, informational interactions are usually mediated by stimuli that involve much smaller quantities of chemicals or energy. The receiver usually has certain highly specialized mechanisms designed to amplify the response to the stimulus. A fundamental property of information is that it can easily be ignored, in contrast to physical or trophic factors.

Consider, for example, that when a lion charges toward an impala the impala will suddenly start running in the opposite direction, but when there is no charging lion the impala will not suddenly start running. Furthermore, we know of no trophic or physical processes that could act on the impala to account for this behavior—the grass is not better away from lions, the gravitational force between the lion and impala is too weak to be a factor and is in any case attractive, the force of electrostatic repulsion is too weak to move such large bodies, and so on. In addition, we believe that it would be advantageous for impala to respond to information about approaching predators. Thus we may conclude that information has been transmitted from the charging lion to the impala.

This was an easy case, but consider one that is more difficult. It has recently been discovered that certain bacteria are "magnetotactic" (Blakemore 1982), which is to say that they swim in the direction of the earth's magnetic field. It has been shown that these bacteria contain iron deposits that are magnets. Because bacteria are very small, it does not take much force to rotate them. All the evidence indicates that the earth's magnetic force acts directly to turn the bacteria into alignment with the earth's field, just as it aligns a compass needle, which is much larger. Indeed, dead bacteria orient just as well as live ones. Thus, there is no evidence in this case of any information transmission between the earth and the bacterium, which in this particular instance is a physical interaction.

The meaning of the information transmitted is determined by comparing the change in the behavior of the receiving organism with knowledge of its requirements and estimates of what behavior would be of adaptive advantage to it in the relevant situations. In the example given of the lion and the impala, it is clearly of adaptive advantage for the impala to avoid lions, which prey on impala. Thus, the information transmitted from the lion to the impala is probably something like: "Danger is rapidly approaching from the left."

Through what channel or stimulus modality is the information transmitted? **Channels** may consist either of chemicals or of various forms of energy such as light, sound, vibrations, electrical or magnetic fields, temperature, or pressure. Some of these may mediate noninformational interactions as well. For instance, light provides metabolic energy to

photosynthetic plants (a trophic interaction), as well as information allowing them to determine day length and consequently the season, in an
informational interaction. The temperature of the environment directly
affects the temperature of the organism (a physical interaction) but it may
also provide information important to the organism's thermal regulation.
Because we rely primarily on vision, it is natural to assume that other
organisms do the same, a bias that scientists must constantly strive to
overcome. An impala might well be using sound or smell, so the researcher
must eliminate such alternatives. For example, considering the speed of
the organism's response and the wind's direction could eliminate the
possibility of chemical transmission. However, distinguishing between
light and sound is more difficult. One could in theory mask the impala's
different sense organs, then observe its behavior in the dark or under loud
masking sounds, or artificially create the sound or light signals suspected
of carrying the information. In practice, more than one channel may be
operating simultaneously.

After the channel is identified, the next point to determine is what is
the **code**. What features of the pattern in the channel are recognized, and
what meaning is attached to them? In physiological terms, what is the
stimulus that caused the observed response? Specific patterns of activity in
a channel acquire meaning as the result of an association between the
pattern and some state of the environment that is relevant to the receiving
organism (Wiley and Richards 1978). A meaningful pattern is often called
a **signal** (by communications engineers) or a **stimulus** (by physiologists).
Some biologists restrict the meaning of "signal" to cases where the
stimulus pattern has been shaped by natural selection to enhance its
information-transmitting function, and use "cue" for patterns that have
not been so shaped (Seeley 1989). This convention is attractive in that
"signal" has stronger connotations of having been deliberately generated
than does "cue." However, it is useful to have a term for meaningful
patterns of both types, and "signal" is well established in other fields to
mean any pattern that is of interest. The term "stimulus" could be used to
include both types of information, although it has connotations associating
it with a particular response. It seems more useful to use "signal" to
represent meaningful patterns of all types and make the important distinction between deliberate and incidental transmission in other terms, as
discussed shortly.

For a given transmission of information to be useful, the receiver must
be able to recognize relevant signals and must possess information about
appropriate actions it should take for the state of the environment as-

sociated with each signal. For example, if light is the channel over which impala recognize predators, are predators recognized by size, shape, color, speed of movement, or their pattern of movement? These are very difficult questions to answer. The general approach is to generate synthetic signals that produce the same behavior in the receiver, then systematically alter these signals to determine which of their features are necessary to cause the response.

It is common to think of the transmission of information as being **communication**. However, this term has implications of deliberate transmission (Beer 1982; Hawkins and Myrberg 1983), whereas most of the informational interactions of interest in sensory ecology involve an inanimate source. How then should we define sensory communication? Should it include all the informational interactions between organisms, or only those in which information is deliberately transmitted? Or perhaps only those in which true information is sent? There is no commonly agreed on definition in biology (Hopkins 1988), but informational interactions between organisms are commonly classified according to whether the interaction has some adaptive advantage to either the transmitter or the receiver, to neither, or to both.

In the case of the transmitting organism, a behavior that is adaptive for other reasons will often generate signals incidentally. Such behaviors should be distinguished from those in which the principal function of the behavior is in fact to transmit information. This distinction is best expressed by considering the function of the behavior rather than the adaptiveness of the signal. The resulting classification scheme is presented in Table 3-1.

The term "true communication" is restricted to cases in which the transmitting organism engages in behavior that is adaptive principally because it generates a signal and the interaction mediated by the signal is adaptive to the receiving organism as well. Thus vocalizations of terrestrial

Table 3-1 Classification of Informational Interactions

Response of the Receiving Organism	Function of the Behavior of the Transmitting Organism	
	Generate the signal	Some other function
Adaptive	True communication	Incidental transmission
Maladaptive	Deception	Misinterpretation

animals are assumed to mediate true communication, because they cost the transmitting animal energy and have no other evident function (Beer 1982). In other words they have no trophic effects and do not carry energy sufficient to exert physical effects, so they must convey information to some receiver, which could evolve to ignore these sounds if the communication proved not to be adaptive to it. The situation is more complex underwater, where some vocalizations may carry sufficient energy to physically injure other animals.

If the behavior of the transmitting organism has primary functions other than communication, the interaction is a case of **incidental transmission**. (If the function is that of transmitting a signal to a third party, such an interaction is of the type called interception by Hawkins and Myrberg [1983], which is a type of incidental transmission.) If the behavior of the transmitting organism has the function of generating the signal but the interaction is not detrimental to the receiving organism, the interaction is of the type called **deception**. If the behavior of the transmitting organism has other functions and the interaction is detrimental to the receiving organism, such an interaction is labeled **misinterpretation**.

In true communication, the association between the state of the environment and the behavior of the transmitting organism generating the signal is one created solely for purposes of communication. In such a case the signal becomes a **symbol** for the state of the environment. In the other three classes of informational interaction between organisms, the type of association is determined by factors outside the interaction between the two parties. For instance, deception occurs when the signal resembles a model. Consequently, true communication is distinguished by the fact that it can be mediated by arbitrary symbols.

3-2 MEASURING INFORMATION CONTENT

The rather late development of a quantitative theory of information independently by Claude E. Shannon and Norbert Wiener in 1948 may be indicative of some of the difficulties in analyzing information. In practice, their theory has not yet proven very fruitful outside the field of communications engineering (telephone, radio, television), but it provides the best way available to measure information. This theory only deals with the capacity to transmit information, saying nothing about the meaning of a message. Scientists are currently working on a more general theory of information, and it has been suggested (Stonier 1990) that information is as fundamental an abstraction as the concept of energy. Whereas energy is the capacity to do work, information is the capacity to organize a system. As this more general view is still under development, this book concentrates on the traditional theory initiated by Shannon and Wiener.

In this theory, information transmission is the removal of uncertainty. The potential information content of any message can be determined by converting it to a standard binary code in which a sequence of binary choices is transmitted. These may be dots and dashes like those in the Morse code, on and off values like those used in most present-day computers, or zeroes and ones. The basic unit of measurement is a single such choice called a **bit** (*b*inary dig*it*). One bit of information is the amount necessary to distinguish between two events of equal probability.

Basic written English consists of about 64 upper- and lowercase characters, a space, and punctuation marks. To represent each of these symbols by binary values requires a series of six values per symbol ($2^6 = 64$). For instance, the letter A, might be represented by 000001, Z by 011010. This suggests that each symbol in English carries about six bits worth of information. We can say with confidence that no more than this amount of information can be contained in an English message, because we can in fact encode any such message in binary form using six binary digits per symbol. However, this estimate overstates the information content of the message. For example, the meaning of most messages could be transmitted using only 26 letters, a space, and a period, which would require only five binary digits per symbol ($2^5 = 32$). A more serious problem is that by using this code we could encode any sequence of these characters, but most of the sequences would not make sense as English and would thus never be sent. The problem is that English is highly redundant and measuring its information content is therefore not easy. The best estimates, based on experiments with people, are that English messages actually contain on average about one bit of information per character (Pierce 1961, 103).

For a message source that produces N different symbols, each with a probability P_i that does not change within a message or between messages, parameter H is defined by

$$H \equiv -\sum_{i=1}^{N} P_i \log_2 P_i$$

(3-1)

Here H is a measure of entropy generalized to cases where the probabilities are not all equal. The parameter H is commonly called the

potential information content per symbol. It actually measures the uncertainty associated with each symbol and thus the amount of information each potentially carries. If the receiver is certain which symbol the source will select, it already has all the information that can be sent; thus, no information will be transmitted (Gatlin 1972, 48). The potential information content per symbol is greatest (H_{max}) when each symbol is equally probable. In this case $P_i = 1/N$ and

$$H_{max} = \log_2 N \qquad (3\text{-}2)$$

If it is known that there are N symbols but their probability is not known, $\log_2 N$ can be considered the maximum possible information content. If there are only two equally probable symbols, then $H_{max} = 1$ bit per symbol. If there is only one symbol, $H = 0$ and no information is transmitted, because there is no uncertainty about it.

Organisms often have multiple receptor cells, which provide parallel pathways for the simultaneous acquisition of information. The different receptors may mediate either separate sensory modalities or different specificities within one sensory modality. If the information from each receptor or set of receptors is independent of the others and is carried in independent pathways, it is useful to identify these pathways as being distinct **stimulus dimensions**.

In this concept each dimension is considered independent of all others in the sense that it is possible to change the value of one without changing any of the others, as in the familiar meaning of dimensions in geometry. There, a plane has ≥ 2 dimensions, because one can move along the north–south axis without moving along the east–west axis or vice versa, and it also has ≤ 2 dimensions, because there is no way to move without moving along at least one of these two axes. Consequently, a plane must have two dimensions. For a sensory example, a light stimulus can include any of a huge number of combinations of different wavelengths. However, the number of independent dimensions that can be perceived is limited by the number of receptors that are tuned to distinct wavelengths. For instance, a protozoan with a single light receptor has only a single dimension to its light perception: intensity. However, humans have three types of cones, providing us with three independent dimensions for our color vision.

To estimate the information capacity of a particular organism or sensory system requires knowledge of the number of its stimulus dimen-

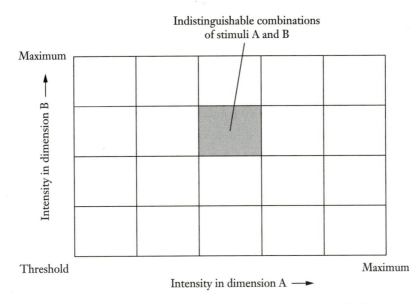

Figure 3-2 An illustration of the concept of stimulus dimensions, which assumes that there are five distinguishable levels of intensity in stimulus A and 4 levels in stimulus B, resulting in twenty distinguishable combinations of these two stimuli.

sions and the resolution of the intensity along each dimension, as shown in Figure 3-2.

If each receptor cell has N_i distinct intensity levels and there are N_r receptor cell types, then (Snyder, Laughlin, and Stavevga 1977; Snyder, Stavevga, and Laughlin 1977)

$$N = N_i{}^{N_r} \tag{3-3}$$

and

$$H_{\max} = N_r \log_2 N_i \tag{3-4}$$

Thus, the information obtainable in an appropriate time window is proportional to the number of receptor types but increases only with the logarithm of the number of distinguishable intensity levels. For example, if we assume that for vision each receptor type can distinguish ten levels of intensity, the number of different colors that can be distinguished becomes

Table 3-2 Channel Capacity of Humans

Sensory System	No. of Receptors	No. of Axons	Capacity (bits/s)
Eyes	2×10^8	2×10^6	107
Ears	3×10^4	2×10^4	105
Skin	1×10^7	1×10^6	105
Smell	7×10^7	1×10^5	10^5
Taste	3×10^7	1×10^3	10^3

Source: Zimmermann, 1978.

10^3, or about 1,000 and the maximum amount of information simultaneously obtained becomes $3 \log_2 10 \cong 10$ bits, which can be seen from $2^{10} = 1,024$.

If the maximum frequency of action potentials in a single neuron is f_{max} then the information that can be transmitted in time t is

$$H_{max} = \log_2 (t f_{max} + 1) \tag{3-5}$$

Some estimates of the rate of information transmission from various sensory systems are presented in Table 3-2.

3-3 THE MAXIMUM RATE OF TRANSMISSION

The rate at which a channel can carry information is limited by two factors, its bandwidth and noise. Bandwidth (B) is a measure of how frequently the channel can be varied. If a channel can switch rapidly from one symbol to another, its transmission capacity is high. The bandwidth is technically defined as the range of frequencies that can be transmitted, but in natural systems the lowest frequency is zero and the range of frequencies equals the highest frequency. For example, with chemical stimuli the bandwidth is determined by how rapidly the concentration varies in a meaningful way that can be detected. In most real channels this value is not well defined (Fontolliet 1986, 46). Experiments have demonstrated that a bandwidth of 7 kHz is required to reliably understand the individual syllables of human speech (Meyer and Neumann 1972, 271). We can communicate over the narrower

Table 3-3 Channel Capacity of Electronic Communications

Channel	Bandwidth	S/N*	Rate (bits/s)
Telephone	3.1 kHz	1,000	3×10^4
AM radio	4.5 kHz	1,000	5×10^4
Television	5 MHz	1,000	5×10^7

*S/N is not well defined, but a value of 1,000 (30 db) is typical.

bandwidths of telephone and AM radio (see Table 3-3) because of the redundancy of human languages.

All channels that carry information have associated with them unpredictable patterns that are not part of the signal. These irrelevant patterns, called noise, can arise from many sources. One fundamental source is statistical fluctuations, such as the number of photons detected at low light levels. The random movement of molecules generates thermal noise. Noise can also originate from transmitters of information other than the one of interest. There is no general way to analyze such secondary transmitters, but a case of well-behaved random noise can be analyzed as follows.

If the total power in a channel of bandwidth B is distributed between the signal S and random noise N, the maximum rate of information transmission R_M is, in bits per second (Young 1971, 81, 87),

$$R_M = B \log_2 (1 + S/N) \qquad (3\text{-}6)$$

If the signal-to-noise ratio is 1, the maximum rate of transmission is equal to the bandwidth of the channel. Some examples of this calculation for commercial communication channels are shown in Table 3-3.

3-4 ENERGY COST OF INFORMATION

When information is obtained there is a decrease in thermodynamic entropy, because the number of different possible states of the receiver is decreased. Thus there is an inherent energy cost in obtaining information which is required by thermodynamics and applies to any organism or device (Volkenstein 1982, 99; Holzmüller 1984, 69). The entropy change for one bit is:

$$\Delta S = k_B \ln 2 \qquad (3\text{-}7)$$

which is equivalent to a free energy change of

$$\Delta G = T\Delta S = k_B \, T \ln 2 \cong 0.7 \, k_B T \qquad (3\text{-}8)$$

where k_B is Boltzmann's constant.

At physiological temperatures this is equal to about 0.6 kcal/mole, which means that the hydrolysis of one ATP molecule could provide the energy to obtain about 20 bits of information. For example, if one percent of the energy metabolism of a person were available for information processing, 10^{20} bits/s could be processed. As we shall see, this number is far in excess of the actual information capacities of organisms, and the energy required by thermodynamics for obtaining information is only a trivial fraction of the energy involved in metabolism. However, the actual costs of constructing and operating practical information processing systems are much higher.

3-5 INFORMATION CONTENT OF ORGANISMS

In order to get a better feeling for the meaning of information and the amount of information involved in various processes, consider the information required to specify the construction of an organism. To define the positions of all the atoms in a person, excluding those of water, would require about 10^{28} bits. If molecules are considered, the number is reduced by two orders of magnitude, that is, a hundredfold (Carlow 1976, 102; Volkenstein 1982, 100). This is still probably a gross overestimate. The information capacity of the human genome, based on nucleotide sequence, is about 10^9 bits, with bacteria being about a thousandfold lower (Carlow 1976, 101). Although some information is transmitted outside the genome, it is probably small compared to what is within the genome. This discrepancy between the apparent information content of an organism and that of its genome can be explained by recognizing that the genome does not specify the positions of all the molecules in an organism. Rather it specifies a process by which new organisms are created out of previously existing ones, in combination with certain resources. An analogy is a computer algorithm (Carlow 1976, 103). A simple program to calculate the square root of 2 can generate an endless sequence of digits. The information capacity of the answer can be as large as one wants, depending only on how long the program is allowed to run. Thus, our present understanding of how organisms reproduce permits only crude estimates about genetic information transmission. We can now begin to see the practical difficulties of measuring the information content of messages.

3-6 INFORMATION-PROCESSING CAPACITIES

Another way of getting a quantitative feeling for information that is closer to sensory ecology is to consider the sensory information-processing capacities of organisms. More data are available for humans than for other organisms. Since information theory was developed specifically to analyze the methods of efficiently providing information to humans, the information content of various kinds of electronic signals is well known. Some examples were shown in Table 3-3.

Consider first the sense of hearing. What rate of information is necessary to satisfy the capacity of our sense of hearing? As shown in Table 3-3, the telephone provides a rate of about 3×10^4 bits/s. With the coding employed, it is sufficient to transmit verbal language but does not provide all the sounds we can perceive in music or in nature. The AM radio is slightly better, but it certainly does not provide all the sensations possible. The more recently introduced compact disk is designed to provide most of the audio information that humans can perceive. Its capacity of about 1 million bits/s can be considered an upper limit on the information rate necessary to satisfy our sense of hearing. Estimates of the amount of information actually extracted by our sense of hearing run much lower, at 10^3 bits/s (see Section 9-7). This illustrates that if electrical engineers could better match their signals to our senses, the signals could be compacted by orders of magnitude.

Vision is our other primary sensory modality. Television has a capacity of about 50 million bits/s, but it certainly does not provide all the information our eyes can take in.

These rates of information transmission are quite large, but do we use all the information? We certainly cannot recall all the information presented. Does this mean that communication engineers are wasteful in presenting us with more information than we can use? Only to an extent.

Table 3-4 The Information Capacity of Humans and Insects

Signal Source	Rate (bits/s)*	References
Reading (partial comprehension)	40	Raisbeck 1964, 47
Speech	30	Zimmermann 1978
Simultaneous chess (several hours)	0.2	Raisbeck 1964, 47
Net lifetime average for humans	0.001	Holzmüller 1984, 79
Insect orientation	0.01–2	Wilson 1962

* These are very rough estimates.

When presented with a large amount of information we direct our attention only to selected parts of it. However, different viewers will select different parts, so the engineer cannot eliminate any.

How much information is actually used is a difficult question. Some estimates are shown in Table 3-4. Raisbeck (1964, 47) has provided an interesting analysis of the ability of chess masters to play simultaneous games, from which he concludes that over a period of many hours the average rate of information transmission that can be retained and used is only about 0.2 bits/s.

With regard to the lifetime memory capacity of humans, Holzmüller (1984, 79) estimates that it is limited to about 1 million bits. Over a lifetime this would correspond to an average rate of storage of about 100 bits/day.

Thus, there is an enormous gap between the information capacities of our sense organs and the capacities of the central nervous system to analyze and retain information. This gap points up the importance of information selection. An organism must select information that is relevant to its needs. In fact, physiological experiments have demonstrated that in the visual system information is highly processed in the retina before it even gets to the main part of the brain.

In summary, there is a general pattern in which information is lost as time goes on after the receipt of a signal. Thus, the answer to the question about the rate at which an organism obtains information via a particular sensory modality depends on where and when it is measured—in the sense organ versus in the brain, or immediately as opposed to some time after reception. Selection also occurs at the first stage of reception. The environment potentially contains much more information than any organism can use. The major focus of sensory ecology is thus to ask precisely what information is used by various organisms.

CHAPTER 4

Stimulus
Transmission

■ ■

A thousand trills and quivering sounds
In airy circles o'er us fly,
Till, wafted by a gentle breeze,
They faint and languish by degrees,
And at a distance die.
—Joseph Addison, An Ode for St. Cecilia's Day, 1699

Information is transmitted from a source to a receiver through a channel which includes some physical agent that carries the information (the stimulus) and all features of the environment that affect the transmission. The speed and range over which transmission can occur depend on the physical properties of the stimulus and the environment. There are three basic mechanisms by which stimuli are transmitted: propagation (light and sound), diffusion (heat and chemicals), and flow (which carries heat and chemical stimuli). This chapter discusses each of these mechanisms in turn and presents the quantitative relationships describing their general properties and the environmental conditions that affect transmission. These basic principles will be applied to specific biological examples in later chapters. More specific information on particular kinds of stimuli is presented in Part 2.

In order to develop some general relationships applicable to a variety of stimuli, certain terms are used here that are not commonly applied to all stimuli. For one, the rate at which a stimulus is produced by its source is termed the **flux** (J). Flux includes the total rate of transmission in all

---

BOX 4-1 Basic Wave Relations

Definitions

t = Time (s)

x = Position along direction of travel of the wave (m)

$d(t)$ = Displacement of particles from equilibrium position (m)

$\dot{d}(t)$ = Speed of displacement (m/s)

$\ddot{d}(t)$ = Acceleration of displacement (m/s^2)

d_m = maximum displacement = amplitude of wave

λ = wavelength of wave (m)

τ = period of wave (s) = $1/f$ = reciprocal of the frequency of the wave

v = speed of propagation of wave (m/s)

Relations

$$d(t) = d_m \sin\left[2\pi\left(\frac{x}{\lambda} - \frac{t}{\tau}\right)\right]$$

$$\dot{d}(t) = -\frac{2\pi}{\tau} d_m \cos\left[2\pi\left(\frac{x}{\lambda} - \frac{t}{\tau}\right)\right]$$

$$\ddot{d}(t) = -\left(\frac{2\pi}{\tau}\right)^2 d_m \sin\left[2\pi\left(\frac{x}{\lambda} - \frac{t}{\tau}\right)\right] = -\left(\frac{2\pi}{\tau}\right)^2 d(t)$$

The *phase* of a wave refers to a particular value of $\left(\frac{x}{\lambda} - \frac{t}{\tau}\right)$, which

---

directions and is expressed in units of energy, mass, or number per unit of time. The density of a stimulus at a particular position is termed its **intensity** (*I*). The intensity of a chemical stimulus is simply its concentration. Intensity may be normalized to a volume, which contains a quantity of stimulus (as with chemical stimuli) or to a surface area through which a quantity of stimulus passes (as with sound or light). Some of the more common units are molecules/cm^3, g/m^3, and W/m^2. The distance from the source of a stimulus will often be referred to as the **range** (*r*).

corresponds to a particular displacement, a particular speed of displacement, and a particular acceleration of displacement.

The *frequency* and *period* of a wave are determined by the agent that generates the wave.

The *speed* of propagation depends on the physical properties of the medium through which the wave travels:

$$v = \sqrt{\frac{T}{\rho}}$$

where

T = Restoring force constant of medium

ρ = Mass density of the medium

The frequency and speed of propagation determine the *wavelength*, as follows:

$$\lambda = v\,\tau = v/f$$

The average *power* (Watts = J/s) transmitted by a wave is given by

$$<\!P\!> = 2\pi^2\rho v \left(\frac{d_m}{\tau}\right)^2 = 2\pi^2\rho v\, d_m^2 f^2$$

The loss of wave energy to friction has no simple theory and depends on the properties of the medium.

4-1 PROPAGATION OF LIGHT AND SOUND

The simplest mechanism of transmission is propagation, in which a stimulus (such as bird song) propels itself away from the source at constant velocity independent of intensity. This mechanism applies to light and sound, which behave like waves. The basic concepts and relations describing waves are presented in Box 4-1.

The propagation of waves is often best described by the movement of **wavefronts** that represent locations all having the same phase of the wave; at other times it is useful to describe propagation by tracing **rays** that represent the path of a small portion of a wavefront. Rays are always perpendicular to the wavefront at the same position.

Initially it is assumed that the environment does not significantly absorb or deflect the stimulus. In this situation, the intensity of the stimulus is attenuated with distance solely according to how the area of the wavefront increases with distance, since the total energy in a wavefront remains constant as the wave travels.

First consider a **plane source**, which is, in theory, a uniform plane of infinite extent that produces the stimulus. In practice, a plane source is a good approximation of a real source if the real source extends laterally in all directions for distances that are large compared to the distance to the receiver. A good example might be plankton in clear water responding to sunlight coming through a sheet of ice. The ice scatters the light in all directions and becomes the effective source. In this situation the intensity (I) does not diminish with distance from the source, and the flux ($\mathcal{J}$) is spread over the area of the plane, independently of the distance. Thus

$$I = \frac{\mathcal{J}}{A}$$

(4-1)

where A is the area of the source. In this situation there is a net movement of the stimulus in only one dimension, which is perpendicular to the source.

Next consider a **line source.** In this case the source is theoretically a line of infinite length, but in practice it need be long only as compared to the distance to the receiver. Here intensity decreases with the distance from the source because the flux is spread out over the surface of a concentric cylinder of increasing radius as the distance increases. If the source has length L and the receiver is distance r away, the area of the cylinder is $2\pi r L$ and $\mathcal{J} = I\, 2\pi r L$, which can be rearranged to

$$I(r) = \frac{1}{2\pi} \frac{\mathcal{J}}{L} \frac{1}{r} \qquad r \ll L$$

(4-2)

Thus, the intensity is proportional to the flux per unit length of the source and inversely proportional to the distance to the receiver. In this situation there is a net movement of the stimulus in two dimensions, both being perpendicular to the source.

The remaining fundamental source geometry is the **point source.** In this situation, the stimulus moves away from the source equally in all three dimensions. The practical limitation is that the extent of the source in all directions must be small compared to the distance to the receiver, as happens with, for instance, a firefly or a singing bird. Here intensity diminishes with distance because the flux is spread out over concentric spheres. The surface area of a sphere is $4\pi r^2$; therefore, $\mathcal{J} = 4\pi r^2 I$, which can be rearranged as

$$I(r) = \mathcal{J}\frac{1}{4\pi}\frac{1}{r^2}$$

(4-3)

In this case, the intensity is proportional to the flux and inversely proportional to the *square* of the distance to the receiver. The shape of the source is not important, as long as the dimensions are all sufficiently small.

These three source geometries are each of major importance. They differ fundamentally from one another in that they involve a net flow of the stimulus in one, two, or three dimensions. Most transmission situations can be modeled by one of them. Each might be relevant to a different range of distances between the source and the receiver. For example, a root fiber with a diameter of 100 μm and a length of 10 cm would look like a plane source to a bacterium 1 μm from the root's surface, like a line source to a worm 1 cm away, and like a point source to a mole 1 meter away.

Similar relationships hold if the stimulus is confined so as to spread in fewer than three dimensions. For example, a stimulus confined to a pipe of a constant diameter can have net propagation in only one dimension, along the pipe's axis, like the case of a plane source in three dimensions, and, as in that case, intensity does not change with distance. Correspondingly, a stimulus confined between two planes can have net propagation in only two dimensions, like the case of a line source transmitting into three dimensions. Again, intensity falls off as the reciprocal of the range from the source.

Up to this point we assumed the **speed of transmission** to have been uniform. However, the speed of some stimuli varies from one environment to another. The speed of sound is equal to the square root of the ratio of the bulk modulus (a measure of elasticity) to the density of the medium. These factors vary greatly between materials and can vary significantly for a given material at different temperatures (Table 4-1). Similar relationships hold for light, where speed is proportional to the square root of the dielectric constant of the medium at the appropriate frequency. (Instead of describing the speed of light in different materials, it is usual to use the

Table 4-1 Velocity of Propagation of Light and Sound
in Air and Water.

Medium*	Temp (°C)	Velocity of Propagation (m/s)	
		Sound	Light (λ = 589 nm)
Vacuum	—	—	$2.99792458 \times 10^{8\dagger}$
Air, dry	0	$331.4^{\dagger}$	$2.9970484 \times 10^{8\ddagger}$
Air dry	20	$343.6^{\dagger}$	$2.9971083 \times 10^{8\ddagger}$
Air, 50% RH	20	$344.1^{\dagger}$	$2.9971087 \times 10^{8\ddagger}$
Water, pure	15	$1,521^{\dagger}$	$2.2477 \times 10^{8\S}$
Water, pure	25	$1,497^{\dagger}$	$2.2492 \times 10^{8\S}$
Seawater	15	$1,555^{\dagger}$	?
Seawater	25	$1,531^{\dagger}$	?

* Atmospheric pressure (760 mm mercury).

‡ Calculated from Levi 1980, 83ff.

† From [Weast, 1985]

§ Calculated from the index of refraction (Weast, 1985, E369).

refractive index, which is the ratio of its speed in the given medium to the speed in a vacuum.) In air, increases in density cause proportional decreases in the speed of light.

The alteration of transmission speed alone would not have much effect on organisms, since it is usually high enough to be effectively instantaneous. However, the wavelike nature of the propagation of light and sound generally leads to the deflection of these stimuli when they pass through environments in which the speed of propagation varies.

When a stimulus crosses a boundary between environments in which the speed of propagation differs, part of it is generally *reflected* back into the first environment, as when light is reflected from the surface of water. The direction of propagation of the reflected stimulus is in the plane containing the original stimulus and perpendicular to the boundary, such that its angle with the boundary is equal to the angle of the incident stimulus with the boundary. In other words, considering the stimulus as a vector, the component of the original stimulus parallel to the boundary is unaffected, but the component perpendicular to the boundary is reversed.

The proportion of the incident energy reflected is governed by the **impedances** (Z_1, Z_2) of the two environments. If the impedances match ($Z_1 = Z_2$), no reflection occurs. For example, light is reflected from water because the impedances of air and water are very different; wet paper is more transparent than dry, because water matches the impedance of the paper fibers better than does air. If there is a mismatch with the wave at normal incidence on the boundary, the intensity of the reflected wave, I_r, is related to the incident wave, I_i, by

$$\frac{I_r}{I_i} = \left(\frac{Z_2 - Z_1}{Z_2 + Z_1}\right)^2 \tag{4-4}$$

In the simple case of plane waves, the impedance of light in transparent materials may be replaced either by the refractive index or the speed. For sound, the acoustic impedance is the product of the speed of sound and the density of the material (Table 9-2).

If it is not absorbed by the second environment, the part of the stimulus that is not reflected continues propagating through the second environment, but its direction is generally altered. This change of direction, called **refraction,** is the mechanism that allows light to be focused by lenses. The new direction is such that, as is seen in Snell's Law,

$$\frac{\sin\theta_1}{\sin\theta_2} = \frac{v_1}{v_2} \tag{4-5}$$

where θ_i is the angle between the direction of propagation in environment i and a line perpendicular to the boundary and v_i is the speed of propagation. For small changes in angle and speed the ratio is approximately equal to the ratio of angles ($\leq 5\%$ error at $|\theta_1 - \theta_2| \leq 5°$). The beam of stimulus, or ray, is closer to the perpendicular in the environment with the slower speed of propagation.

If there is a gradient of the propagation speed, the rays *curve* toward the direction of lower speed. Figure 4-1 shows that rays may be turned around almost 180°, although the natural environment rarely has sufficient differences in the speed of propagation to bend rays through such large angles. A more common situation in nature is for rays to be confined in a layer of minimum propagation speed (Figure 4-2). Some examples are discussed in Chapter 9 on sound.

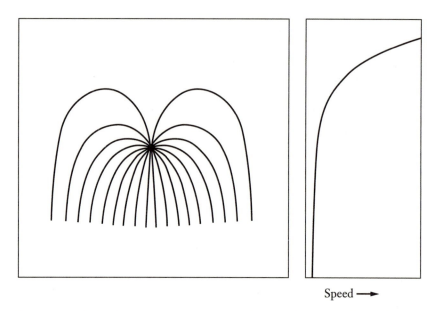

Speed ➝

Figure 4-1 Refraction of rays in an exponential gradient of speed. The paths of rays leaving a source at 20° intervals are shown for a fixed time interval. As shown on the right, the speed of propagation increases upward. Note that the rays that travel the farthest nearly catch up with those traveling the shortest distance, because the former travel faster over most of their path.

If the stimulus passes from an environment of lower to higher propagation speed and approaches the boundary at an angle close enough to parallel to the surface, it will be completely reflected. Such a condition, called **total internal reflection,** is why looking at a water surface at a shallow angle, one can see only light reflected from above, and nothing under the surface.

When a stimulus with wavelike properties (sound or light) passes an edge between materials with dissimilar transmission properties so that part of the wave passes through one material and the other part through the second, a fraction of the stimulus is diverted into new directions. This phenomenon is called **diffraction.** The proportion diverted into various directions has a complicated dependence on the geometry and size of the edge relative to the wavelength of the stimulus. Generally, diffraction's effects are inconsequential when the wavelength is very small compared to the sizes of the objects with different transmission properties. The

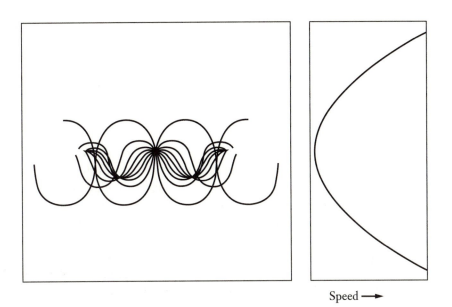

Speed ⟶

Figure 4-2 Refraction of rays in a parabolic gradient of speed. As shown on the right, the source is near a minimum of its propagation velocity. The paths of rays leaving a source at 20° intervals are shown for a fixed time interval. Note that the rays are confined to a range of distances around the position of the velocity's minimum. The rays traveling at steep angles to the horizontal travel farther, but these also travel faster horizontally, because they spend more time at positions of high propagation velocity. There are positions with high intensity (many rays close together) and low intensity (rays far apart) distributed around the source in a complicated pattern.

wavelength of visible light is about half a micrometer, that of audible sound is on the order of a meter. Consequently, how light is diffrated is important only within sense organs, but how sound is diffracted is often important in the process of its transmission to organisms. This is the main reason barriers often do not block sound thoroughly, and for example why you can hear through an ear turned completely away from a sound source.

Propagating stimuli are also subject to **scattering.** Part of their energy is diverted into more or less random directions and is lost from the initial beam. Scattering can be caused by reflection, refraction, and diffraction in a heterogeneous environment. For instance, air bubbles in water scatter light to create whitewater, and water drops in air scatter light to create fog and clouds. Even without heterogeneities in the environment, however, light is scattered by interaction with the electrons in molecules. At wavelengths much longer than those that are absorbed (as is the case for visible light in air or water), the scattering power is proportional to the density of matter and inversely proportional to the fourth power of the wavelength. Thus, blue light (short wavelength) is scattered more strongly than red (long wavelength). This process is what makes the sky blue and sunsets red.

Reflecting surfaces are often described with reference to one of two extreme situations. The term **diffuse reflection** indicates that a surface scatters light in many directions. **Specular reflection** indicates that light is reflected in a regular manner from a smooth surface as in a mirror.

The remaining effect the environment can have on propagating stimuli is to cause **absorption,** a process in which the energy of a stimulus is converted to other forms, usually heat. A given thickness of material will absorb a specific fraction of the stimulus passing through it. This situation leads to an exponential decay of intensity in a parallel beam:

$$I(r) = I(0)e^{-\alpha r} \tag{4-6}$$

where $I(0)$ is the intensity of the beam at the point from which the distance r is measured and α is the absorption coefficient. The latter is the reciprocal of the distance in which the environment absorbs all but $1/e$ ($\cong 37\%$) of the stimulus.

Attenuation is the reduction of intensity from all causes, whether by absorption or scattering. In a homogeneous environment, attenuation leads to a similar exponential decay of intensity. The symbol α is also commonly used to represent an attenuation coefficient.

It is easier to think about absorption or attenuation in terms of the distance over which a significant attenuation occurs. We can define an **attenuation length** (L_α) as $1/\alpha$, the distance over which intensity decreases by $(1 - 1/e) \cong 63\%$. In this case the attenuation law becomes

$$\frac{I(r)}{I(0)} = e^{-\alpha r} = e^{-r/L_\alpha} \tag{4-7}$$

Alternatively, for quick mental calculations, the half-attenuation length ($L_{0.5A}$) is useful. It is defined by

$$\frac{I(r)}{I(0)} = 2^{-r/L_{0.5A}} \tag{4-8}$$

The relation between the two attenuation lengths is

$$L_{0.5A} = L_\alpha \ln(2)/\alpha \tag{4-9}$$

An important characteristic of the exponential decay law for attenuation is that attenuation has little effect over distances that are small compared to the attenuation length but leads to extremely rapid loss of intensity over distances much larger than the attenuation length. For example, increasing a distance tenfold decreases the intensity by a hundredfold, through geometrical spreading from a point source. If these distances are, for example, 0.01 and 0.1 of the attenuation length, intensity will decrease by only 10 percent (0.990 to 0.905), but if the distances are 1.0 to 10 times the attenuation length, intensity will decrease by a factor of over 8,000 (0.368 to 4.54×10^{-5}). Consequently, the decrease in intensity with distance is dominated by geometrical spreading at distances that are small compared to the attenuation length, but it is dominated by attenuation at larger distances. This effect can be seen later in Figures 9-17 and 12-6 and Table 13-1.

The value of α generally depends on the wavelength or frequency of the stimulus as well as on the nature of the environment. As seen in Table 4-2, it can vary over an enormous range, from micrometers to megameters.

The various types of propagation are illustrated in Figure 4-3, by tracing a set of typical rays emanating from a source into two dimensions.

4-2   DIFFUSION OF CHEMICALS AND TEMPERATURE

Diffusion is a transmission mechanism that differs from propagation in that the rate of transmission is dependent on the intensity distribution. Specifically, the rate of transmission is proportional to the gradient (the rate of change with change in position) of intensity. Thus

$$\mathcal{J}_r = -AD \frac{dI}{dr} \tag{4-10}$$

where D is the diffusivity, $\mathcal{J}_r$ is the flux in the direction of r and through the area A, which is perpendicular to the direction r; and the minus sign indicates that the net flux is directed down the gradient, dI/dr, which is assumed to be uniform over A. This relationship derives from the statistics

Table 4-2 Attenuation Length of Light and Sound in Air and Water.

Stimulus	Attenuation Length $(1/\alpha)$, m		
	Air*	Pure Water	Mud
Light			
300 nm	$>10^3$	1.5†	$<10^{-3}$
400 nm	$>10^4$	17†	$<10^{-3}$
500 nm	$>10^4$	40†	$<10^{-3}$
700 nm	$>10^4$	1.7†	$<10^{-3}$
1,000 nm	$>10^3$	0.03†	$<10^{-3}$
Sound			
20 Hz	$2 \times 10^{5\ddagger}$	$5 \times 10^{10\S}$	$2 \times 10^{4\,//}$
1,000 Hz	$1 \times 10^{3\ddagger}$	$2 \times 10^{7\S}$	$5 \times 10^{3\,//}$
50,000 Hz	$3^\ddagger$	$3 \times 10^{3\S}$	$1 \times 10^{3\,//}$

* Atmospheric pressure (760 mm mercury) at 50% RH and 20 C.

† Levi 1980, 980.

‡ *CRC Handbook of Chemistry and Physics*, 66 ed.

§ From equation 9-7

//Sündermann 1986, 340.

of the random movement of molecules (see Berg 1983 for a discussion of this phenomenon). It applies to the **diffusion of chemicals** through a stationary medium, in which case D is the diffusion constant for a specific chemical in a given medium and I is the concentration (strictly speaking the activity) of the chemical. This relationship also applies to the **conduction of heat** through a stationary medium, in which case D is the thermal diffusivity (that is, the thermal conductivity divided by the product of the density and the specific heat capacity) of the medium and I is the temperature of the medium. Typical values for the diffusivities of air and water are given in Table 4-3.

The diffusion equation can be solved to give an analytical solution for certain simple geometries, as given for many by Carslaw and Jaeger (1959).

An **instantaneous source** is assumed to release a quantity, Q, of a stimulus at time zero, as for instance the release of alarm pheromone by an ant in still air. The stimulus then spreads through the environment according to the relationship just supplied as well as various geometric con-

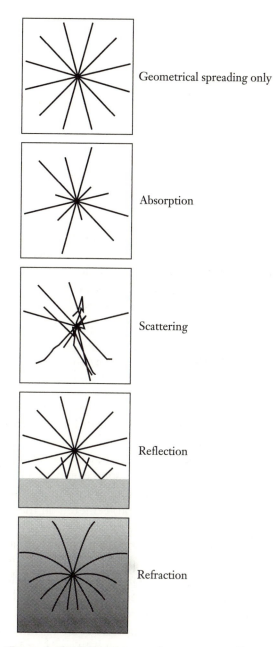

Figure 4-3 Ray paths for different types of propagation. All the paths shown are for the same time interval. The absorption and scattering lengths are equal to the path lengths of the unaltered rays. At any position the density of the rays is indicative of the intensity.

Table 4-3 Diffusivities for Heat and Chemicals in Air and Water.

	Diffusivity (m²/s)	
Medium	Thermal	Chemical*
Air	2.2×10^{-5}	2×10^{-5}
Water	1.4×10^{-7}	2×10^{-9}

Source: Derived from Campbell 1977, 149.
* Values typical of small molecules. For large molecules, D decreases significantly with increasing molecular size.

straints. For a line or plane source Q is the quantity released per unit length or area, respectively. If the ground or some other surface reflects the stimulus, a source on the surface can be modeled by setting Q as equal to twice the actual emission (Sutton 1953, 139; Bossert and Wilson 1963). The general solution for an instantaneous source is

$$I(r,t) = \frac{Q}{(4\pi Dt)^{d/2}} \; \exp\left(\frac{-r^2}{4Dt}\right)$$

(4-11)

where r is the range from the source. The parameter d is the dimensionality of the stimulus flow. Thus, $d = 1$ for diffusion occurring in only one dimension, as in a pipe, or in three dimensions away from a plane source; $d = 2$ for diffusion occurring in two dimensions away from a point, or into three dimensions away from a line source; and $d = 3$ for diffusion in three dimensions from a point source. The profile at any time is a Gaussian curve (meaning a normal distribution) centered on the starting position, with a standard deviation of $(2 D t)^{1/2}$.

For a point source, the intensity at r has a maximum value at a time of $t = r^2/6D$, and the root mean square distance of the stimulus from the source is $r_{rms} = (6Dt)^{1/2}$. In the case of a plane source, half of the stimulus is assumed to diffuse away from each of the two sides from each unit area. If stimulus is released from only one side, as from a solid surface, the situation can be modeled by assuming that twice as much is released altogether. (These relationships can be derived from Carslaw and Jaeger [1959, Section 10.2] or Crank [1975].) The intensity at various times and distances is shown in Figures 4-4 and 4-5.

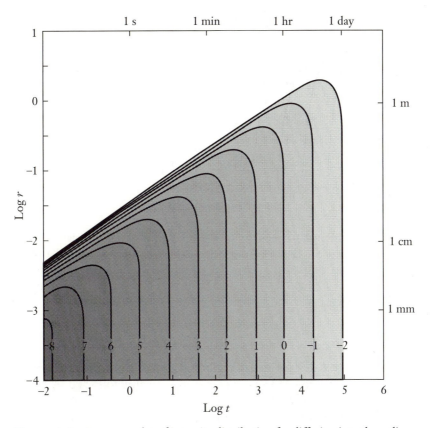

Figure 4-4 A contour plot of intensity distribution for diffusion into three dimensions from a point source pulse. Here r is the number of meters in the distance to the source, and t is the number of seconds in the time since the pulse. The contour values (–2 to 8) are equal to log I/Q in units of seconds and meters. The diffusivity was assumed to be 2×10^{-5} m²/s, which is appropriate for thermal or chemical diffusion in air. Other diffusivities can be estimated by adjusting the time in proportion to the difference in diffusivities.

This plot makes clear the asymmetry in diffusion from a pulse. For instance, at a range of 1 cm (a horizontal cross section of the plot), the intensity increases a million-fold (from 10^{-2} to 10^4) during the period $10^{-1.2}$ (=0.063) to $10^{-0.6}$ (=0.25) seconds after the pulse, but it decays over this range during the period of ten seconds to one day. Similarly, the spatial distribution of intensity at a particular instant in time (a vertical cross section of the plot), at say one second, increases over the same intensity range between $10^{-1.35}$ (=45 mm) to $10^{-2.5}$ (=3 mm) but is nearly constant closer to the source. Thus, in both time and space there is a rapid increase in intensity followed by a slow decline (in time) or plateau (in space).

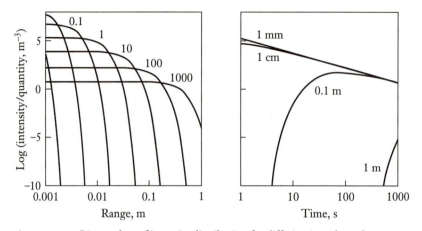

Figure 4-5 Linear plots of intensity distribution for diffusion into three dimensions from a point source pulse. These plots depict the same data as the previous plot but make clearer the variations with time and distance. *Left*: Intensity distribution as a function of distance for times of 0.001 to 1,000 seconds, in tenfold increments. *Right*: Intensity distribution as a function of time for distances of 1 mm to 1 m, in tenfold increments.

At all distances, the pattern is similar on the appropriate time scale. At a given position there is a rapid increase in intensity, followed by a slow decline. For example, 1 cm from a point source in stationary air the intensity would increase nearly a millionfold between 0.1 and 1 second after the release, but it would take several hours to decline over this range. At a given time there is an area near the source that has a nearly uniform intensity, but with a steep decline at the edge of this area. This edge moves away from the source at a rate proportional to the square root of the time. This general pattern holds for all the basic source geometries. For lower dimensionalities, relative changes in intensity occur over larger spans of both time and distance.

These plots can be used for other diffusivities, by multiplying the time by the ratio of the diffusivities. For small molecules diffusing in water ($D = 2 \times 10^{-9}$), time is slowed by a factor of 10^4. The intensity would peak at 1 cm from the source in about 10^4 s or three hours, rather than at 1 s, as for diffusion in air.

In many cases of biological interest there is some threshold intensity, I_{Th}, that provides information to an organism. Assume in addition that diffusion is occurring in three dimensions from a point (Bossert and Wilson 1963). The range of the threshold intensity is then given by

$$r_{Tb}(t) = \sqrt{4Dt \ln\left(\frac{Q}{I_{Tb}\,(4\pi Dt)^{3/2}}\right)}$$

$$\text{for } t \le (4\pi D)^{-1}\,(Q/I_{Tb})^{2/3} \tag{4-12}$$

At times greater than $(4\pi D)^{-1}\,(Q/I_{Tb})^{2/3}$ the intensity is below threshold everywhere. The maximum range is

$$r_{\max} = \left(\frac{3}{2\pi e}\right)^{1/2}\left(\frac{Q}{I_{Tb}}\right)^{1/3} \cong 0.42\left(\frac{Q}{I_{Tb}}\right)^{1/3} \tag{4-13}$$

at time

$$t_{\max} = \frac{1}{4\pi De}\left(\frac{Q}{I_{Tb}}\right)^{2/3} \cong 0.029\,\frac{1}{D}\left(\frac{Q}{I_{Tb}}\right)^{2/3} \tag{4-14}$$

where e is the base of natural logarithms. The range subsequently decreases to zero at time

$$t_{fade\ out} = \frac{1}{4\pi D}\left(\frac{Q}{I_{Tb}}\right)^{2/3} = e t_{\max} \cong 2.72 t_{\max} \tag{4-15}$$

Thus, the fadeout time is a constant 2.7 times the time to reach the maximum range.

It is important to note that in all these relations, the quantity released and the threshold intensity appear only as the ratio of the two parameters (Bossert and Wilson 1963). This ratio, Q/I_{Tb}, has dimensions of volume and may be thought of as the volume in which the quantity released would, if uniformly distributed, be at threshold intensity. The ratio might be given a name such as **transmission volume**, symbolized by V_T. The equations could then be simplified to

$$r_{\max} \cong 0.42 V_T^{1/3}$$

$$t_{\max} \cong 0.029 D^{-1} V_T^{2/3}$$

$$t_{fadeout} \cong 0.078 D^{-1} V_T^{2/3} \tag{4-16}$$

The actual volume of the active space, V_{AS}, is at its maximum

$$V_{AS} \equiv (4/3)\pi r_{max}^3 \cong 0.31 V_T \qquad (4\text{-}17)$$

Thus, the actual volume of the active space is not very different from the transmission volume, which can be simply calculated from the physiological parameters, Q and I_{Th}.

Note that the maximum range is independent of the diffusion rate and consequently is approximately the same in both air and water. In contrast, the time scale is inversely proportional to the diffusion constant, so that the kinetics will be slowed 10,000 times for diffusion in water.

A **continuous source** is assumed to start producing a stimulus at zero time and continue at steady rate J. A good example might be waste products released by a bacterial cell. For a continuous point source resulting in transmission in three dimensions, the distribution is

$$I(r,t) = \frac{J}{4\pi Dr} \, erfc\left(\frac{r}{(4Dt)^{\frac{1}{2}}}\right) \qquad (4\text{-}18)$$

where $erfc$, the error function complement, is the area under the normal curve out to infinity. The distribution of intensity with this type of source is shown in Figure 4-6. At sufficiently long times $erfc$ goes to 1, and the relation is

$$I(r,\infty) = \frac{J}{4\pi Dr} \qquad (4\text{-}19)$$

Intensity is inversely proportional to distance. The range at which the intensity equals a threshold intensity is

$$r_{max} = \frac{J}{4\pi DI_{Th}} \qquad (4\text{-}20)$$

In this case of continuous release with a fixed threshold, the rate of release and the threshold value enter only as their ratio, analogous to the case of instantaneous release. The equations can be simplified by representing this ratio by a new parameter for the transmission volume rate, $\dot{V}_t \equiv J/I_{Th}$. The maximum range then becomes

$$r_{max} = 0.08 \, \frac{\dot{V}_T}{D} \qquad (4\text{-}21)$$

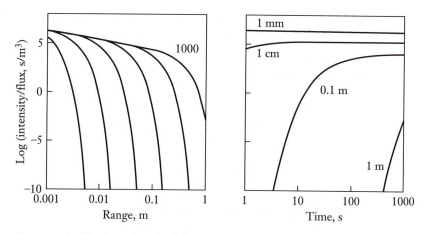

Figure 4-6 The intensity of a diffusing stimulus around a continuous point source. *Left*: Intensity distribution as a function of distance is given for times of 0.01 to 1,000 seconds in tenfold increments. *Right*: Intensity distribution as a function of time for distances of 1 mm to 1 m, in tenfold increments.

For a continuous line source (transmission in two dimensions), the general solution involves more specialized functions, which can be found in Carslaw and Jaeger (1959, Section 10.4). At long times ($t \gg r^2/4D$) the distribution becomes

$$I(r,t) \approx \frac{\mathcal{J}/L}{4\pi D}\left[\ln\left(\frac{4Dt}{r^2}\right) - 0.577\right] \tag{4-22}$$

where $\mathcal{J}/L$ is the flux per unit length and the number is Euler's constant. Alternatively, if the intensity (I_1) is known at some distance (r_1) after a steady state is established, then

$$I_2 = I_1 + \frac{\mathcal{J}/L}{2\pi D}\ln\left(\frac{r_1}{r_2}\right) \tag{4-23}$$

For a continuous plane source (transmission in one dimension), the solution is

$$I(r,t) = \frac{\mathcal{J}}{A}\left\{\sqrt{\frac{t}{\pi D}}\,\exp\left(\frac{-r^2}{4Dt}\right) - \frac{|r|}{2D}\,erfc\left[\frac{|r|}{\sqrt{4Dt}}\right]\right\} \tag{4-24}$$

where $\mathcal{J}/A$ is the flux per unit area (Carslaw and Jaeger 1959, Section 10.4).

The remaining temporal type of stimulus source to be considered is the **periodic source,** which is assumed to produce stimulus in a repeating pattern. For instance, certain moths release pheromones in this manner. A sinusoidal pattern is assumed; any other periodic pattern can be treated as a sum of the sinusoidal components. If the source produces an intensity in a *plane* that varies as $I(0,t) = I_{ave} + I_{1/2} \sin(2\pi f t)$, where f is the frequency of the periodicity, then the intensity distribution is

$$I(r,t) = I_{ave} + I_{1/2} \exp\left(-\frac{r}{r_d}\right) \sin\left(2\pi f t - \frac{r}{r_d}\right)$$
(4-25)

where $r_d = (D/\pi f)^{1/2}$ is the **damping distance** ([Campbell 1977, 16] and top panel of Figure 4-7). This distance is that over which the amplitude of the variations decays to $1/e$, or 37 percent of its starting value. Note that high frequencies damp faster than low ones. Consequently, at a sufficient distance from the source, any periodic pattern is well represented by a sinusoidal variation having the fundamental frequency of its periodicity. The distribution of intensity is shown in the middle panel of Figure 4-7. The overall pattern can be interpreted as an intensity wave of wavelength $(4\pi D/f)^{1/2}$ that propagates with velocity $(4\pi Df)^{1/2}$. However, the damping is such that over a distance of one wavelength the amplitude is attenuated by a factor of 0.002 (Carslaw and Jaeger 1959, 66). Thus, the waves do not propagate very far before damping out. This points up the severe limitations on using diffusion to transmit information rapidly. This model has been used a great deal to describe temperature distribution in the ground, thus providing important information on the history of climatic temperature variations (Lachenbruch and Marshall 1986). In addition, soil-dwelling microorganisms may use this pattern of temperature changes to locate the surface (see Chapter 6).

Periodic release from a *point source* is more complicated. If stimulus is released and absorbed continuously at a point according to $\mathcal{J}_0 \cos(2\pi f t)$, the distribution of intensity will be

$$I(r,t) = \frac{\mathcal{J}_0}{4\pi D\, r} \exp\left(-\frac{r}{r_d}\right) \sin\left(2\pi f t - \frac{r}{r_d}\right)$$
(4-26)

which is similar to that of a plane source, except for the additional damping of $1/r$ (Carslaw and Jaeger 1959, 263). Thus, the point source attenuates even faster than the plane source. A periodic line source would attenuate

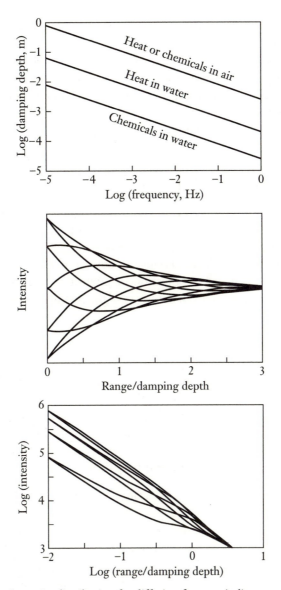

Figure 4-7 Intensity distribution for diffusion from periodic sources. *Top*: Damping depth as a function of frequency for heat or small chemicals diffusing through air or water. *Middle*: Intensity distribution from a plane source, with the full range of intensities included. The different lines represent the temperature distribution at different times. *Bottom*: Intensity distribution for a point source diffusing into three dimensions. The distribution is dominated by spherical spreading, and the intensity gradient never reverses slope.

at a rate intermediate between that of a periodic point source and a plane source. It in fact involves special functions.

However, the alternate release and absorption of a stimulus from a point is not very likely. A more useful model is the one for periodic release, $\mathcal{J}(t) = \mathcal{J}_{ave} (1 + \cos(2\pi f t))$, where $\mathcal{J}_{ave}$ is the average rate of release. Bossert (1968) has determined the steady-state distribution for this type of point source. In the notation used here, he found that

$$I(r,T) = \frac{\mathcal{J}_{ave}}{4\pi r D}\left[1-exp\left(-\frac{r}{r_d}\right)\cos\left(2\pi f t - \frac{r}{r_d}\right)\right] \tag{4-27}$$

Bossert's result has been divided by 2 to make it appropriate for an isolated source. This distribution is shown at the bottom of Figure 4-7. The intensity has a sinusoidal component but is dominated by the $1/r$ decay with distance and the gradient never reverses. Following Bossert, if an organism is on the threshold of detection of a modulation of intensity when $I(t) = I_{ave} + I_{1/2} \cos(2\pi f t)$, the maximum range of detection of the modulation is

$$r_{max} = r_d \ln\frac{I_{ave}}{I_{1/2}} \tag{4-28}$$

If $I_{1/2}/I_{ave} = 0.05$, the maximum range is only three damping distances, or about the characteristic diffusion length (see following section). As seen in Table 4-4, this range is at most a few centimeters for periods up to a minute.

It can be seen from the equations above that the dimensionless ratio $r^2/4Dt$, or its square root, appears frequently. It is consequently useful to define a **characteristic diffusion length,** or

$$r_D \equiv (4Dt)^{1/2} \tag{4-29}$$

which can be interpreted as a rough measure of the distance over which the intensity changes in time t. For example, it is equal to $\sqrt{2}$ (= 1.41) times the standard deviation of the distribution of intensity from an instantaneous source. Alternatively, a characteristic diffusion time can be defined as

$$t_D \equiv \frac{r^2}{4D} \tag{4-30}$$

Typical values for these parameters are presented in Table 4-4. It is clear from these values that diffusion is a relatively slow process, one not suitable for transmission over long distances.

Table 4-4 Characteristic Diffusion Lengths.

| | | Characteristic Diffusion Length, r_D (m) | |
| | | Water | |
Time Interval	Air	Heat	Chemical
1 second	0.01	0.0007	0.0001
1 minute	0.07	0.006	0.0007
1 hour	0.5	0.04	0.005
1 day	2.6	0.1	0.03
1 year	50	4.2	0.5

If the time interval is the period of a sinusoidal source, the damping depth is less than r_D by a factor of $2\sqrt{\pi} = 3.545$.

4-3 FLOW OF AIR OR WATER

As seen above, the transmission of stimuli by diffusion is a slow process with a very limited information-carrying capacity. If it were the only transmission mechanism the use of chemical stimuli would be restricted to short ranges. However, in addition to spreading by diffusion, chemical stimuli are often carried by flows of air or water.

In discussing this mechanism of transmission we will use the set of Cartesian coordinates x, y, and z. The origin will be placed at the source, and oriented with x increasing downstream and z perpendicular to the surface, if any. The velocity of the flow will be v. The term *fluid* can apply to either air or water.

UNIFORM FLOW: PLUMES AND TRAILS

Uniform flow refers to the condition in which the velocity, v, of the medium with respect to the source is the same everywhere within the region of interest, which is to say there is no turbulence or nearby surface restricting flow. Either the source or the medium can be considered stationary. Thus, the analysis applies to both a stationary source creating a **plume** in a uniform wind or current and to a source moving through a stationary medium leaving a **trail** of stimulus.

As is discussed in the next section, turbulence usually disrupts uniform flow in the open atmosphere or ocean, for spaces > 1 cm. Consequently, the equations developed in this section apply only to natural situations such as the flow of air or water through soil or through distances of less than about 1 centimeter. In spite of this limitation, these relations provide necessary starting points for thinking about what happens in turbulent flow, which cannot be predicted.

In the case of an instantaneous release of a stimulus, it will diffuse from its center as previously predicted for dispersion in three dimensions, but the center will move with the flow, away from the source. Explicit expressions of this phenomenon are given by Sutton (1953, 134ff) for point, line, and plane sources. If the stimulus is released continuously at constant rate J from a point, the distribution is

$$I(r,\theta) = \frac{J}{4\pi r D} \exp\left[\frac{-(1 - \cos\theta)r\,v}{2D}\right]$$

(4-31)

where r is the straight-line distance from the source to the position of interest and θ is the angle between this line and the downstream direction (the x axis). Note that $(1-\cos\theta)r = r-x$, and the latter expression is used in some forms of this equation. A plot of the shape of the resulting plume or trail is shown in Figure 4-8. Along the downstream direction ($\theta = 0$), the exponential term becomes 1; the intensity falls as $1/r$ and, surprisingly, is independent of the velocity. Thus, a remarkable feature of this transmission mechanism is that the flow does not extend the range downstream but simply reduces the range in all other directions. Consequently, a uniform flow *reduces* the active space of a continuously emitting source, although it might greatly speed up its transmission downstream.

The intensity declines to a level I_{Th} at a distance of $r_{max} = J/4\pi D I_{Th} = 0.08\,\dot{V}_T/D$ downstream from the source. How rapid a movement is necessary to distort the sphere of pure diffusion to an elongated plume? The magnitude of the elongation is fundamentally determined by the ratio of the rate of spread caused by diffusion to the rate of spread due to flow. The rate of spread via diffusion is inversely proportional to the distance from the source. Thus there is always a region around a small source in which diffusion dominates flow. In the plume the range perpendicular to the direction of flow is reduced by half when $v \cong 1.4\,D/r$. In the atmosphere this equals typical wind speeds only at distances in the range of 10 to 100 μm. In water the range dominated by molecular diffusion would be even smaller. For a flow fast enough to form an elongated plume, the intensity

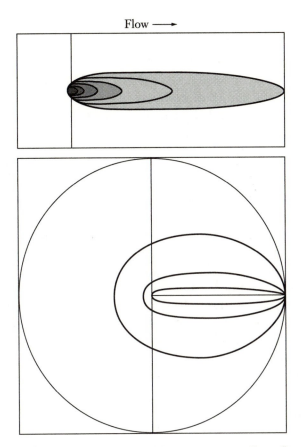

Figure 4-8 The active space for a constant source in a uniform flow. *Top*: Contours of equal intensity, with the contours differing by a factor of two. *Bottom*: A single intensity contour for different flow speeds. In each case, the source is on the thin vertical line The shapes depend only on the dimensionless ratio vr_{max}/D, where $r_{max} = \mathcal{J}/4\pi DI_{Tb}$. In the upper panel the contours are plotted for ratios of 6.25 to 200. In the lower panel, contours are plotted for flow speeds of 0 (which is circular), 5, 50, 5,000, and > 10,000 (the line downstream from the source).

If r_{max} were 100 m and the diffusivity was appropriate for air (2×10^{-5} m^2/s), the wind speeds depicted in the lower panel would be 0, 1×10^{-6}, 1×10^{-5}, and 1×10^{-3} m/s. Natural wind speeds of > 0.1 m/s would be contained within the horizontal line to the right of the source. More realistically, if r_{max} were 1 cm, a range over which wind might be uniform, the contours depicted would correspond to winds of 0, 0.01, 0.1, and 10 m/s, which includes the range of natural wind speeds.

is above the threshold out from the plume axis a maximum distance of approximately $(\mathcal{J}/8.54vI_{Th})^{1/2} = 0.34\dot{V}_T/v)^{1/2}$, which occurs a distance approximately $\mathcal{J}/34DI_{Th} = 0.029\dot{V}_T/D$ downstream of the source (Roberts 1923). Under common conditions, a plume would be remarkably thin *in the absence of turbulence*. With a wind of 1 m/s and a maximum range of detection 100 m downwind of the source, a plume would have an effective maximum width of only 10 cm.

These mathematical relations were derived by Roberts (1923; see also Sutton 1953, 136), who assumed that the results could be applied to a turbulent flow by interpreting D as a "coefficient of eddy-diffusion." It will be seen later that this is a poor assumption for problems of stimulus transport. Thus, this mathematical result should be applied only to uniform flows, and D should be interpreted as the usual molecular diffusivity. This restriction severely limits the application of these formulas so that they are best applied to the trails of small organisms (Bossert and Wilson 1963) or small sedimenting particles. Roberts (1923) and Sutton (1953, 137) also treat a line source with a solution that involves Bessel functions and will not be given here. A plane source perpendicular to a uniform flow gives rise to a uniform concentration.

As mentioned with respect to stimulus diffusion in a stationary medium, the effect of an impervious boundary such as the ground adjacent to the source is to double the effective concentration, assuming the boundary acts as a reflector. If there is a finite separation (z_0) between the source and the boundary, the effective strength of the source varies from Q at $x \ll z_0$ to $2Q$ at $x \gg z_0$. An exact solution is readily obtained (Sutton 1953, 139) by including an image source (Figure 12-3). If the boundary absorbs or destroys the stimulus, the effective release rate will be reduced.

The distribution applicable to a periodic source has been derived by Bossert (1968). As with the calculations of Roberts discussed above, his intention was to apply the results to turbulent flow, using an appropriate value for the diffusion constant. As will be seen in the next section, his is a very poor approximation. However, the result is a rigorous description of molecular diffusion in a uniform flow. Rearranging Bossert's equation 10 and dividing by 2 to represent an isolated source, one obtains the following for a periodic source in a uniform flow:

$$I(r,\theta,t,) = I_{ave}(r,\theta) \, [1+M(r) \cos(2\pi ft-\phi(r)], \text{ where}$$

$$I_{ave}(r,\theta) = \frac{\mathcal{J}}{4\pi rD} \exp[-(1-\cos\theta)\frac{vr}{2D}], \text{ time average;}$$

$$M(r) = \exp[(1-F\cos\beta)vr/2wD], \text{ degree of modulation;}$$

$$\phi(r) = (v\,r/2wD)F\sin\beta, \text{ phase shift;}$$

$$\beta = (1/2)\arctan\frac{8\pi f D}{v^2}$$

$$F = \left[1 + \left(\frac{8\pi f D}{v^2}\right)^2\right]^{\frac{1}{4}} \tag{4-32}$$

If the velocity goes to 0, this expression becomes equivalent to that previously given for a periodic source without flow. If the frequency goes to infinity, the expression takes on the form previously given for a constant source in a uniform flow. The relative modulation depends only on the distance from the source, not on the distance from the axis of the plume. This distribution is shown in Figure 4-9.

Flow ⟶

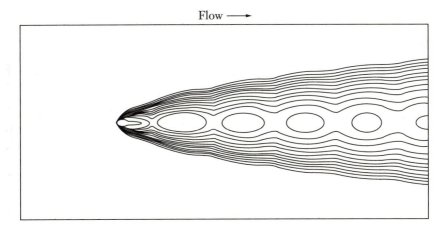

Figure 4-9 The active space for a periodic source in a uniform flow. Contours of equal intensity are drawn for intensities differing tenfold at a particular instant in time. Over time the maxima drift downstream, to the right.

TURBULENT FLOW

Turbulent flow differs from the other types of stimulus transmission in being fundamentally unpredictable. The most we can do is discuss its limits and make some necessarily vague generalizations. Turbulence is created when a flow is hindered in one location relative to another one. This hindrance generates torque on the fluid, which then tends to generate vortices, eddies, waves, and the like. Another source of turbulence is temperature differences in the fluid which causes density differences in it that in turn give rise to differences in buoyancy, thus creating torques around horizontal axes. Such differences are the ultimate cause of almost all natural currents. If viscous forces are sufficiently strong compared to inertial forces in the fluid, the energy of rotation becomes converted quickly into heat and the flow remains laminar. Otherwise, vortices and related disturbances develop, producing turbulence.

This balance of forces is measured by the **Reynolds number,** defined as

$$R_e \equiv \frac{vL}{v} \qquad (4\text{-}33)$$

where v is the velocity of the fluid, L is some length characterizing the flow, and v is the kinematic viscosity (1.5×10^{-5} m^2/s for air and 1.0×10^{-6} m/s for water). The value of R_e can be interpreted either as the ratio of inertial forces to viscous forces or as the ratio of the time required for transport via turbulence to the time for transport via diffusion (Tennekes and Lumley 1972, 10). If the Reynolds number is above 10,000, turbulent flow develops. If it is below 2,000, the flow is laminar. At intermediate values, turbulence may or may not develop. These limits are depicted graphically for air and water in Figure 4-10.

In fully developed turbulent flow, the vortices that are initially generated produce smaller vortices, which in turn produce still-smaller vortices. The energy contained in the larger vortices is thus continually transferred to the smaller ones. This concept is portrayed nicely in the verse by F. A. Gifford (Campbell 1977):

> Great whirls have little whirls
> That feed on their velocity;
> And little whirls have lesser whirls,
> And so on to viscosity.

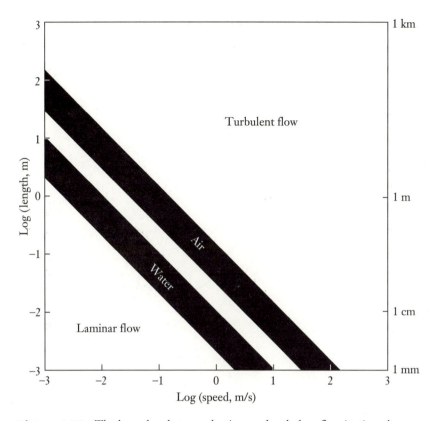

Figure 4-10 The boundary between laminar and turbulent flow in air and water. The boundaries are defined by Reynolds numbers between 2,000 and 10,000 in the black regions, in which turbulence may or may not develop.

In any range of vortex size where energy is not being added or lost, an inertial subrange is said to exist. In this range viscosity has little effect and the density of vortices of different relative sizes can be predicted (Tolmazin 1985, 74). Vortices' sizes are usually measured in terms of their wave number, k, where eddy size $L = 2\pi/k$. The power spectral density is then $S(k) \approx 1.4\ \varepsilon^{2/3}\ k^{-5/3}$ where ε is the rate at which kinetic energy is passed from one size vortex to another. In simple cases this rate can be estimated as v^3/L where L is the size of the largest eddies, or the width of flow (Tennekes and Lumley 1972, 20). This relation can be seen in the energy spectrum for the ocean given in Figure 4-11. In vortices below a certain size, the viscous forces domi-

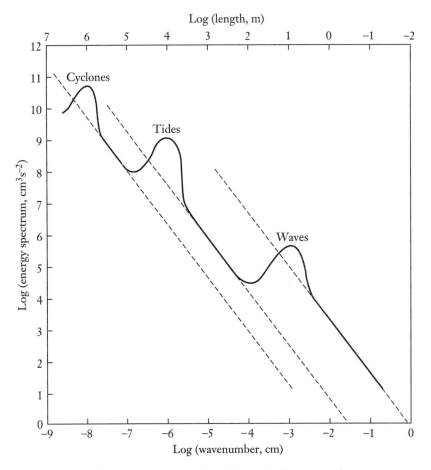

Figure 4-11   The energy spectrum of turbulence in the ocean. The energy spectrum of turbulent flows in the ocean is shown as a function of the size of the associated vortex, with the vortex size represented by wave number. The dashed lines have a slope of –5/3, as predicted by theory in size regions where external agents are not adding energy to the system. As shown, energy is added at certain size scales by waves, tides, and cyclones. (After Tolmazin 1985, Figure 4.4.)

nate and the energy is dissipated as heat. The limit where this begins to occur is given by the **Kolmogorov scale,** expressed as $L_K = (\nu^3/\varepsilon)^{1/4}$. For the ocean, L_K is approximately 1 cm (Tolmazin 1985, 74).

   The **limiting vortex size** in the atmosphere is estimated to be several millimeters or centimeters (Sutton 1953, 102). This can be seen by applying the Reynolds number criterion. A typical wind velocity taken as 1 m/s, used to determine the length scales that give a Reynolds number of 2,000 with the viscosity of air, gives a length of 3 cm. Similarly, if the typical current velocity in the ocean is 0.1 m/s, the corresponding length scale is 2 cm.

This lower limit on vortex size means that there is always a layer of flow next to any solid that is laminar rather than turbulent. This **laminar, or viscous, sublayer** (Tennekes and Lumley 1972, 160) is to be distinguished from the **turbulent boundary layer** (Sutton 1953, 76). This unstirred layer is important because the only mechanism of transport through it is molecular diffusion or self-propulsion. Diffusion through the laminar sublayer is sometimes the rate-limiting step in stimulus transport. Experiments on transport from leaves indicate that the thickness of the unstirred layer can be estimated approximately by using $\delta = 0.4 \, (L/v)^{1/2}$, where L is the linear dimension in cm of the surface in the wind direction and v is the wind velocity in cm/s (Nobel 1974, 305). For surface dimensions of centimeters and a wind velocity between 0.1 and 10 m/s, the unstirred layer varies in thickness from 1 to 0.1 mm. Diffusion through it would occur in a fraction of a second.

Another consequence of having a lower limit of vortex size is that, when a stimulus is released from a source that is effectively smaller than the smallest vortex, the plume can expand only by the process of molecular diffusion (Aylor 1976). The plume's centerline may meander because of larger vortices, but the instantaneous diameter of the plume will be restricted by diffusion, producing a filament (Figure 4-12). On this small size scale, the equations for uniform flow (see pages 71–74) apply. Once diffusion has increased the diameter of the filament to the size of the smallest vortices, turbulence will begin to tear the filament apart. This consideration needs more attention, for micrometeorologists generally use tracer sources that are larger than the smallest eddies, whereas biological sources and receivers are often smaller.

In describing turbulence in the natural environment one comes up against a property the meteorologists call **intermittency.** If you try to determine the average of some property such as wind speed or direction, you find that the result depends on the period over which you average. The problem in studying turbulent flow in nature is that there are forces creating turbulence over an enormous range of time and distance scales. In the atmosphere, wind obstructed by sand grains leads to millisecond fluctuations, bushes and rocks to fluctuations in seconds, hills and thermals in minutes and hours, daily temperature cycles in days, seasonal temperature cycles in years, climatic changes in hundreds or thousands of years.

Figure 4-12   A filament of pheromone near its source. The female moth of the woolybear caterpillar releases an aerosol of pheromone, which forms a thin filament just visible below the moth. (Courtesy of Wendell L. Roelofs and Joe Ogrodnick, New York State Agricultural Experiment Station, Geneva, N. Y.)

This complexity is illustrated by a frequency spectrum of climatic variability shown in Figure 4-13. The quixotic nature of turbulence can be readily appreciated by anyone who has tried to avoid the smoke around a campfire. There is in fact no time period over which one can average to get reproducible results; that is, there is no "stationary" average (Pasquill 1962, 17). This situation makes working with turbulent flows extremely difficult. Nonetheless, there are certain recognizable patterns, such as thunderstorms and weather fronts. Migrating birds often bet their lives on certain statistically predictable features of these patterns. Because insects do manage to find the source of an airborne pheromone, there must be some useful information in it.

Most attempts to describe the movement of chemicals in a turbulent flow have assumed there to be a continuous distribution such as that produced by an effective diffusion rate (Roberts 1923) or have assumed that the plume has a binormal cross section (Munn 1966, 119). These models have frequently been used in studies of air pollution, but using

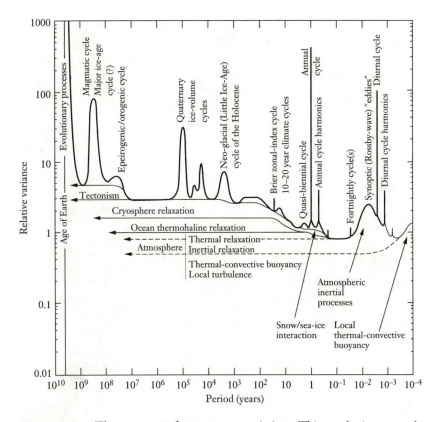

Figure 4-13 The spectrum of temperature variations. This graph gives a rough idea of some of the causes of different frequencies in the variance of weather. (From Mitchell 1976.)

them to describe the dispersal of airborne pheromones involves serious oversimplification (Jones 1983; Elkinton and Cardé 1984). The problem is that in turbulent flow the distribution of concentration in a plume is much different from the continuous distribution generated by diffusion. A good example is shown in Wright (1958) and Figure 4-14. This distinction may not be important in studies of air pollution, but in studies of behavior it is very important. If the limiting vortex size is 1 millimeter and a typical wind speed is 1 m/s, we can expect fluctuations down to milliseconds. Since

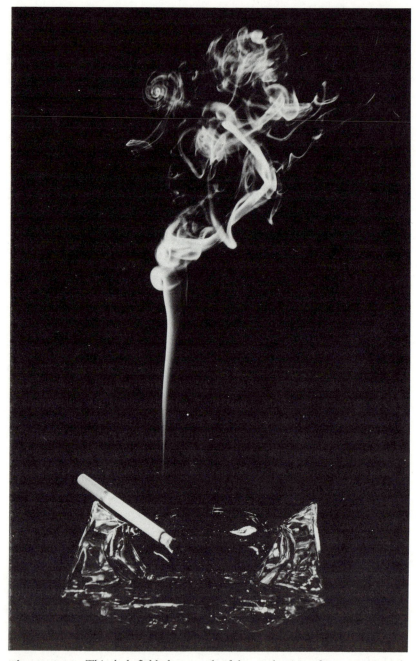

Figure 4-14 This dark-field photograph of the smoke rising from a cigarette illustrates the filamentary nature of the distribution of concentration in a plume.
(Courtesy of Vincent Mallette, Georgia Institute of Technology.)

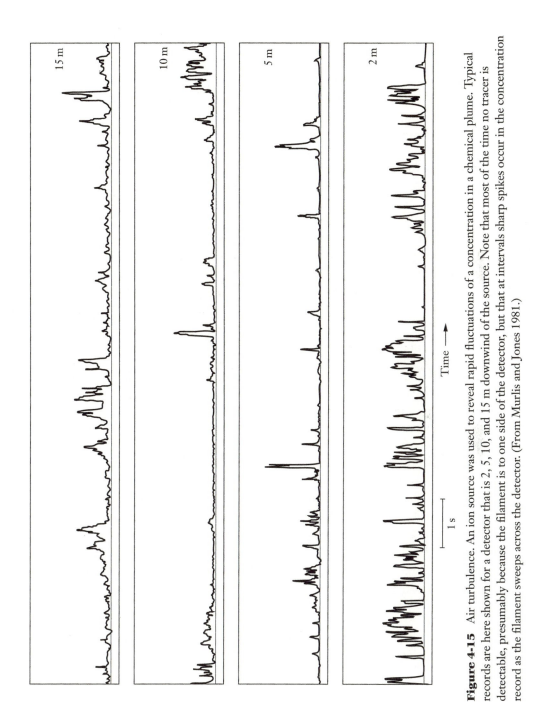

Figure 4-15 Air turbulence. An ion source was used to reveal rapid fluctuations of a concentration in a chemical plume. Typical records are here shown for a detector that is 2, 5, 10, and 15 m downwind of the source. Note that most of the time no tracer is detectable, presumably because the filament is to one side of the detector, but that at intervals sharp spikes occur in the concentration record as the filament sweeps across the detector. (From Murlis and Jones 1981.)

biological receptors can respond to changes in this range, these fluctuations are likely to be important.

That **fluctuations** on a subsecond time scale do occur has been demonstrated dramatically by Murlis and Jones (1981; Jones 1983), who used ions as tracers, because they could be detected rapidly. Under ordinary atmospheric conditions on an open heath with a wind speed of 5 m/s and neutral stability, they sampled between 2 and 15 m downwind of the source over periods of 48 min with 0.01 s time resolution. As shown in Figure 4-15, the signal at all positions was extremely spiky. Bursts of spikes lasted on average 0.3 s and returned with an average period of about 1.2 s. About 80 percent of the time no signal was present. The mean concentration declined with the distance from the source, but the maximum intensities observed were similar at all the distances sampled. The average cross-sectional diameter of the plume, which was about 0.8 m at 10 m, increased in proportion to the distance from the source. But within the plume the concentration was certainly not continuous. As shown in Figure 4-16, similar observations have been made in water (Atema 1985). This fact must be considered in studying what information an organism can gain from a plume and what the best strategies are for finding its source.

This problem is illustrated by some calculations by Bossert (1968). He argued that the channel capacity for transmitting information by diffusion of chemical stimuli in the atmosphere with no wind is limited to about 0.1 bits/s at a range of 1 m and about 100 bits/s at a range of 1 cm. His argument seems well founded, and these low rates are consistent with what was seen above in the discussion on diffusion (see Section 4-2). However, he goes on to argue—mistakenly—that a turbulent current can increase the information-carrying capacity enormously by ten-thousandfold. He—along with many others, including this author—failed to appreciate the important differences between diffusion and turbulent transport discussed earlier. Diffusion always leads to a continuous distribution in which the intensity changes only slowly. In contrast, turbulence leads to an intermingling of fluid containing no stimulus with fluids that do. In consequence, the receiving organism sees a very noisy signal. Bossert estimated that 10 m downwind from a periodic source a receiver could detect the variations at frequencies as high as about 10 Hz and 100 bits/s could be transmitted. However, in their experiments Murlis and Jones (1981) found that the signal 10 m downwind from a *steady* source consisted of pulses on the scale of seconds. Often, periods of several seconds elapsed when no signal was received, even though the source was transmitting steadily. Thus, the estimates just given of channel capacity must be overly optimistic by at least two orders of magnitude, and probably more. Indeed, it seems possible that turbulent flow might actually reduce channel capacity compared to that of pure diffusion.

These problems are fundamentally difficult, and turbulence must be the most longstanding unsolved problem in physics. Aylor (Aylor 1976; Aylor et al. 1976) has attempted to estimate the fluctuations relevant to moth sex pheromones in a forest, but this work required assumptions thought not to be valid.

Rough practical estimates of the width of a ground-level plume in open country under conditions of neutral stability and wind speeds from 2 to 10 m/s have been furnished by Pasquill (1962, 133 ff). If the concentration is averaged over several minutes and its edge is defined as that point at which the concentration is 0.1 of the peak at the same distance from the source, the width at 100 m downwind is about 40 m. Under a variety of stability conditions, the plume's width varies in proportion to the standard deviation in wind direction over the sampling period. The concentration distribution across the wind is nearly Gaussian (normal) on average, but at any given time it certainly could depart from this distribution. The width varies with distance x within the range of 50 to 1,000 m as x^p, where p = 0.6, 0.8, or 0.9, respectively, for stable, neutral, or unstable temperature distributions. Note that this description of a plume defines its edge with respect to the concentration at the centerline an equal distance downwind. This does not describe a contour having an equal concentration and thus is of little use in most biological applications.

The dramatic effect of turbulence on the size and shape of a plume can be illustrated by considering parameters typical of moth sex attractants. A typical emission rate by a female is $\mathcal{J} = 1 \times 10^{10}$ molecules/s (see pages 135–140). A typical threshold concentration is $I_{Tb} = 1 \times 10^4$ molecules/cm^3 (see Table 7-5). The ratio gives a transmission volume rate of $\dot{V}_T = 1$ m^3/s. Using relations from page 72 suggests a maximum range of detection downwind in a uniform flow of 4 km, independent of wind speed, and a maximum width of only 34 cm, assuming a wind speed of 1 m/s. In contrast, detailed sampling (Elkinton and Cardé 1984) suggests that a typical plume is more cone shaped, with its edges at least 20° from the axis and extending at least 80 m downwind. Let us model a plume as a sector of a sphere with its source at the center, its radius equal to the maximum range, and its angle determined by variability in wind direction. With a radius (range) of 100 m and a half angle of 20°, the plume's volume is 1×10^5 m^3, or about ten times what the plume would be in a uniform wind. This volume is possible because the effective plume is composed mostly of

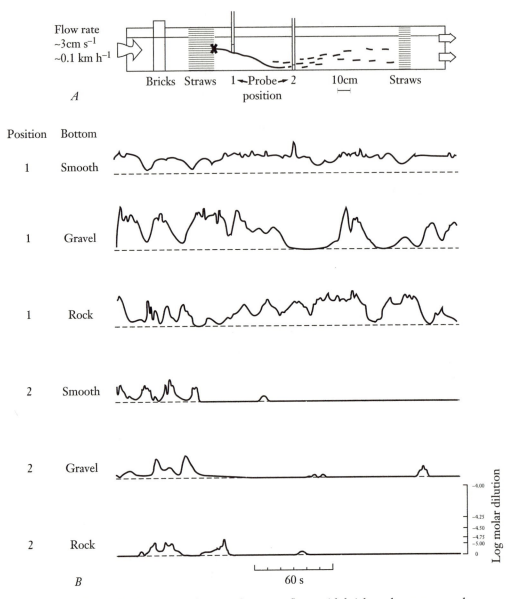

Figure 4-16   Water turbulence. (*A*) Side view of seawater flume with bricks and straws to produce laminar flow. The flume's bed was either smooth or covered with gravel or rock. Dye is released at point X and measured with a probe having approximately 1 cm³ effective volume, at positions 1 or 2. (*B*) Records obtained at the two positions over the different bottoms. The commonly observed peaks are of 1 to 3 seconds' duration and several times the background. (From Atema 1985.)

volumes of fresh air intermixed with regions of high concentration. A range of 100 m is much more realistic than one of 4 km (Wall and Perry 1987). Our ability to predict turbulent flows remains so poor that this simple sector model is probably as good as any now available.

4-4 STATIONARY STIMULI

Some stimuli do not move. Gravitational, magnetic, and electric fields spread at the speed of light, which is effectively instantaneous as far as organisms are concerned. However, the only gravitational and magnetic fields with which organisms are known to have any significant interaction are those produced by the earth. These fields are steady as far as organisms are concerned. Although they do not provide information about changes in an organism's environment as most stimuli do, they may provide valuable information about the organism's orientation and location on the earth's surface.

The force of **gravity** varies in principle with the subject's distance from the center of the earth. However, organisms move such a small distance vertically compared to the size of the earth that this variation is insignificant. Thus, the only useful information provided by gravity is the direction toward the center of the earth.

The **magnetic** field of the earth is somewhat more complex, since it is not directed radially toward the center. Rather, its lines of force emanate from one pole, travel around the earth, and go into the other pole. It is doubtful that the magnetic field's intensity variations are detectable by organisms. There is evidence, however, that some organisms like birds and bees can make use of the horizontal component of the force to determine their north–south direction. Long-distance migrators might also be able to use the inclination of the magnetic field's force with respect to gravity or the horizon to determine their latitude.

Another type of stationary stimulus is the **tactile** stimulus provided by touching a solid object. This stimulus does not move unless the object moves, and information is rarely transmitted over any appreciable distance by the movement of solid objects. More frequently, the organism moves toward an object and acquires information from contact with it. Various types of information may be obtained, such as size, shape, mass, texture, and taste. The type of information thus acquired depends on very specific details of the nature of the contact. Consequently, this type of information is very difficult to analyze, and relatively little is known about it.

CHAPTER 5

Signal
Detection

■ ■

Nothing is great or little otherwise than by comparison.
—Jonathan Swift, *Gulliver's Travels*, 1726

Given the transmission of a stimulus to an organism, the next problem is that of signal detection. Is the intensity of the stimulus sufficient? Does noise interfere with detection? What features of the stimulus are detected? What strategies are useful to the receiving organism? A large body of theory about signal detection has been developed, primarily for efficient design of electronic communications. This chapter presents some basic ideas useful to sensory ecology. Here the term *signal* will be used in the general sense of a meaningful stimulus from any source.

Signal detection can be said to occur when the output of a sensory system (the **subjective intensity**) exceeds some threshold value. The **objective intensity** impinging on the receiver is determined by the intensity of the signal transmitted by the sender and the degree to which the signal is attenuated in the environment between the sender and receiver. The subjective intensity, which controls behavior, is determined by the objective intensity, the sensitivity of the receptor, and the signal-processing systems in the receiving organism. The threshold is usually set by the receiver at an appropriate level in regard to the level of noise in the system.

5-1 INTENSITY NOISE

There are many sources of noise that affect subjective intensity and in various situations different noise sources dominate. In general, it is useful

89

to distinguish three types of noise: receptor, channel, and environmental. **Receptor noise** is present in all receptor systems as a result of the thermal motion of molecules. Man can reduce the noise in fabricated infrared detectors by cooling them, but no other organism is known to employ such a strategy. Instead, receptor noise is reduced by averaging over a sufficiently large number of molecules or other elements. Examples of receptor noise are the hiss in radio output when one is tuned between stations and the dark current of a photomultiplier tube. In humans we might imagine that visual phantoms and hallucinations arise from noise in receptors or signal-processing parts of the nervous system.

The term **channel noise** is used to describe noise caused by fluctuations in the signal. For example, photon noise results from statistical fluctuations in the number of photons impinging on a detector. Our vision in dim light is limited by this type of noise. At a distance, sound intensity is modulated randomly by fluctuations in density and velocity of the intervening medium, whether it is air or water (see Section 9-4). The ability of bacteria to detect chemical gradients is limited by random fluctuations in the rate at which molecules diffuse into and out of their vicinity (see Sections 7-3 and 17-2). In many such cases channel noise, as well as receptor noise, is nearly random and can be treated by statistical analysis.

In contrast, **environmental noise** is created by competing sources, whose output is usually not random. Some examples are the difficulty of detecting the song of a particular bird in the presence of other bird songs, or a flying moth's mistaking a streetlight for the moon, which it uses as a directional reference. Environmental noise usually limits the detectability of temperature and sound stimuli (see Chapters 6 and 9). In most cases environmental noise is usually specific to a given situation, so that no general analysis can be made.

5-2 DETECTION THEORY

Theories of signal detection have been extensively developed for use in electronic communications and radar (Selin 1965; Poor 1988). It may be noted that detection theory is closely related to statistical hypothesis testing, which is more familiar to biologists.

In general, a channel has noise even in the absence of a signal. As depicted in Figure 5-1, any of a range of intensities may be observed, with varying probability. In many cases the probability distribution is likely to be approximately normal (Gaussian). When a signal is present, the distribution is shifted and may or may not retain the same shape. The problem is to decide whether or not a signal is indeed present when a given intensity is observed. In simple cases the problem reduces to one of determining a threshold intensity upon which to base the decision. If the observed intensity is below the threshold, one concludes that a signal is not present. Higher intensities are interpreted as meaning a signal is present. If there is no overlap in the two probability distributions, an intermediate threshold intensity can be chosen as a threshold which produces completely reliable decisions and noise poses no problem. The degree of overlap is usually measured by the **signal-to-noise** (*S/N*) ratio. In Figure 5-1, the signal (*S*) is a measure of the distance between the mean intensities with and without the signal, and the noise (*N*) is a measure of the width of the probability distributions.

Detection theory addresses the question of the best strategy to adopt when the two distributions do overlap. If the probability distributions in Figure 5-1 are known, the obvious strategy is to set the threshold, I_{Tb}, at the intensity level where the two distributions intersect. By doing so, for any

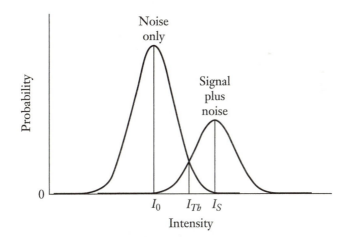

Figure 5-1 The probability of observing a particular intensity in the presence and absence of a signal. The areas under the two curves must sum to 1, since the signal is either present or not. The difference between I_S and I_0 is the strength of the signal, and the width of the distributions is a measure of the noise. The intensity at I_{Tb}, where the curves intersect, is the optimal threshold. Intensities above I_{Tb} are more likely to occur when a signal is present, intensities below it more likely when one is absent.

observed intensity the more probable case of the presence or absence of the signal is chosen, which maximizes the overall probability of reaching correct decisions. If the distributions overlap, it is not possible to be completely reliable. Reliability can be improved to any desired level by making additional observations, if possible. This strategy is analogous to that of increasing the integration time of a receptor, as described shortly. A more sophisticated strategy would be to repeat the observations only when ambiguous, intermediate intensities are observed. For example, humans often sniff several times after detecting a hint of an important smell.

Usually, a situation is not so well-defined that an optimal threshold intensity can be chosen this easily. In particular, the probability distributions may not be known. Even if there is knowledge about the noise in the channel, it is also necessary to know the prior probability that a signal is present, which would influence the relative height of the two probability distributions and hence their point of intersection. Another complication is that the cost of missing a signal may be different from the cost of falsely concluding that it is present. For example, the cost of missing an alarm call may be death, whereas the cost of falsely sensing an alarm is only wasted time and energy. In such cases it is desirable to minimize the cost rather than the number of errors.

The simplest quantitative way to incorporate all these considerations is to consider a channel that has a mean intensity of I_0 in the absence of a signal and I_S in its presence, and has noise that has a normal distribution with standard deviation I_{SD}, whether or not a signal is present. Furthermore, assume that the prior probability of a signal's being present is P_S and that the cost of an error minus any cost of a correct determination is c_0 when no signal is present and c_S when one is present. Then the threshold that minimizes overall cost is calculated as (Poor 1988, 17)

$$I_{Tb} = \frac{I_0 + I_S}{2} + \frac{I_{SD}^2}{I_S - I_0} \ln\left(\frac{(1 - P_S)\, c_0}{P_S\, c_S}\right) \tag{5-1}$$

Note that when the probabilities and costs are equal for the two situations $P_S = 0.5$ and $c_0 = c_S$, the second term goes to zero and, as anticipated, the threshold becomes the midpoint, $(I_0 + I_S)/2$, between the two distributions, which, by assumption, have an equal width. In the absence of knowledge about the prior probability of a signal's being present, the threshold that minimizes the cost with the worst possible prior probability is also the midpoint (Poor 1988, 28).

The trade-off between the two types of errors is shown in Figure 5-2. For a given signal-to-noise ratio, the receiver can choose a threshold that will give any combination of "false-alarm" and "miss" errors that falls on the appropriate curve. Both types of error can be reduced only by increasing the signal-to-noise ratio. In the absence of signal ($S/N = 0$) the only strategy available is pure guessing, but the receiver could guess the signal to be always present, always absent, half and half, or some other combination. All these possibilities lie on the diagonal in Figure 5-2. If a clear signal ($S/N \gg 1$) is available, correct decisions can always be made, and the performance will correspond to that in the upper-left-hand corner. For ambiguous signals ($S/N \approx 1$), performance will be confined to the appropriate curve in the upper left half of the figure, but the receiver will be able to operate at any point along that curve.

For a more specific example, consider a channel in which events occur randomly in time. Such events might be the arrival of photons or molecules. If I is the average rate at which such events are received over a time interval t, the number of events is $N_t = I\,t$. Then the root-mean-square (rms) deviation, or the standard deviation, in N_t is $N_t^{1/2}$, and the relative rms deviation is $N_t^{1/2}/N_t = N_t^{-1/2}$. Consequently, the relative intensity noise is expressed as

$$\frac{I_{SD}}{I} = (I\,t)^{-1/2} \tag{5-2}$$

and the spread in the distribution of intensity can be reduced to an arbitrarily small value by increasing either the integration time or the stimulus flux to increase the number of events observed. Note, however, that the noise is reduced only in proportion to the square root of the stimulus intensity or sample time. Increasing sample size is the basic reason our visual processes are relatively slow and is why nocturnal animals have large eyes (see Sections 8-1 and 8-3). As another example, Berg and Purcell (1977) concluded that the relative error in determining concentration, C_0, of a substance with diffusion constant D using a receptor that effectively sampled a volume of radius r in a time t is $\Delta C_{rms}/C_0 = 0.437\,(\text{D}\,t\,C_0\,r)^{-1/2}$, which also involves the reciprocal square root of time and intensity (C_0).

In Section 3-2, it was pointed out that the number of distinguishable intensities, N_i, in a receptor cell is an important determinant of its information capacity. This number can be related to intensity and noise in the following way (Snyder et al. 1977).

The stimulus that maximizes information transmission is random in nature. Assume it has a standard deviation of S_{SD}. Then if a receptor cell

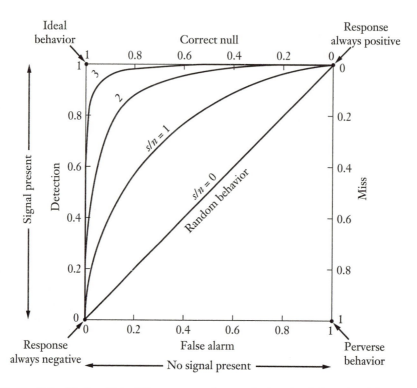

Figure 5-2 Trade-offs in different types of errors during signal detection. The axes are conditional probabilities. For example the bottom axis is the probability of a false alarm occurring, given the condition that no signal is present. For any particular signal-to-noise ratio (*S/N*) the probability of error depends on the threshold employed, which determines the bias of the detector for reporting either "signal" or "no signal." A detector that is given a particular *S/N*, which is often determined externally, can select a threshold that would produce errors corresponding to any point on the curve appropriate to that signal-to-noise ratio (0, 1, 2, and 3 are shown). The most appropriate point to select on the curve depends on the relative costs of the different types of errors. If the costs are equal, the best performance will be on the part of the curve closest to the upper-left-hand corner, which gives equal error probabilities whether or not a signal is actually present. A reduction in errors for both conditions of signal and no signal can be achieved only with an increase in the *S/N* ratio. In this model a normal distribution of noise was assumed, with the same width whether or not a signal was present, as in the previous figure; other distributions would change the shapes of the curves. The signal-to-noise ratio used here was determined by using the difference in mean intensity with and without the signal as the signal and the standard deviation of the intensity as the noise. With *S/N* = 0 the performance is 50 percent correct; with *S/N* = 1 the best possible performance is about 70 percent correct; and with *S/N* = 3 accuracy of 90 percent is possible.

has a noise level of N_{SD}, one can estimate the number of distinguishable intensity levels there will be. Assume that for them to be distinguished two intensity levels must differ by $2N_{SD}$, corresponding to approximately 95 percent reliability. The range of intensities is approximately twice the standard deviation of signal plus noise. Thus, dividing this by the required difference gives the number of distinguishable intensity levels:

$$N_i \cong \frac{2(S_{SD}^2 + N_{SD}^2)^{1/2}}{2N_{SD}} = \left(1 + \frac{S_{SD}^2}{N_{SD}^2}\right)^{1/2} \tag{5-3}$$

Thus, it can be seen that the number of distinguishable levels is closely related to the ratio of the variances of the signal and the noise.

5-3 SIGNAL PROCESSING

The previous analysis assumed that noise and signal intensities and distributions were fixed. In real life, noise intensity is likely to vary, and signal intensity almost always varies with its distance from the source. Another consideration is that information concerning changes in the presence or absence of a signal is usually more important than information concerning its continued presence at a constant level. Organisms almost universally adjust to these conditions, by using a process of sensory adaptation in which changes in intensity are enhanced and steady intensities are suppressed. An example is shown in Figure 5-3.

Similar considerations apply to variations of intensity with position as well as in relation to time. One of the first steps in visual-signal processing by the retina is to enhance differences in intensity with respect to position. This general process might be called **contrast enhancement**. The following discussion of signal processing usually refers to temporal changes, but the same considerations could just as well apply to spatial patterns.

A simple model of adaptation is the process of mathematical differentiation, in which the rate of change in intensity becomes the transformed stimulus (Foster and Smyth 1980). In practice, differentiation can be accomplished simply by taking the difference between intensities at points differing in time or space by an appropriate amount. The basic requirement is for some simple form of memory and a mechanism of comparison. The differentiation, which can be performed by a simple electrical circuit of resistors and capacitors, can certainly also be done by biological cells.

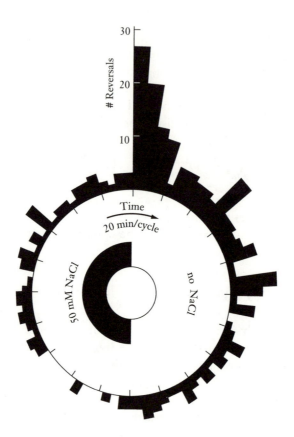

Figure 5-3 An example of sensory adaptation. Individual nematodes were exposed to two concentrations of an attractive chemical, NaCl, in an alternating cycle of 20 min, and the number of "reversal bouts," which cause a nematode to change its direction of locomotion, was recorded for each fifteen-second interval. It can be seen that immediately after the decrease in concentration there is a large increase in the probability of a reversal bout, a probability that then declines rapidly in the first minute, reaching the basal level in a few minutes as the animals adapt to the new concentration. After an increase in concentration there is a reduction in probability that lasts several minutes, until the animals adapt to this concentration. Adaptation like this is important to efficient movement along a stimulus gradient (see Section 17-2). (From Dusenbery 1980b, p. 330.)

The major question here is to identify the optimal parameters. For instance, how large a time difference should be used? If it is too large, the response will be slow; if too small, the time over which the signal can be averaged will be small and thus the two samples will be noisy. In addition, using a small time interval will mean that the difference is small. The small difference, combined with the large noise in the samples, will mean that the output will be noisy. Thus, a signal-processing system must make compromises between conflicting requirements. More-sophisticated signal-processing strategies are possible. This section introduces some of the more important concepts, with a few examples.

TIME AND FREQUENCY

In order to discuss signal processing it is convenient to introduce the concept of describing a stimulus pattern in terms of the frequencies present in it. A general mathematical result is that any function that is of practical interest, such as a pattern of intensity $I(t)$ with $0 < t < t_0$, can be described by a **Fourier series** (see, for example, Ackerman 1979, 595–99), as follows:

$$I(t) = (a_0 / 2) + a_1 \cos(\omega t) + b_1 \sin(\omega t) + a_2 \cos(2\omega t) + b_2 \sin(2\omega t) + \ldots$$

or

$$I(t) = (a_0 / 2) + \sum_{n=1}^{\infty} [a_n \cos(n\omega\, t) + b_n \sin(n\omega\, t)] \tag{5-4}$$

where

$$a_n = (2/t_0) \int_0^{t_0} I(t) \cos(n\omega\, t) dt$$

$$b_n = (2/t_0) \int_0^{t_0} I(t) \sin(n\omega\, t) dt$$

$$\omega = 2\pi/t_0$$

This relation* can be understood (Jeffrey 1979, 600–603) from the fact that the cos and sin terms represent orthogonal[†] functions (analogous to orthogonal axes) and the coefficients a_n and b_n are projections of the

---

* For readers not familiar with calculus, integrals of the form $\int_a^b F(x)\, dx$ can be interpreted as the area under a plot of the function $F(x)dx$ between the values $x=a$ and $x=b$.

[†] $\int \sin n\, t \sin m\, t\, dt = \int \sin n\, t \cos m\, t\, dt = \int \cos n\, t \cos m\, t\, dt = 0$ for $n \neq m$.

complete function $I(t)$ on each of the orthogonal functions, which are also described as the cross-correlations between the functions.

The sum of sin and cos terms of the same frequency can be replaced with a single sin term including a **phase** parameter, φ_n:

$$I(t) = (a_0 / 2) + \sum_{n=1}^{\infty} [c_n \sin (n\omega t + \varphi_n)] \tag{5-5}$$

where

$$c_n = \sqrt{a_n^2 + b_n^2}$$

and

$$\varphi_n = \arctan (a_n / b_n)$$

It is frequently useful to convert stimulus patterns back and forth between the time domain, $I(t)$, and the frequency domain, c_n's and φ_n's. The same information can be contained in either description. However, some transformations, such as that made by our sense of hearing, convert $I(t)$ to a frequency spectrum consisting of c_n's, but the phase information, φ_n's, is lost (see Section 9-2). The human sense of vision has also been described in terms of Fourier series using spatial coordinates in place of time. Thus the spatial resolution of the visual system is described in terms of its sensitivity to different spatial frequencies.

The terms in the infinite series oscillate at increasingly higher frequencies, and the more terms included in the series the better the match to the original function (see Figure 5-4). This result demonstrates that an arbitrary pattern contains components of many different frequencies; in other words, it has a wide bandwidth.

FILTERING

A basic form of signal processing is a **filter** that continuously transforms a signal from one form to another. The simplest filters, which remove high or low frequencies from the signal, are called **low-pass** or **high-pass filters**

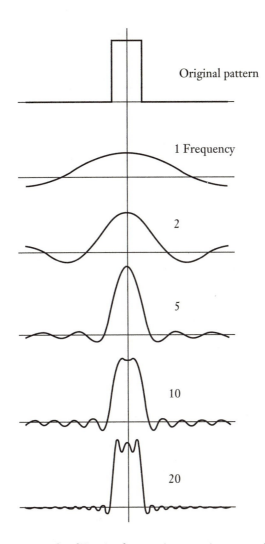

Figure 5-4 An example of Fourier frequencies summing to match a pattern. The terms in the Fourier series were weighted to sum up to the rectangular pulse shown at the top. The lower curves show the shapes produced when increasing numbers of terms are included in the summation; the first curve includes only the first term, the bottom curve the first twenty terms. The curves approximate the rectangular pulse more and more accurately as more terms are employed. This comparison indicates how infinite series can lead to an accurate representation of the original rectangular pulse and shows how the first few terms can be used as an approximation for some purposes.

respectively. All real detectors have an upper frequency limit that is determined by how fast they can respond to a change in input. Consequently, they act as low-pass filters. Low-pass filtering is desirable to the extent that it effectively increases the time interval over which signals are averaged, which can be beneficial in increasing the signal-to-noise ratio, as discussed in the previous section, but it has the disadvantage that response is slowed.

If the signal of interest is known to have components that exist in a narrow band of frequencies, a **band-pass filter** that reduces the frequency components outside the band can improve the signal-to-noise ratio. For instance, crickets have receptors that are tuned to the frequency of the song of their species (Hutchings and Lewis 1983). Specifically, the calling song of *Gryllus campestris* has a carrier frequency of 4 kHz, and this species is most sensitive to sounds at this frequency. At other frequencies, sounds must be more intense to be detected. Consequently, noise at other frequencies has less impact on detection of the calling song. In the simplest case, when noise consists of completely random changes in intensity all frequencies will be present equally. (This situation is called white noise, in analogy with the fact that the color white contains all colors represented equally.) A cricket exposed to white noise filters out all frequencies except those in a narrow band. Doing so reduces the noise intensity, improving the signal-to-noise ratio.

A simple example of a low-pass filter is a **running average**, in which intensities from a local interval are averaged together to produce the output. This interval then slides along the intensity pattern to produce the output pattern. All real detectors act this way to some extent, because of their limited frequency response. A running average can be elaborated into a band-pass filter by making the output the difference between the averages of two adjacent intervals. In other words, in this case the filter has both positive and negative lobes. (These effects can be seen below in Figures 5-6 and 5-8.)

The behavior of a signal-processing mechanism that is linear* can be completely predicted by knowing its response, $R_i(t)$, to an impulse signal occurring at $t = 0$, which consists of a pulse of sufficiently short duration that the response depends on only the product of the amplitude and duration of the pulse. The response, $R(t)$, to any stimulus pattern, $I(t)$, can be represented mathematically by the convolution integral of the stimulus and the impulse response (Hancock 1961, 126):

---

* Linear means that filtering the sum of two different inputs produces an output that is the same as the sum of the outputs of the two individually filtered inputs. Real filters are never completely linear, because they have limited response ranges. However, many real filters approximate linearity for small signals, which are usually the most important ones.

$$R(t) = \int_{-\infty}^{t} I(\tau)\, R_i\,(t - \tau)\, d\tau$$

(5-6)

This integral simply represents a running average with the possibility that it can be formed by weighting various parts of the interval differently, according to the shape of $R_i(t)$. For a normal running average $R_i(t)$ would have a constant value within a certain time interval and be zero outside that interval.

Another way of understanding this integral is to consider that the stimulus pattern $I(t)$ can be represented by a series of impulses, as shown in Figure 5-5. Since the filter is assumed to be linear, its output at any given time is simply the sum of the remaining responses to each of the previous impulses. Practical filters have impulse responses that decay to zero over some time interval t_i, and only a stimulus intensity this far into the past will influence the output at time t. Thus, the limits of integration could be taken as $t - t_i$ to t, rather than as the infinite limits given in equation 5-6.

If the stimulus is composed of the sum of a signal $S(t)$ to be detected and noise $N(t)$, the output can be represented as the sum of two integrals involving the signal and noise independently:

$$R(t) = \int_{-\infty}^{t} S(\tau)\, R_i\,(t - \tau)\, d\tau + \int_{-\infty}^{t} N(\tau)\, R_i\,(t - \tau)\, d\tau$$

(5-7)

The optimum technique for detection, a **matched filter**, is obtained when the impulse response matches the signal to be detected:

$$R_i(t) = S(t_0 - t)$$

(5-8)

where the signal occurs at time t_0. Thus, (Schwartz 1963, 149)

$$R(t) = \int_{-\infty}^{t} S(\tau)\, S(t_0 - t + \tau)\, d\tau + \int_{-\infty}^{t} N(\tau)\, S(t_0 - t + \tau)\, d\tau$$

(5-9)

The first integral of the pair can be interpreted as comparing a copy of the expected signal shifted in time to the actual signal. At time $t = t_0$ there will be a match, at which point both terms within the integral will have the same sign for all values of τ and will be large for the same values of τ. Consequently, their product will always be positive and the integral will sum to a large value. In contrast, the second integral involves the noise, which would generally not

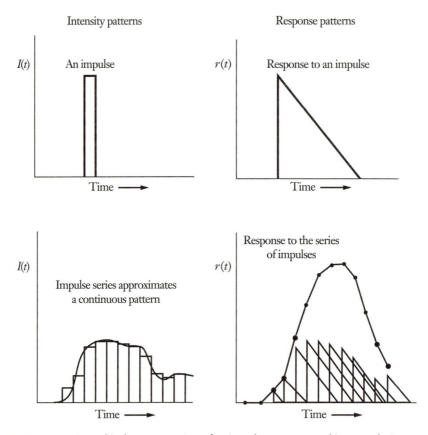

Figure 5-5   A graphical representation of an impulse response and its convolution. *Left*: Two intensity patterns. *Right*: Responses to these patterns. The top pair of graphs shows the intensity pattern of a single impulse represented by the rectangle and the response to it, in this case a triangle with instantaneous rise and decay at a fixed rate assumed. The bottom graphs show that a continuous intensity pattern can be approximated by a series of impulses of the appropriate size, and then (assuming that the receiver is linear) the response to the pattern can be accurately determined at any time simply by summing the remaining effect of each previous impulse. This is the procedure carried out by the convolution integral of the stimulus and impulse response in equation 5-6.

be correlated with the expected signal. The two terms in the product will often have opposite signs, and one will be small when the other is large. Thus, the second integral will generally sum to a small value. In this way the processed signal will be enhanced when the input stimulus contains a pattern resembling the signal; for example, see Figure 5-6.

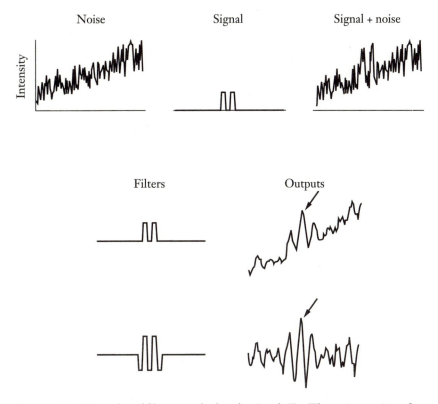

Figure 5-6 Examples of filters matched to the signal. *Top*: The noise consists of 100 random numbers ($0 \leftrightarrow 2$), representing high-frequency noise, superimposed on a drift ($0 \rightarrow 2$), representing low-frequency noise. The signal to be detected consists of two pulses ($0 \leftrightarrow 1$). The intensity received, that is, the objective intensity, is the sum of the signal and the noise. In this example the signal is marginally discernible to the eye. A filter with two positive lobes matching the signal (*middle row*) produces an output that enhances the signal but increases with the drift in intensity. A similar filter, with equal positive and negative lobes (*bottom row*) enhances the signal and eliminates the drift in intensity. The averaging within each lobe reduces the high-frequency noise, while taking the difference between adjacent intervals with equal weight removes the drift from low-frequency noise.

This strategy can be illustrated with bacterial chemotaxis (see Section 17-2), where the impulse response has been measured (Block et al. 1982). It has a positive peak lasting about 1 s, followed by a negative peak of about 3 s duration (Figure 5-7, upper right). The areas of both peaks are ap-

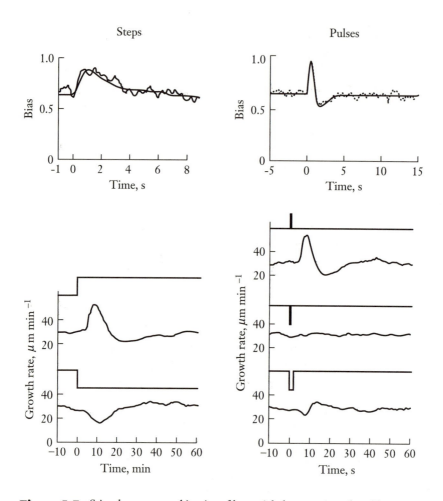

Figure 5-7 Stimulus response kinetics of bacterial chemotaxis and mold phototropism. Behavioral responses of two very different microorganisms to step increases in intensity (*left*) and to brief pulses (*right*). *Top*: The response is the bias in the direction of turning of the flagella of cells of the bacterium *Escherichia coli*, and the stimulus is the concentration of an amino acid (Segall 1986). *Bottom*: The response is the rate of growth of sporangiophores of the mold *Phycomyces blakesleeanus*, and the stimulus is light intensity (Galland 1987). Both organisms respond in a manner suggestive of the filter in Figure 5-8.

proximately equal. This response pattern indicates that the signal-processing mechanism sums up the stimulation over the past 1 s which reduces noise, and compares it by subtraction to the average stimulation over the previous 3 s, which provides sensitivity to change. The equality of the positive and negative peaks means there is no response to a constant stimulus. The Fourier transform of the impulse response is a band-pass peaking at 0.25 Hz.

A similar pattern of signal processing is demonstrated in a mold. Sporangiophores of *Phycomycetes blakesleeanus* grow toward light, a phototropism that has been studied in great detail (Galland and Lipson 1987; Presti and Galland 1987). Of interest here is the fact that the kinetic pattern of the response to a change in intensity is similar to that of bacteria, but with a slower time scale (see Figure 5-7, bottom). This type of filter is probably a good match for many stimulus situations, because it provides band-pass characteristics to reduce noise and emphasizes changes in intensity that usually carry more-important information than do steady levels of intensity (Foster and Smyth 1980). Its effects are illustrated in Figure 5-8.

For unknown signal patterns or general signal processing the optimum filter design is not clear. A general property of Fourier transforms is that there is an inverse relationship between the spread of a function in time or space and the width of its Fourier transform in the frequency domain (bandwidth). For some examples see Figure 5-9. Furthermore, the bandwidth of a pulse of constant frequency is approximately the reciprocal of its duration (Pye 1983).

Marr and Hildreth (1980) have argued that the optimal spatial filter for vision should be one having minimum width in both the spatial and frequency domains. The function that uniquely satisfies these requirements is the Gaussian function. To use this function as a detection device, take the second derivative and define detection as occurring when the output crosses zero. This process leads to a function that has a "Mexican hat" shape (Figure 5-10) that is similar to a difference in two Gaussian distributions (Ratliff 1978) with widths in the ratio of 1.6 : 1. The effect of convoluting this pattern with rectangular pulse signals is shown in Figure 5-10.

Another basic question concerning detection is to determine at what distance from the target it occurs—in other words, what is the range of detection? As discussed, this depends on the output of the source, the attenuation in the environment, the noise present, and the characteristics of the receiver. The general aspects of attenuation were discussed in Chapter 4, on signal transmission. More specific aspects of all three considerations are discussed in the chapters on specific stimuli in Part 2.

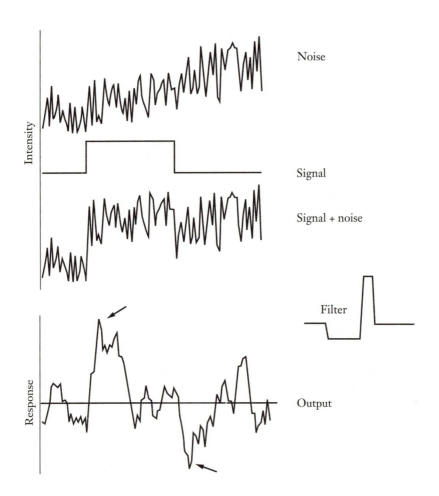

Figure 5-8 The effect of an asymmetrical filter. The noise (*top*) is the same as in Figure 5-6. The signal is a single pulse (0 ↔ 1). The intensity received (the objective intensity) is the sum of the signal and the noise. In this example the signal is marginally discernible to the eye. The input is filtered by taking three times the average of five adjacent points (positive lobe of filter) and subtracting the average of fifteen points from an adjacent interval (negative lobe of filter). This filtering pattern, which is appropriate for many organisms (see Figure 5-7), yields the output response (or subjective intensity) (*bottom*). Averaging within each lobe reduces high-frequency fluctuations and taking the difference between the two lobes of equal weight removes the drift of low-frequency noise. More importantly, the changes in the signal are represented by a positive peak (*top arrow*) for the increase in signal intensity and a negative peak (*bottom arrow*) for the decrease in signal intensity. Thus, contrast information is emphasized and changes in intensity can be easily detected as levels of the output response.

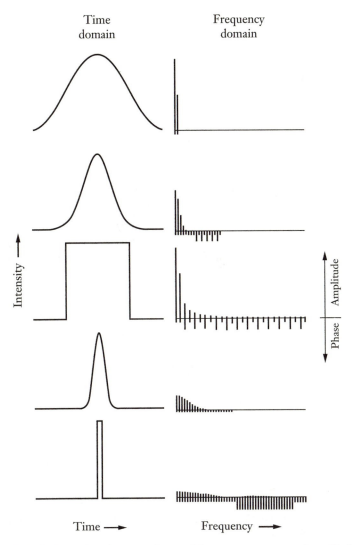

Figure 5-9 Relationships between time and frequency representations. Each row contains a graphical representation of the same pattern, presented in the time domain on the left and the frequency domain on the right. In the frequency domain the intensity of each frequency component is indicated by the height of the line above the axis, with the phase indicated below the axis. The frequency plot starts with 0 frequency (*left*) and extends to fifty times the reciprocal of the period in the time domain. A wide pulse in the time domain has a narrow frequency distribution (*top*), while a pulse of short duration has a wide range of frequency components (*below*). Rapid changes in intensity, as in rectangular pulses, are associated with high-frequency components.

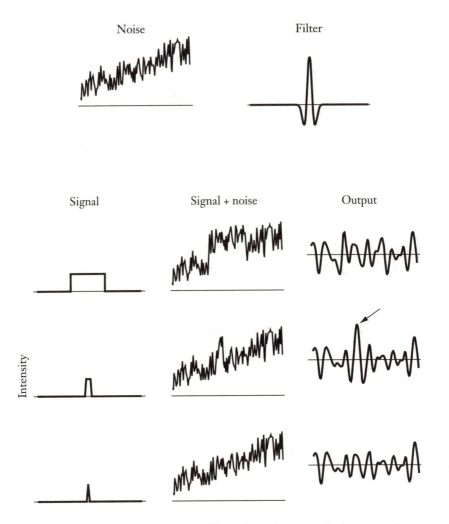

Figure 5-10 Difference of Gaussians filter. The noise pattern is the same as in previous figures. The filter is the difference between two Gaussian distributions, which produces the "Mexican hat" shape. The signals are single pulses of the same height but of three different durations. This filter yields a strong response only to a signal of the appropriate width (*arrow*).

PART 2

Stimulus Properties

■ ■ ■

These six chapters consider the properties of various stimulus modalities. Among the concerns here are the factors limiting physiological sensitivity and precision, as well as the strength of signals that are of interest and of background noise. Attempts are made to estimate the range and rate of the effective transmission of information, using various stimulus channels.

CHAPTER 6

Temperature

■ ■

> Truly, to enjoy bodily warmth,
> some small part of you must be cold,
> for there is no quality in this world
> that is not what it is merely by contrast.
> —Herman Melville, *Moby Dick*, 1851

Thermal sensation is important to most organisms as a means for them to control their temperature. Even bacteria migrate, in a thermal gradient, toward the temperature to which they have been acclimated (Maeda et al. 1976). Large animals can maintain a body temperature different from that of their environment, and most regulate their body temperature within a narrow range by various physiological and behavioral responses. Such regulation is aided by an ability to obtain information about the temperature of the environment.

Some animals use external temperatures as signals to obtain information about other aspects of their environment. For instance, certain nematodes may use thermal stimuli to locate a particular level in the soil (Dusenbery 1989). And many parasites of warm-blooded animals use thermal stimuli to help them guide their attack on the host (Cohen et al. 1980). It has further been suggested that wasp parasites, which oviposit through tree bark onto bark beetles, locate the potential host's position from outside the bark by temperature (Richerson and Borden 1972). Fish may solve the problem of determining water current direction in open water by using information from the fine structure of temperature differences (Westerberg 1984).

Studying responses of surface-dwelling animals to temperature is complicated by humidity changes and convection currents that are caused by temperature differences (Madge 1961). Consequently, the focus in this chapter is on simple organisms that have extremely sensitive responses to temperature and that inhabit soil, where the temperature patterns are predictable.

In addition, it is known that certain snakes use patterns of infrared radiation (having a wavelength longer than that of visible light) to distinguish objects in their environment of differing temperature. Although this is an indirect acquisition of temperature information, it is included here because the mechanism for detecting the radiation involves a temperature difference in the receptor organ. This subject is discussed in Section 6-4.

6-1 PHYSICAL LIMITATIONS ON THERMORECEPTION

The sense organs of many organisms have a level of sensitivity that operates near the limits of what is physically possible. What are the limits for thermal sensitivity? Calculations based on general physical principles demonstrate that temperature fluctuations will be 10^{-6} °C or less in a typical thermal receptor (Dusenbery 1988). As we will see, this thermal noise is quite small compared to signals in the environment.

Another potential limitation on thermal sensitivity is that thermal changes of a relevant size and time course may not penetrate to an interior receptor. Temperature changes generated at a surface decay exponentially with depth, as was explained in Section 4-2. If the thermal diffusivity is that of water and temperature fluctuates with a period of 1 s, the damping depth is 200 μm (see Table 4-4). Since sensory information is rarely acquired faster than on a one-second time scale, thermal signals can clearly penetrate to the receptors in small organisms. In humans the thermal receptors in the skin are several hundred micrometers from the surface, and their depth may limit the speed and sensitivity with which thermal changes can be detected (Hensel 1982, 14, 59).

6-2 LIMITS OF THERMOTAXIS

In analogy with the analysis of chemotaxis by Berg and Purcell (1977), the minimum thermal gradient that can be detected can be determined by equating the temperature change signal with the receptor noise, as determined in Section 9-3. If an organism is moving with velocity v along thermal gradient G for time t, the change in true temperature will be

$$\Delta T \approx v\, t\, G \tag{6-1}$$

For receptors that are less than 300 μm in extent and for integration times of a second or more, the thermal noise can be expressed as follows (Dusenbery 1988):

$$\Delta T \approx T \left(\frac{k_B}{c_v}\right)^{1/2} \left(\frac{c_v}{4\pi\, k\, t}\right)^{3/4} \tag{6-2}$$

where k is the thermal conductivity (6.2×10^{-3} J s^{-1} cm^{-1} K^{-1} for water), c_v is the heat capacity per unit volume (4.2 J cm^{-3} K^{-1} for water), and T is temperature in Kelvin. Equating the signal with the noise and solving for the minimum detectable gradient,

$$G_{Tb} \approx T \left(\frac{k_B}{c_v}\right)^{1/2} \left(\frac{c_v}{4\pi k}\right)^{3/4} t^{-7/4}\, v^{-1} \tag{6-3}$$

At standard temperature, using the thermal properties of water,

$$G_{Tb} \approx 4.2 \times 10^{-7}\, t^{-7/4}\, v^{-1} \tag{6-4}$$

in units of °C and s.

There is a relatively strong dependence on the time available for integrating the signal. If, instead of measuring the temporal change in a receptor the organism measures the difference between two receptors separated by a distance L, the equivalent formula is

$$G_{Tb} \approx 4.2 \times 10^{-7}\, t^{-3/4}\, L^{-1} \tag{6-5}$$

in units of °C and s. In this case the effect of time is less strong.

Using these relationships, it has been estimated that typical bacteria could respond to gradients as small as 3×10^{-4} °C/cm (Dusenbery 1988). Bacteria with this degree of sensitivity in soil would almost always be in a thermal gradient to which they could respond, as discussed in the next section. However, the only detailed report thus far of thermotaxis in bacteria employed a gradient of about 20 °C/cm (Maeda et al. 1976). Thus, it is not clear whether the thermosensory abilities of bacteria come any-where near the theoretical limit.

Pseudoplasmodia of the cellular slime mold *Dictyostelium discoideum* have been shown to migrate in thermal gradients away from a temperature several degrees below that of their acclimation temperature (Whitaker and Poff 1980). They are able to respond to gradients as small as 0.05 °C/cm (Bonner et al. 1950; Poff and Skokut 1977). Unlike bacteria, the slime mold appears to employ a tropotactic mechanism (see Table 17-1) in which the spatial gradient is detected across the pseudoplasmodium. A tempera-ture difference of 5×10^{-4} °C or less is thus detected across a 100 μm-wide pseudoplasmodium. If their sensitivity were limited only by thermal fluc-tuations in their receptors, slime molds could respond to gradients as shallow as 1×10^{-7} °C/cm (Dusenbery 1988). This is nearly a million times less than the threshold that has actually been determined. In nature the gradients would usually be much larger (see Table 6-1, below).

Like bacteria, nematodes are capable of responding to temperature by klinokinesis (see Table 17-1) (Dusenbery 1980; Goode and Dusenbery 1985; Dusenbery 1988). And it has been estimated that in principle they could have a threshold as low as that of the slime mold (Dusenbery 1988). The most sensitive response to a thermal gradient that has been reported for any organism, 0.001 °C/cm, is for a nematode (Pline et al. 1988). But even this limit is a thousand times higher than what is theoretically possible. Nematodes have also been shown to respond to temporal temp-erature changes of about 10^{-4} °C/s within a few seconds (Dusenbery 1988). Thus, they must be able to detect a change of 0.001 °C. Again, this is much less sensitive than the noise expected in a thermal receptor.

In mammals thermoreceptors have been studied a great deal, and they do not seem to be as sensitive as the responses previously discussed (Hensel 1981; 1982). The minimum sensible temperature change is greater than 0.1 °C (Hensel 1981, 20). This finding is in reasonable agreement with Ivanov's determination that the precision of human body temperature regulation is 0.05 to 0.08 °C (Ivanov 1984). Although a temperature signal becomes somewhat attenuated before reaching the receptors embedded several hundred micrometers below the surface in human skin, it would still be about half the magnitude of that at the surface (Hensel 1982, 14).

In summary, no organism has been shown to have a sensitivity to temperature stimuli that is anywhere close to the physical limits. The following discussion of thermal signals in the environment explains why.

6-3 THERMAL SIGNALS UNDERGROUND

Sunlight falling on land surfaces causes relatively large daily changes in temperature, with changes of tens of degrees in the air and the soil close to the surface being common. In most environments, complicated patterns of thermal gradients are established by the movement of shadows cast by clouds or vegetation, and by evaporative cooling, or air movements. Analyzing this thermal environment would be very complicated. In contrast, the subsoil environment can be reasonably well modeled, by assuming the conduction of heat through a homogeneous medium.

Temperature variations at the surface penetrate into the soil primarily as a sinusoidal wave (see Section 4-2). In order to determine the sizes of temperature gradients, take the derivative of equation 4-19:

$$\frac{dT}{dz} = -\frac{T_{1/2}}{z_d} \exp\left(-\frac{z}{z_d}\right)\left[\cos\left(2\pi ft - \frac{z}{z_d}\right) + \sin\left(2\pi ft - \frac{z}{z_d}\right)\right] \quad (6\text{-}6)$$

where r is replaced by z for the soil depth and I is replaced by T for temperature. The term $\cos(\alpha) + \sin(\beta)$ is limited to the range ± 1.414, and its average magnitude is close to unity. Consequently, this term in the equation above can be ignored for our purposes. Thus,

$$\frac{dT}{dz} \cong -\frac{T_{1/2}}{z_d} \exp\left(-\frac{z}{z_d}\right) \quad (6\text{-}7)$$

For typical temperate regions, daily variations have values of $z_d = 10$ cm and $T_{1/2} = 15\ °C$. For annual variations, $z_d = 200$ cm and $T_{1/2} = 10\ °C$ are appropriate (Campbell 1977, 16; Nobel 1974, 375). Using these parameters, some values for the average gradient at various depths in soil are presented in Table 6-1.

Above 30 cm the gradient is dominated by the daily variation and averages more than 0.1 °C/cm. Twice a day the gradient will briefly go to zero. Below 40 cm the daily gradient is usually insignificant compared to the annual. The gradient caused by the annual variation becomes quite small only at depths of several meters. This gradient will go to zero twice a year. Another source of heat is the interior of the earth. The average rate of conduction of heat to the earth's surface is 1.4×10^{-6} cal cm^{-2} s^{-1} (Garland 1971, 341). This rate of flow through average soil produces a temperature gradient of about 6×10^{-4} °C/cm. The temperature profile is

Table 6-1 Temperature Gradients in Soil.

Depth	Average Gradient Magnitude (°C/cm)	
	Daily	Annual
0	1.5	0.05
1 cm	1.35	0.05
10 cm	0.56	0.05
36 cm	0.042	0.042
1 m	7×10^{-5}	0.03
10 m	6×10^{-44}	3×10^{-4}

also influenced by climatic changes, in temperature to depths of hundreds or even thousands of meters (Garland 1971, 335).

Organisms may make use of these gradients to aid them in migration. The gradients could be used, for instance, by small animals as collimating stimuli (page 397), since the gradients extend over many centimeters. Computer modeling suggests that the migrating pseudoplasmodia of slime molds can use information in the dynamic pattern of the soil temperature to move to the surface in order to release spores (Dusenbery 1988) and that nematodes can use patterns of temperature changes to move toward a particular level in the soil in order to search for roots (Dusenbery 1989). These temperature changes also represent environmental noise that may limit the detection of temperature signals from other sources. For example, it has been suggested that nematodes may locate plant roots by being attracted to the metabolic heat that roots generate (El-Sherif and Mai, 1969). However, calculations (Dusenbery 1987) indicate that the thermal gradients generated would be on the order of 10^{-3} °C/cm, which organisms could respond to, but this is much smaller than the gradients created by other causes. Thus, we must conclude that the environment is too noisy for this type of information transmission.

6-4 INFRARED RADIATION

In any material not at absolute zero of temperature, the atoms and molecules are in constant motion. This irregular movement of the molecules, all of which carry electric charges, generates electromagnetic radiation (light), which propagates away from the material. The higher the temperature of the material, the faster the molecular motion and the more radiation given off. One can see this radiation as light from a flame, or from heated metal, or from the sun.

The amount of radiation leaving the surface of a given material depends on the efficiency with which the motions of the molecules are coupled with electromagnetic waves. A black surface looks black because it efficiently absorbs all the colors of visible light, and it also radiates light with maximum efficiency. This type of radiation is thus often called **blackbody radiation**. The characteristics of radiation from an ideal blackbody can be predicted from physical theory, although any real material will radiate somewhat less energy.

The efficiency of radiation of a material is measured by its emissivity (Campbell 1977, 49), which has a value of 1 for an ideal blackbody and zero for an ideal white body. Natural materials have emissivities ranging from about 0.05 to 0.95. All natural materials have high emissivity for far-infrared radiation.

The total energy radiated by a black body is proportional to the fourth power of its absolute temperature. (The constant of proportionality, called the Stefan-Boltzmann constant, is equal to 5.67×10^{-8} W m^{-2} K^{-4} [Weast 1985, F-195].) Thus, a body at 20 °C (293 K) emits about 400 watts per square meter, and a one-degree temperature difference generates about a 1 percent change in the total energy radiated. Objects in the environment have temperatures that frequently differ by several degrees, and the intensity of their infrared radiation can provide information about them. Military technology frequently exploits this strategy to guide missiles to warm targets. People do not normally notice the large energy loss involved, because all objects in the environment are usually at nearly the same temperature and are both radiating and absorbing similar amounts of radiation. When one is exposed to a particularly hot or cold object, the change in energy balance becomes noticeable.

The spectrum of the radiation emitted can also be predicted. The energy flux of radiation from a black body in a unit interval of frequency is calculated as

$$e\,(f,T) = \frac{2\pi h f^{3}}{c^{2}\,[exp\,(hf/k_{\mathrm{B}}T) - 1]} \tag{6-8}$$

where f is frequency, T is absolute temperature, h is Planck's constant, and k_{B} is Boltzmann's constant (Davidson 1962). This distribution is plotted in

Figure 6-1. The frequency of maximum emission per frequency interval is simply

$$f_{max} = 5.87 \times 10^{10} \, T \tag{6-9}$$

in units of K and Hz.

At 20 °C (293 K), the peak frequency is 1.72×10^{13} Hz, which corresponds to a wavelength of 17 μm, in the infrared region of the spectrum. For the peak to reach the edge of the visible spectrum ($\lambda = 700$ nm or $f = 4.3 \times 10^{14}$ Hz), the temperature would have to increase to 7,000 °C, which is somewhat hotter than the sun.

From these descriptions it becomes apparent that infrared light carries useful information about objects in the environment. The problem for organisms trying to obtain some of this information is to devise sensitive detection mechanisms. The organisms that have been most successful at this are certain snakes that use infrared signals for prey localization. The pit organ of rattlesnakes has been studied in some detail (Bullock and Diecke 1956; Hartline 1974). This structure consists of a membrane with an area of 3 to 4 mm² but only 10 to 15 μm thick that is suspended in an air-filled cavity that insulates it from the rest of the animal. The endings of some 3,500 neurons are embedded in this membrane. Each receptor has branches that extend over an area with a diameter of about 40 μm. This system can thus be considered similar to a square array of detectors having about 60 elements on a side and little overlap of the receptive fields of adjacent elements.

Infrared radiation from a given direction passes through an opening and falls on part of the membrane. This part is then heated rapidly, because of its low thermal mass, which is caused by the extreme thinness of the membrane and its being insulated by suspension in air. There, neurons of ordinary thermal sensitivity respond both to temporal changes in temperature and to differences between the local temperature and the average temperature of the membrane. This organ thus provides the snake with a coarse, but high-contrast, infrared representation of its environment. The system is analogous to a very crude pinhole camera. The receptive field of individual neurons is in the range of 45 to 60° wide (Newman and Hartline 1982).

Various experiments indicate that a temperature change of the membrane of about 0.003 °C in 0.1 s is detected. If the detection were limited by thermal fluctuations in the receptor, a limit of 7×10^{-8} °C would be predicted (Dusenbery 1988). Thus, the receptor is nowhere near as sensi-

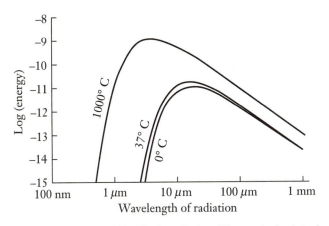

Figure 6-1 The spectrum of blackbody radiation. The vertical axis is the logarithm of the energy in watts per square meter of surface area per Hz spread in frequency. The short-wavelength side of the distribution is very steep, and an increase in temperature shifts the curve to shorter wavelengths. Thus a detector that is sensitive only to wavelengths below the peak will be relatively sensitive to the temperature of the source.

tive as it might be. However, it appears to be sufficiently sensitive to discriminate objects whose surface temperatures differ by only 0.1 °C (Bullock and Diecke 1956). It seems likely that objects exhibiting such small temperature differences would almost always be present at the surface of the ground. There is probably no adaptive advantage to having a more sensitive receptor. Estimates indicate a sensitivity to radiation of approximately 10^{-4} W/cm^2 for radiation with a wavelength of 1 to 10 μm (Hartline 1974). Neurons in the brain respond to movements of a man's head 2 m away. Estimates suggest that a mouse could be detected at a range of 30 cm (Buning 1983). In snakes, strikes can be guided solely by the pit organs with an angular accuracy of about 5° (Newman and Hartline 1982). This sense is probably most useful for hunting in underground burrows, where the background temperature differences would be minimized.

To summarize, most environments have thermal signals that are large enough to be easily detected. It is likely that many more organisms make use of this source of information than we now realize.

CHAPTER 7

Chemicals

■ ■

Smells are surer than sounds or sights
To make your heart-strings crack.
—Rudyard Kipling, *Lichtenberg*, 1934

Chemicals are probably the original stimuli, since they can participate directly in biochemical reactions without needing a sensory transduction step. This may be the reason that chemicals seem to be the most universal of stimuli. Indeed, it is possible that all organisms make use of chemical stimuli; many bacteria certainly do. Even certain viruses infect their host cell by a mechanical mechanism triggered by chemical recognition of the host cell's surface. However, the ubiquity of chemical stimuli is hard for humans to appreciate, because we pay so much more attention to sight and sound. Although humans tend to ignore smells, these stimuli seem in fact to be more directly linked to the emotions than are sound and light. This linkage probably reflects the antiquity of the use of chemical stimuli to control basic functions in our primitive ancestors.

This chapter reviews basic information on what is known about the use of chemical stimuli. The first section discusses the terminology that has been developed to bring some order to discussions of the diversity of chemical interactions. The second section then considers the specificity that is available among chemical stimuli to estimate the rate at which information can be obtained. The remaining sections describe the sensitivity of which organisms are capable, the nature and quantity of the chemical stimuli present in the environment, and the mechanisms for the transmission of chemicals to estimate the ranges at which chemical stimuli can carry information.

7-1 TERMINOLOGY PERTAINING TO CHEMICAL STIMULI

The variety of chemical interactions that occur between organisms and their environment has prompted efforts to classify chemical stimuli according to the type of interaction. Unfortunately, many researchers have not distinguished between chemicals that carry information and those with trophic, including toxic, effects (Whittaker and Feeny 1971; Nordlund 1981). In part, this failure to make a distinction is understandable, because it is easier to demonstrate that an organism responds to a chemical than to determine whether it mediates informational or trophic effects. However, this failure leads to confusion about adaptive pressures on the organisms involved, and it becomes desirable to make this distinction clear.

Because most chemical stimuli are released by organisms, the traditional classification scheme organizes chemicals according to whether the transmitter or the receiver is the one to benefit from the transfer of the chemical (Whittaker and Feeny 1971; Nordlund 1981). This distinction often parallels one based on whether the chemical carries information or trophic effects, because organisms can easily evolve to ignore information, but not trophic effects. For example, repellents are likely to be toxic to the receiver and have benefits for the producer, while attractants are likely to be informational and to benefit the receiver.

The exceptions are less common. If being attracted to a chemical stimulus is detrimental to an organism it can easily evolve to ignore it, unless it is mistaken for another stimulus having important benefits. A transmitting organism may benefit from releasing a chemical lure that deceives the receiver into taking actions unfavorable to itself. But the receiving organism will continue to respond only if the unfavorable situation is rare, compared to the related beneficial stimulus situation.

In order to clarify the distinction between informational and trophic effects while remaining as consistent as possible with traditional terminology, the classification scheme in Table 7-1 is proposed. A similar scheme is used by Hölldobler and Wilson (1990, 227).

A **semiochemical** carries information from another organism. The literal meaning of this term restricts its use to chemicals that act as a signal.

In analogy with the term *hormone*, the suffix *-mone* is used only to denote chemicals that carry information and have no trophic effects. A **pheromone** is a semiochemical that acts between individuals of the *same* species. There is usually no question of toxic effects acting within a given

Table 7-1 Classes of Chemical Interaction.

Relationship of Transmitter and Receiver	Type of Interaction	
	Trophic	Informational
Environment and Organism	(O_2, water, salt)	**Chemical Stimuli***
Interorganismal	(Food, milk, toxins)	**Semiochemicals**
Intraspecific	(Milk, nuptial gifts)	**Pheromones**
Interspecific	——— Allelochemics ———	
Benefit to Receiver	(Food)	**Kairomones** (Food scents)
Benefit to Sender	(Toxins, antibiotics)	**Allomones** (lures)
Benefit to Both	(Flower nectar)	**Synomones** (Floral scents)
Intra-organismal	(Blood sugar)	**Hormones**

* The interaction terms in boldface are general class names. The terms in parentheses give specific examples.

species, but nutritional transfers (trophic effects) often occur from parent to offspring and sometimes also from male to female. A simple basis for discriminating between informational and nutritional transfers is the quantity of the chemical transferred. For a nutritional transfer to be significant, a relatively large amount of the chemical in question must reach the recipient, whereas true pheromones usually act in very low quantities. The differences are generally obvious.

A **kairomone** is a semiochemical that acts between individuals of different species with the receiver benefiting, as is the usual case. An example is the scent of prey to a predator. In many chemical interactions between different species it is difficult to distinguish informational effects from trophic ones, because toxic effects are the likely trophic interaction and toxins may be potent. In the absence of sufficient knowledge to make the distinction, chemicals mediating interactions between species may be called **allelochemics** (Whittaker and Feeny 1971). A common situation involves a chemical from one species repelling a second species. In many cases it will not be evident whether the chemical is avoided because it itself is toxic or because it indicates a situation unfavorable for other reasons. In the absence of more information such chemicals should simply be called **repellents**, with nothing implicated about their mode of action.

In Table 7-1, no distinction is made between olfactory, taste, or gustatory stimuli, because no universal distinctions are possible (Hsiao 1985; Laverack 1988).

7-2 SPECIFICITY OF CHEMICAL STIMULI

Chemical stimuli are distinguished from all others in the variety of independent signals that may be simultaneously present in the environment. This variety results from the complex ways in which carbon atoms can combine to form chemicals and from the fact that cells can generate **receptor** proteins with such a high degree of sensitivity and specificity that the simplest changes in the structure of a chemical may drastically alter its effectiveness in activating a receptor. Specificity as well as sensitivity may be further enhanced by specialized receptor cells, which may in turn be part of a specialized receptor organ. It is often convenient to use the word *receptor* without specifying a given level of organization such as a molecule, cell, or organ.

Receptors vary significantly in their specificity. Highly specialized ones such as those for detecting pheromones are very specific. Any chemical change, including an interchange of stereoisomers, generally reduces their sensitivity from tenfold to a millionfold, although related compounds may act as synergists or inhibitors of the main stimulus (Kaissling 1971; Tamaki 1985). Some receptors are specialized for chemicals that occur at relatively high concentrations, such as CO_2, but others seem less highly "tuned" to a specific chemical. There is, for example, a receptor in locusts for volatile chemicals released by green vegetation. This receptor responds to a variety of approximately six-carbon carboxylic acids and alcohols (Kaissling 1971).

Most organisms have a large variety of receptor molecules, but we do not know how many. This is an important question, because each receptor type may represent a separate stimulus dimension (see Section 3-2) if information from each type of receptor is kept separate in the nervous system. In order to estimate the information-carrying capacity of chemical stimuli it is necessary also to have estimates of the number of intensity steps resolved along each dimension.

For chemical stimulation, we can make the following estimate concerning the resolution of intensity. Although individual receptor cells typically have an instantaneous response range of only about two orders of magnitude (Atema 1985), adaptation and multiple receptors with differing affinities can provide a much wider response range for sensation, which is about 200,000 in humans (Wilson and Bossert 1963). To discriminate a concentration (intensity) difference generally requires having a difference of 10 to 30 percent in fish (Blaxter 1988) and 50 percent in humans (Wilson and Bossert 1963). Consequently, there are about thirty distinguishable steps in intensity along each dimension for humans ($1.5^{30} \approx 2 \times 10^5$), which would correspond to about five bits of intensity information ($2^5 = 32$).

The number of independent dimensions in the human sense of smell remains a major question. John Amoore initiated a long-term program to identify the primary odors, using panels of individuals with specific deficiencies (**anosmias**) in their sense of smell (Amoore 1967; 1969; 1971). Such individuals typically have thresholds for smelling particular chemicals that are a hundred and sometimes a million times higher than those of normal individuals but have normal thresholds for other chemicals. These individuals are presumed to lack a particular type of receptor and a corresponding stimulus dimension. The odor that is most strongly affected is assumed to be one detectable only by the missing receptor, which is identified by this so-called **primary odor**. Eight primary odors have been identified in this way (Table 7-2).

Anosmias are surprisingly common, constituting a few percent of the population in many cases, as in color blindness. Such abnormalities should probably not be regarded as defects as much as a normal variation in any population. If the initial research trends continue, nearly everyone would be found to have at least one "defect"; fewer than 15 percent of the population would be free of all eight of the specific anosmias identified, if they are distributed independently within the population. The best guess as to the total number of primary odors is about 30 (Amoore 1975). If these thirty are each independently perceived in only ten concentration steps, the total number of possible odor sensations becomes astronomical: on the order of 10^{30}, which corresponds to 100 bits of information ($2^{100} \approx 10^{30}$).

Another approach to determining the number of stimulus dimensions is anatomical. The chemosensory axons in arthropods and vertebrates terminate in distinctive concentrations of neurons called **glomeruli** (Ache 1988). Evidence from insects suggests that each glomerulus corresponds to a distinctive chemosensory dimension (MacLeod 1971; Hildebrand and Montague 1986). The sphinx moth (*Manduca sexta*) has about 60 such glomeruli, suggesting that there may be sixty dimensions to its chemosen-

Table 7-2   The Primary Odors Identified for Humans.

Characteristic Chemical	Formula	Normal Threshold*	Source of Odor[†]	Anosmics[‡] (%)	Reference
Isovaleric acid	$C_5H_{10}O_2$	140	Sweat	2	Amoore 1969; 1971
l-carvone	$C_{10}H_{14}O$	43	Mint	8	Pelosi and Viti 1978
1-pyrroline	C_4H_7N	22	Semen	20	Amoore et al. 1975
1,8-cineole	$C_{10}H_{18}O$	21	Camphor	30	Pelosi and Pisanelli 1981
Isobutyraldehyde	C_4H_8O	2	Malt	36	Amoore et al. 1976
ω-Penta decalactone	$C_{15}H_{28}O_2$	2	Musk	9	Amoore et al. 1977
Trimethylamine	C_3H_9N	0.5	Fish	7	Amoore and Forrester 1976
5α-androst-16-en-3-one	$C_{19}H_{28}O$	0.2	Urine	50	Amoore et al. 1977

[*] In ppb in an aqueous solution that produces a perceptible odor in the gas phase.
[†] A natural odor source that has the same perceived quality as the characteristic chemical. The odor of malt derives from the fermentation of essential amino acids. The odor of semen derives from the oxidation of amines. 5a-androst-16-en-3-one is a mating pheromone in pigs that is also produced by oxidation of the corresponding alcohol from human urine. Carvone and cineole are found in many plants.
[‡] The percentage of anosmic individuals found in the sampled population.

sation. If each of these glomeruli is independently perceived in a minimum of two concentration steps, the number of distinct sensations would be $2^{60} \approx 10^{18}$. In vertebrates the number of glomeruli is about 2,000 (Allison and Warwick 1949), suggesting that these are 2,000 dimensions and at least as many bits of information, depending on the intensity resolution along each dimension.

    Quite recently a family of genes has been identified with properties suggesting that it codes for olfactory receptor proteins (Buck and Axel 1991). Present estimates are that there are at least 100 different genes in this family. This information provides another estimate of the number of stimulus dimensions. The power of molecular-genetic techniques should lead to rapid improvements in our understanding of how chemical stimuli are perceived. In any case, it is clear that there is enough potential specificity for chemical stimuli to transmit simultaneously a large quantity of information.

The rate at which information is presented by chemical stimuli is even harder to estimate, because chemoreceptors often adapt rapidly and sense organs are usually designed to expose the receptors in pulses, such as sniffs or flicks of antennae. As a best guess, one might assume that a new set of sensations is acquired every 10 to 100 seconds. If so, this would suggest that the maximum average rate at which we can acquire information from chemical stimuli is in the range of 1 to 10^4 bits per second—not a very precise determination.

7-3 SENSITIVITY FOR DETECTION OF CHEMICAL STIMULI

The usual mechanism by which biological systems detect chemical stimuli is by binding of the stimulus molecules to a protein receptor coupled to biochemical processes within a cell. In principle, a chemoreceptor protein can detect the presence of a single stimulus molecule. However, if the molecule stays bound to the receptor no further information can be obtained by it. In order to continuously monitor the concentration of a chemical stimulus, the molecules must somehow be removed or released. The simplest situation is reversible binding, in which molecules rapidly bind and dissociate from the receptor protein. The intensity, or concentration, of the molecules is determined by the fraction of time a receptor has a signal molecule bound to it.

The accuracy with which concentration can be measured is dependent on the time interval over which the occupancy of the receptor is averaged, as explained in Chapter 5. Most receptor cells in fact have thousands of receptor proteins on their surface, which aids in the averaging process. Quantitative analysis (Berg and Purcell 1977) indicates that in the region around a receptor where diffusion dominates over any bulk flow the accuracy of measurement is limited by the rate at which signal molecules diffuse into the vicinity of the receptor. This limitation occurs because once a signal molecule has been detected and released the nature of diffusion is such that the molecule is likely to remain in the vicinity for some time and probably bind to the receptor again. Any such binding after the first one provides little new information, since it is already known that that molecule is present.

Assuming a receptor with effective radius R that measures the concentration of a stimulus chemical with diffusion constant D for time t and the true concentration is C, Berg and Purcell provide several arguments that, ignoring factors of odor unity, the relative error will be given approximately by

$$\frac{\Delta C}{C} \approx (tDCR)^{-1/2} \tag{7-1}$$

where C is in units of molecules per cubic centimeter. For a 1 M solution, $C = 6.0 \times 10^{20}$ cm^{-3}. For a pure gas at standard temperature and pressure, $C = 2.5 \times 10^{19}$ cm^{-3}. The relative noise is inversely proportional to the square root of the rate of diffusion, the time over which the measurement is averaged, the concentration, and the effective size of the receptor. Table 7-3 presents the predicted concentration limits for airborne and water-borne stimuli for $\Delta C/C = 1$ (signal-to-noise ratio of 1 for detection) for receptors of size 1 µm (bacterium), 1 mm, and 1 cm (an arthropod antenna), assuming an integration time of 1 s, which is appropriate for bacteria (Berg and Purcell 1977) and insects (Kaissling 1971).

It can be seen that in water a bacterium might detect concentrations as low as 10^{-12} M. A bacterial sex pheromone appears to function at concentrations near this limit (Suzuki et al. 1984), but the time interval over which it acts is probably much longer than the one second assumed. Chemotactic responses, which have the appropriate time interval, have a lower limit of 10^{-7} M (Adler 1969; Mesibov et al. 1973). At the other extreme, a lobster might detect 10^{-17} M. The lowest concentrations observed to cause responses in aquatic animals appear to be about 10^{-13} M (Derby and Atema 1988). In air, an insect with large antennae might detect a concentration of a few molecules per cm^3, but the lowest-observed responses are about a thousand molecules per cm^3.

It is not clear why chemical receptors are not as sensitive as predicted by theory. One possibility is that receptor proteins are not specific enough to respond to such very low concentrations of appropriate chemicals, while not responding to much higher concentrations of the other chemicals present in the environment. In effect, sensitivity may be limited by receptor noise.

A great deal of research has been performed on the attraction of male moths to females via a pheromone released by the females. The large antennae carried by the males testify to the importance of the information carried by the pheromone, and the males' receptors seem to be highly

Table 7-3 Concentration Limits.*

Type of Concentration Situation	Radius	Threshold Concentration		Example or Reference
		Mol/cm^3	Moles/l	
Predicted				
Air	1 μm	5×10^4	——	Receptor cell
Air	1 mm	50	——	Insect
Air	1 cm	5	——	Mammal
Water	1 μm	5×10^8	10^{-12}	Bacteria, nematode
Water	1 cm	5×10^4	10^{-17}	Lobster
Observed				
Bacteria		5×10^{14}	10^{-7}	Adler 1969; Mesibov et al. 1973
Bacteria[†]		2×10^{10}	4×10^{-11}	Suzuki et al. 1984
Bacteria[†]		5×10^8	8×10^{-13}	Stephens 1986
Algal sex attractant		4×10^9	6×10^{-12}	Boland et al. 1982
Nematode		4×10^{12}	6×10^{-9}	Balan 1985
Lobster		5×10^7	10^{-14}	Atema 1985
Ants		$10^{10} - 10^{15}$	——	Kaissling 1971
Bees		$10^9 - 10^{11}$	——	Kaissling 1971
Moth sex attractants		$10^3 - 10^6$	——	Mankin et al. 1980; Kaissling 1971
Dogs		10^6	——	Moulton 1977; Marshall et al. 1981
Humans		$10^9 - 10^{11}$	——	Kaissling 1971

* Predicted values are based on the following assumptions:
  $D = 0.2$ cm^2/s in air and 2×10^{-5} cm^2/s in water.
  The integration time is 1 s.
  The relative error, $\Delta C/C$, = 1 at threshold.
[†] Indicates that the response is much slower than the 1 s assumed in the calculations.

optimized. The chemicals in the pheromone have been identified for many moth species, and careful determinations of the thresholds have been made. The lowest thresholds are in the vicinity of 10^3 to 10^4 molecules/cm^3 (Kaissling 1971; Mankin et al. 1980). This limit is about a hundredfold higher than would be predicted by an effective receptor of radius 1 mm and an integration time of 1 s. Detailed experiments (Kaissling 1971) with the male silkworm moth indicate that, at the threshold with 50 percent of the animals' wing fluttering, a total of 310 molecules reaches the twenty-five thousand receptor cells. Thus, even when losses in transporting of the stimulus from the air to the internal receptor cells are considered, the threshold is two orders of magnitude above the theoretical limit of one molecule. This limitation is probably due to noise in the receptors.

The improved effectiveness of larger receptors is a consequence of more rapid diffusion of the stimulus to their larger surface area. The constraint of slow diffusion could also be alleviated by stirring the medium around the receptor. Berg and Purcell (1977) calculate that the energy cost of such stirring is prohibitive for receptors less than approximately 10 mm in size. Stirring can benefit macroscopic organisms, however, which may explain the antenna flicks commonly seen in lobsters and perhaps other arthropods (Schmitt and Ache 1979) as well as sniffing in mammals. Flying insects probably also benefit from the airflow over their antennae. Kaissling (1971) has maintained that the antennae of many insects are efficient filters for the molecules in the airstream flowing over or through them, although others argue that this flow is unimportant (Mayer and Mankin 1985). It has been established that copepods increase their reaction time as well as the range of their chemoreceptors by generating appropriate currents flowing from the environment over their receptors (Andrews 1983).

Dogs are often considered to have a good sense of smell. Indeed, a beagle has twenty-five times the area of olfactory epithelium, which carries the receptor cells, that a human has (Albone and Shirley 1984, 245). Nonetheless, there is little reliable evidence, and estimates of the differences in sensitivity between dogs and humans have ranged from millionfold or billionfold down to insignificant (Moulton et al. 1960). Precise comparisons between dogs and humans are difficult to make because of the large range of individual variation. Nonetheless, some careful comparisons have been made with two odorants (Moulton and Marshall 1976; Marshall et al. 1981; Marshall and Moulton 1981), with the result that German shepherds were about 1,000 to 10,000 times more sensitive than humans. Dog's superiority in distinguishing mixtures of odors may be even more important than their sensitivity (Moulton et al. 1960). In spite of their abilities, dogs can apparently locate prey by smell only within a few meters. Their famous ability to track other animals probably depends also on having their noses close to the ground (Albone and Shirley 1984, 275).

7-4 NATURE AND QUANTITY OF CHEMICAL STIMULI

In order to carry information through the environment chemicals must ordinarily be somewhat soluble in water, because the receptor cells are covered by at least a thin layer of water. If the chemicals are to be carried in air, which is useful for terrestrial organisms, they must also be volatile.

In aquatic environments, signal chemicals are generally much more polar than airborne chemicals. Chemicals that elicit feeding in aquatic animals (Carr 1988) are usually amino acids or nucleotides that are highly soluble, present in high concentrations in prey organisms and in low concentrations in the general environment, because of the action of bacteria (Figure 7-1).

The chemicals used to obtain information range in size from ammonia, NH_3, with a molecular mass of 17, to proteins with molecular masses in the thousands. Chemicals that must be volatile are limited in size to about 20 carbon atoms, or a molecular mass of about 300.

WASTE PRODUCTS

The chemicals available in the greatest quantity to carry information from organisms are the waste products they excrete. These are usually simple molecules. For example, carbon dioxide is used to locate hosts in soil by insect larvae (Doane et al. 1975) and nematodes (Klinger 1963; 1965), and in air by mosquitoes (Gillies 1980) and black flies (Sutcliffe 1986). Such chemicals have continued usefulness to parasites because under selective pressure the host cannot easily stop producing them. The great disadvantage of CO_2 is that it has little specificity—there are many sources of it, both plants and animals. The open atmosphere contains about 0.03 percent CO_2. Its production in soil is typically on the order of 1 gram per hour per square meter of surface area (Wesseling 1962) and its concentration is typically 0.2 to 1 percent (Black 1968, 159). Mammals and birds expire air that contains about 5% CO_2 (Altman et al. 1958, 285). The high level of CO_2 in the environment makes it easy to develop receptors that are sufficiently sensitive so that the natural concentration is always above threshold (Pline and Dusenbery 1987). The challenge in extracting information is to detect small changes in CO_2 concentrations, which may be why mosquitoes and flies appear to respond to other host chemicals at a longer range than they do CO_2 (Gillies 1980; Vale and Hall 1985). Calculations analogous to those on thermotaxis (see Section

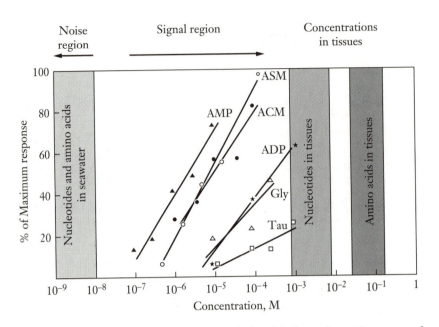

Figure 7-1 Concentrations of some chemicals found in the marine environment and response thresholds. Data are for the shrimp *Palaemonetes pugio*. and adenosine 5'-monophosphate (AMP), adenosine 5'-diphosphate (ADP), glycine (gly), taurine (tau), an artificial crab mixture (ACM), and an artificial shrimp mixture (ASM). (From Carr 1988.)

6-2) indicate that a nematode may be potentially able to respond to gradients that are too small to be easily tested in the laboratory but could in principle provide useful directional information a meter from the roots of a host plant (Dusenbery 1987).

Early in the history of evolution, the use of other waste products probably forced changes in waste disposal. For instance, the common pattern among motile animals of disposing of their waste in discrete packets as opposed to continuous release can be viewed as an adaptation to avoid leaving an obvious chemical trail that parasites or predators could follow. In addition, some organisms may condense their waste into non-volatile or insoluble products because it reduces their "apparency."

LEAKING CHEMICALS

Some molecules that are not waste products are inevitably released, at reduced levels, during gas exchange or leakage through the body wall. For example, in the ocean the background concentration of most amino acids is about 10^{-8} M (Atema 1985), with most of these presumably being leaked out of organisms unintentionally, since they are valuable to all organisms. A related observation is that ants recognize dead nest mates by the presence of oleic acid, and remove from the nest anything releasing this chemical (Hölldobler and Wilson 1990, 297).

There is evidence for some of the specific chemicals that bacteria-eating organisms use to locate their food. For instance, slime molds probably use folic acid ($C_{19}H_{19}N_7O_6$) (Gerisch 1982). Microbivorous nematodes may use cyclic AMP ($C_{10}H_{12}N_5O_6P$) (Ward 1973), pyridine (C_5H_5N) (Dusenbery 1976), or amyl acetate ($C_7H_{14}O_2$) (Balan 1985).

Plants are known to give off a wide variety of chemicals, the significance of which will be discussed here, although in many cases the release may serve a function rather than being an accidental leaking. Many volatile chemicals such as terpenes are released to the atmosphere, and many nonvolatiles wash off in rainwater or are released by roots (Finch 1980). Some of these chemicals are common to many plants, but others are unique (Visser 1986).

In a survey of fifty-four plant species, including crops, shrubs, herbs, and trees (Evans et al. 1982), 68 percent emitted isoprene (C_5H_8) at ≥ 0.01 µg carbon per gram of dry leaf per hour and 37 percent at ≥ 1 µg/g h. Monoterpenes ($C_{10}H_{16}$) were detected from 10 percent of the species, with a maximum value of 3 µg/g h. These included pinene, camphene, limonene, and 1,8-cineole ($C_{10}H_{18}O$). Natural nonmethane hydrocarbon emission rates averaged over the United States for a year are estimated to be 450 µg/m² h (Lamb et al. 1987). The most prevalent compounds are isoprene (15 percent) primarily from deciduous trees, and a-pinene (16 percent), mostly from coniferous trees. The so-called green odor of plants comes from various isomers of hexanol, hexenol, hexanal, and hexenal (Visser 1986). The volatile molecules are generally in a size range of 100 to 200 daltons (Finch 1980). Many of these are toxic defensive chemicals, and deciding whether their release is beneficial to the plant and, if beneficial, whether the benefit is waste removal or defense is a delicate question. Some chemicals may also be released by plants for communication purposes (Baldwin and Schultz 1983; Rhoades 1985) or sounding (see Section 12-6). In any case, insects are thought to use many of these chemicals to locate appropriate hosts (Miller and Strickler 1984). A summary of some findings in this regard is presented in Table 7-4.

Mammals give off a variety of chemicals in their breath, urine, and sweat (Schlunegger 1972; Savina et al. 1975) that are not obvious waste

Table 7-4 Kairomones That Attract Phytophagous
Insects to Their Hosts.

Insect	Plant*	Chemicals[†]
Beetles		
Anthonomus grandis	Cotton	α-pinene, limonene, β-caryophyllene, β-bisabolol
Hylobius abietis	Pine	α-pinene, 3-carene, α-terpineol
Ips grandicollis	Pine	Geraniol, limonene, myrcene, methyl chavicol
Dendroctonus pseudotsugae	Douglas fir	α-pinene, limonene, camphene
Flies		
Delia brassicae	Cabbage	Allylisothiocyanate, hexyl acetate
Delia antiqua	Onion	Propanethiol, dipropyl disulfide, 2-phenylethanol
Psila rosae	Carrot	*trans*-methylisoeugenol, *trans*-asarone
Moths		
Acrolepiopsis assectella	Leek	Propylthiosulfinate, methylthiosulfinate
Choristoneura fumiferana	Spruce	d-α-pinene, l-β-pinene

* Identified in Chapman (1986).
[†] Identified in Visser (1986).

products. The total volatile effluvia from humans contains at least several hundred different compounds (Ellin et al. 1974). Breath contains at least 100 identified organic chemicals, in the concentration range of 0.06 to 10 ng/l, with larger quantities of propanone (120 ng/l), isoprene (33 ng/l), and acetonitrile (24 ng/l) (Albone and Shirley 1984, 236). Urine and sweat contain a similar set of volatile chemicals, plus an array of steroids (Oertel and Treiber 1969; Albone and Shirley 1984, 32–33, 176–78). Although steroids are usually minor components, many of them are odorous or are converted to odorous chemicals by exposure to the environment. For example, humans excrete about 1 mg per day of 5α-androst-16-en-3α-ol ($C_{19}H_{28}O$) (Gower 1972); In the environment this steroid is oxidized to the corresponding ketone, and the quantity excreted could be diluted to 5,000 liters and still be detectable to humans having normal sensitivity (Table 7-2). Another example is that tsetse flies apparently locate host cattle by using a combination of CO_2, acetone (C_3H_6O), and 1-octen-3-ol ($C_8H_{16}O$) detected in air downwind (Vale and Hall 1985).

PHEROMONES

The use of waste products or leaking chemicals to obtain information would lead to their classification as kairomones, assuming the information transmitted benefited only the receiver. In contrast, pheromones are released because they benefit the sender as well as the receiver. Pheromones are typically more complex molecules than hormones, because the sender has an interest in synthesizing and releasing them, and because specificity is important and for large molecules there is a much greater number of possible structures. Wilson and Bossert (1963) predicted that pheromones would need at least 5 carbon atoms to provide sufficient diversity, and this has proven correct. For terrestrial animals the long-range sex attractants are usually in the range of 200 to 300 daltons, or 15 to 20 carbon atoms, whereas alarm substances, which must act quickly, are in the range of 100 to 200 daltons and 6 to 15 carbons (Wilson 1970). Among some 120 species of moth (Lepidoptera), eighty compounds were identified as being components of long-range sex pheromones. They ranged in size from 10 to 21 carbon atoms and were mostly straight-chain hydrocarbons or oxidized hydrocarbons, with a mean molecular mass of 240 (Tamaki 1985). Flies (Diptera) almost all use as sex attractants hydrocarbons with 7 to 40 carbon atoms (Tamaki 1985). The alarm pheromones in insects are more diverse and are frequently terpenes (Blum 1985). Insect trail pheromones, which last for hours or days, are often more polar, with carboxylic acids being common (Haynes and Birch 1985).

Pheromones that are transmitted only through water are generally more polar and may be as large as proteins. Peptides are an obvious possibility for waterborne pheromones, which makes it interesting to ask how large they should be to provide sufficient specificity. Given that twenty amino acids compose proteins, there are 20^N different peptides N amino acids long. Thus, there are 160,000 possible tetrapeptides. However, all organisms have many kinds of proteins that can break down into a great variety of peptides, a process that could produce a given structure by chance. A rough estimate of the possible number of structures produced in this way by one organism is on the order of 10^5, assuming that a typical organism produces one thousand independent proteins that are each on average one hundred amino acids long. One might then guess that, to provide a reasonable degree of specificity, peptides should be used that have one hundred to one thousand times more possible structures. A hexapeptide meets this criterion with about 6×10^7 possible structures and can be considered the minimal useful size for a pheromone.

The properties of some of the better-characterized pheromones are presented in Table 7-5. Hölldobler and Wilson (1990, 263–269) list most of the pheromones that have been identified in ants.

Steroids, best known as internal information-carrying molecules (hormones), are also used as pheromones to communicate between individuals. Some examples are their use in induction of structures for sexual reproduction in a water mold (Raper 1970) and as a mating pheromone in pigs (Gower 1972). The widespread use of steroids for carrying information may be due to the fact that the fused rings of the steroids' structure provide a rigid framework upon which numerous modifications can be made and detected. Of the eight primary odors established for humans (Table 7-2), the lowest threshold is for the one steroid. On a molar basis it is ten-times more potent than the next most potent primary stimulus, despite its lower volatility.

Moth sex attractants are typically released at rates on the order of one nanogram per minute (Baker 1981, 1985; Wall and Perry 1987), corresponding to about 10^{10} molecules per second, although some insects release as much as a thousand times less (Mayer and Mankin 1985). Thomas Baker (1985) has pointed out that the moth sex attraction system is characterized by much lower emission rates than seems necessary, which he explains as providing the advantage of sexual selection for the most able males. Another clear advantage is that communicating with a weak signal makes it harder for others to tune in, which may keep predators and parasites from using the pheromone to locate the female themselves. A number of other types of pheromones are, in fact, exploited by parasites and carnivorous insects to locate prey (Albone and Shirley 1984, 91; Aldrich 1985; Haynes and Birch 1985). Another possible factor is that high sensitivity may be a by-product of high specificity. A highly specific receptor protein is likely to make many contacts with the molecule that activates it, which is likely to produce a high binding affinity for the molecule.

It has recently been discovered that certain moths release pheromones in **pulses**, at approximately one-second intervals, and a variety of possible functions have been proposed (Conner et al. 1980). A simple model (Dusenbery 1989a) suggests that a likely function is to increase the range of attraction for a given amount of pheromone. The argument is simply that the male needs only one whiff of pheromone above its threshold to know that there is a female upwind. Thus, it is advantageous to generate high peak concentrations. Quantitative analysis suggests that the effect is most important for short-range attraction in which diffusion has less effect

Table 7-5 A Sampling of Identified Pheromones.

Organism	Function*	Formula	Structure	Threshold[†]	Reference
Bacterium	Induction	$C_{13}H_{21}O_4$	Ring and tail	10^{-12}	Stephens 1986
Bacterium	Induction	$C_{45}H_{68}N_8O_{10}S$	Octapeptide	10^{-10}	Suzuki et al. 1984
Green algae	Induction	30,000 MW	Glyco-protein	10^{-16}	Maier and Müller 1986
Brown algae	Gamete attraction	C_8H_{12}	Linear	?	Maier and Müller 1986
Brown algae	Gamete attraction	$C_{11}H_{16}$	Ring	10^{-11}	Maier and Müller 1986
Water mold	Gamete attraction	$C_{15}H_{24}O_2$	Ring	?	Raper 1970
Water mold	Induction	$C_{29}H_{42}O_5$	Steroid	10^{-9}	Raper 1970
Slime mold	Aggregation	$C_{10}H_{12}N_5O_6P$	Cyclic AMP	10^{-9}	Devreotes and Steck 1979
Sea anemone	Alarm	$C_7H_{16}NO_4$	Linear	10^{-10}	Carr 1988
Ant	Alarm	$C_8H_{16}O$	Branched	10^9	Bradshaw and Howse 1984
Ant	Alarm	$C_{11}H_{24}$	Linear	10^7	Hölldobler and Wilson 1990, 261
Ant	Trail	$C_6H_9NO_2$	Ring	?	Bradshaw and Howse 1984
Ant	Trail	$C_6H_{12}O_2$	Linear	?	Bradshaw and Howse 1984
Ant	Trail	$C_{15}H_{24}$	Branched	?	Bradshaw and Howse 1984
Caterpillar	Trail	$C_{27}H_{46}O_2$	Steroid	?	Peterson 1988
Silk moth	Male attraction	$C_{16}H_{30}O$	Linear	10^3	Kaissling 1971
Meal moth	Male attraction	$C_{16}H_{28}O_2$	Acetate	10^4	Mankin et al. 1980
Gypsy moth	Male attraction	$C_{19}H_{38}O$	Branched	10^3	Elkington et al. 1984
Beetle	Male attraction	$C_{17}H_{33}O$	Branched	10^6	Shapas and Burkholder 1978
Fish	Induction	$C_{21}H_{32}O_3$	Steroid	$<10^{-9}$	Dulka et al. 1987
Pig	Aphrodisiac	$C_{19}H_{28}O$	Steroid	?	Gower 1972

* Induction is activation of the preparation for sexual reproduction in one sex due to the presence of the other sex.
[†] Threshold concentrations are given in molarity for pheromones normally transmitted through water (values << 1) or in the number of molecules per cm^3 (values > 1) for pheromones normally transmitted through air.

in smoothing out the peaks. In such cases the effective range can be increased tenfold by pulsing (Figure 7-2).

It is clear that most sex-attractant pheromones in moths are actually blends rather than single chemicals (Linn et al. 1984; Tamaki 1985; Linn et al. 1986; Linn et al. 1987). This provides many more opportunities for distinguishing the pheromones of one species from those of another. However, given that there are millions of possible chemical structures for thousands of species, the need for using blends is not clear. Some understanding is gained by realizing that most of the chemicals used in the pheromone blends of moths are closely related to one another and many can be generated by relatively small changes in biosynthetic steps (Steck et al. 1982; Bjostad et al. 1985). Thus, the use of blends can be understood as having evolved because it is much easier to change a blend by regulating enzyme activity than to synthesize a novel structure by creating a new enzyme. There is in fact evidence that different blends do form a reproductive isolating mechanism (Cardé et al. 1977; Tamaki 1985; Löfstedt et al. 1986).

Another potential reason for using blends of chemicals in a pheromone is that a blend might provide information about the distance to the source. With turbulence, concentration is so variable (see Figure 4-15) that the concentration of a single component is unlikely to provide useful information on the range to the source. It has been suggested (Tamaki 1985) that blends with differing thresholds for each component could provide distance information. Although this strategy does not provide any information that is not available from the concentration of a single component, sensory adaptation may limit the information that can be obtained from the concentration of a single component.

Another possible strategy would be to generate a pheromone blend with chemicals that differed in their diffusion rates and were well mixed when released. A receiver far away should experience an increase in the more rapidly diffusing chemical both before and after receiving a pulse of the slower-diffusing ones. The difference between how the two chemicals are received should decrease as the source is approached. Although this seems like it might be a useful strategy, there is no evidence that it is actually employed. Indeed, the chemicals in the blends of moth pheromones are generally so similar in size that there would be little difference in their diffusion rates. However, there could be related strategies based on differing rates of evaporation from released aerosol droplets or perhaps from releases from different sites on the female. A range-estimating strategy based on differences in diffusion rates would

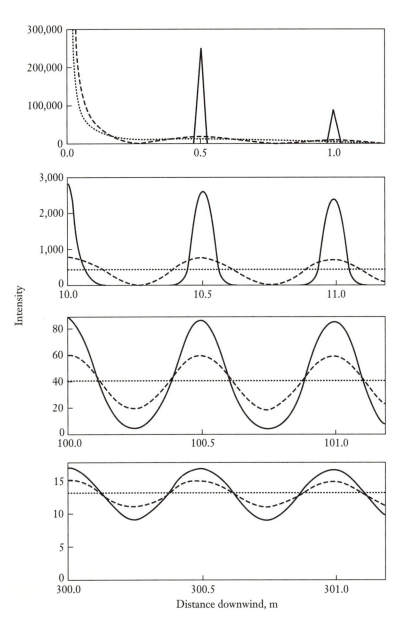

Figure 7-2 The range of a pulsed chemical release. A simple model assuming uniform flow was used to illustrate the potential effect of a pulsed release on the range of an attraction. Intensity (s/m^3) is concentration divided by the rate of release, shown here for various distances downwind. The solid line is for a series of instantaneous pulses, the dashed line is for sinusoidal release, the dotted line for constant release. (After Dusenbery 1989a.)

probably work better in water, where the diffusion rates vary more with molecular size.

Still another strategy for determining distance would be to use blends of chemicals with differing stabilities. For example, if a given chemical were oxidized in the atmosphere at an appropriate rate, its concentration relative to a stable compound in the blend would provide distance information. Alternatively, the release of a single compound would be sufficient if both the original and the oxidized form were detected. Although this strategy has not been demonstrated, it seems the most likely means for determining distance. In aquatic environments the opportunity for chemical alteration is greater than in air (Atema 1985), so that this mechanism might be common there.

7-5 TRANSMISSION OF CHEMICAL STIMULI

Information is transmitted slowly via chemicals. Ultimately, a stimulus must diffuse through an unstirred layer of water covering the receptor cells. If this layer is as thin as 30 μm, the characteristic diffusion time (see Section 4-2) is 0.1 s. Often, chemoreception is even slower. In the silkworm moth, for instance, reception and decay take nearly a second (Kaissling 1971). This consideration explains why olfactory receptors in vertebrates have among the smallest unmyelinated axons found in the body and consequently the slowest conduction velocity, producing in large animals a delay of as much as $\approx$ 20 ms (Moulton 1978).

Chemical stimuli are often transported by wind or water currents over distances that are much greater than if diffusion were the only mechanism. In nature, such flows are almost always turbulent in the size scales relevant to most organisms (see Section 4-4). Turbulence mixes together currents that both have and do not have the stimulus chemical, in complicated patterns that are not predictable. As a result, the receiving organism often experiences sharp pulses of stimuli (Figures 4-15 and 4-16), and it is difficult to send signals by pulsing the stimulus, although the advantage of doing so to obtain greater range with the same amount of pheromone remains, as explained in the previous section.

When currents flow through a stationary but porous substrate, such as wind through vegetation, the stimulus chemical may adsorb and dissociate from the substrate in a way that could smooth out the distribution of stimulus. In such a situation, movement of the stimulus would be analogous to the process of chromatography. There have in fact been some clear cases of such reversible adsorption (Perry and Wall 1986).

Atema (1988) has emphasized that in general chemical stimuli exist in patches, which may vary in size from micrometers to kilometers. The smallest patches exist in water around microorganisms or in small bits of organic matter (Alldredge and Cohen 1987). Since transport at such a small scale occurs entirely by diffusion, these patches are spherical—or hemispherical, if on a surface—and have highly predictable distributions. At larger scales, turbulence becomes important and much predictability is lost. In addition, currents distort the patch into a plume, which may permit much faster transmission at long ranges downstream. In aquatic environments, where vertical mixing is usually small, patches may exist as "pancakes" just a few meters thick and perhaps one kilometer in diameter. Fish may then search vertically for a layer with the appropriate chemical stimuli (Døving et al. 1985).

7-6 RANGE OF ACTION OF CHEMICALS

The range over which chemicals can carry information depends primarily on the transmission mechanism, the time available, and the ratio of the amount released to the threshold. Diffusion through water is a slow process (see Table 4-3), and the range of transmission in it is relatively short. For example, the characteristic diffusion length $(4 Dt)^{1/2}$ for a period of one day is about one centimeter in water and one meter in air.

The influence of the ratio of the released amount to the detection threshold (see Section 4-2) is demonstrated for diffusion in Table 7-6. Cases are known that extend the range of ratios over ten orders of magnitude (Wilson 1970), suggesting a range of communication distances of at least a thousandfold, simply by varying the sensitivity of the receiver as compared to the amount of stimulus released. If dispersal by currents is considered, the range of distances would probably be even larger.

Many organisms broadcast chemical *alarm pheromones* to help protect their related neighbors from danger. An alarm signal must act rapidly, but then fade away over an appropriate time interval. By their nature, chemicals are appropriate agents for transmitting alarm signals that do not have to travel far. For example, chemical alarm signals are common among ants, where they function through a few centimeters of air and persist for only a few minutes because of their rapid diffusion. On these small length scales, the situation can be modeled fairly accurately, assuming pure diffusion, as in Table 7-6 (Wilson 1970). In aquatic environments, a sea anemone releases a polar, seven-carbon compound that has a threshold of

Table 7-6 Calculated Range of Chemical Signaling.

Organism	Information	Amount Released (No. of mol)	Threshold Concentration (mol/cm^3)	Predicted Range* (cm)	Reference
Alga	Sex attraction	10^{11}	10^{11}	0.4	Maier and Müller 1986
Ant	Alarm	4×10^{15}	10^{12}	6	Wilson 1970
Moth	Sex attraction	10^{13}	10^{4}	400	Typical values

* The predicted range assumes instantaneous release and pure diffusion without flow, which is realistic only for distances of a few centimeters.

3×10^{-10} M. Termination occurs by rapid dilution of the signal in the turbulent environment. Other aquatic signals are terminated by photosensitivity or chemical reactivity (Carr 1988).

The attractant *sex pheromones* of moths probably act over the longest range of any chemical stimulus. However, evidence for the precise range of attraction remains poor (Wall and Perry 1987). Many claims to demonstrate a range of attraction have ignored the distance covered by the search phase (see Section 14-1) before the pheromone was detected. A critical summary suggests that the range of attraction for several species is at least several hundred meters, and probably more than one kilometer for some of the largest (Wall and Perry 1987). Once a male has arrived in the vicinity of a female, the male often produces a pheromone of its own that elicits the cooperation of the female in mating. These courtship pheromones are short range ones that act over a few centimeters or require contact (Baker 1985).

The attraction of insects to food or hosts is poorly documented, but it seems to be limited to a few tens of meters outdoors (Finch 1980) and about one meter in "still" air (Hidaka 1972; Hagstrum and Davis 1982).

7-7 CHEMICAL TRAILS

Chemical stimuli are distinctive in that they persist in the environment. Consequently, they are often used to mark territory or trails. The trail pheromones of ants are similar to the pheromones they use for other purposes but are normally discharged as dilute aqueous solutions, which presumably slows their evaporation. The trails may persist from a few minutes to days (Bradshaw and Howse 1984).

An interesting question about chemical trails is whether they contain information distinguishing one direction from the other. At first one might assume that the chemical gradient along the axis of a trail could be followed, but this is probably thousands of times smaller than the gradient across a trail and it is doubtful that such a gradient could be followed. Evidence from insects indicates that they cannot determine the direction in which a trail was laid (Baker 1985; Haynes and Birch 1985; Hölldobler and Wilson 1990, 271). In contrast, snails seem to be able to determine the direction of a trail (Wells and Buckley 1972) by reference to unknown cues, and snakes are able to determine the direction another snake has taken, by using the pattern of pheromone deposition (see pages 268–269).

In other cases, stimuli inadvertently left behind are used by predators to locate prey. A well-known example is the ability of dogs to track other animals based purely on the scent remaining. This seems like magic to humans, but let us try to analyze the situation quantitatively. In spite of stories to the contrary, it appears that in practice a track can be followed only for a few hours, and even less in hot, sunny conditions (Gibbons 1986). The dog's basic strategy is to move back and forth across the trail, sniffing with its nose close to the ground (Gibbons 1986). It is not clear whether the dog relies on sniffing air containing only volatilized chemicals or whether, like snakes, it takes in particles as well. It is also difficult to estimate the nature and quantity of the chemicals lost from an animal that potentially become available on the track, which thwarts any rigorous analysis. However, the feasibility can be illustrated by the following considerations. The amount of 5α-androst-16-en-3α-ol excreted by a human could, when oxidized, be diluted to 5,000 l per day and still be detectable by the human nose (see page 134). If this solution were distributed along a 10 km-long daily track, there would be half a liter of odorous solution for every meter of track. This suggests that there is sufficient odorous excretion from an animal that even the human nose could potentially follow a trail—if the nose were close enough to the substrate.

Chemical trails may also be left by inanimate sources such as debris sinking in water. Shrimp and other planktonic animals have been shown to follow such vertical food trails (Hamner and Hamner 1977). The shrimp normally swim horizontally but, upon contacting the scent trail of a sinking piece of food, they swim downward along the trail to the source.

They apparently adopt the strategy of assuming the trail was created by a sinking (rather than an ascending) object and swim downward whether or not that is the direction of the source. This again suggests that there are few clues to the direction of a chemical trail.

CHAPTER 8

Light

■ ■

In every object there is inexhaustible meaning;
the eye sees in it what the eye brings means of seeing.
To Newton and to Newton's Dog Diamond,
what a different pair of Universes.
—Thomas Carlyle, *The French Revolution*, 1837

Light is one of the stimuli most utilized for obtaining information. Nearly all organisms live where light is available at least some of the time, because sunlight is the ultimate source of energy for all organisms in almost all ecosystems. Some bacteria, and most plants, obtain information from light, as do all but a few arthropods and vertebrates. Light is a distinctive stimulus in that its intensity, direction, frequency, and polarization can all convey useful information. This stimulus is the easiest for us to comprehend, because vision is the sense we pay most attention to and because few animals surpass us in their ability to extract information from light. Most animals obtain information about objects in their environment from reflected sunlight, but a few generate their own light, for sounding their environment (see Chapter 12) or communication with other animals (Chapter 13).

8-1 LIGHT RECEPTORS

Light-reception organs come in a wide range of sizes and designs. In single-cell organisms they are relatively simple. For example, photosynthetic bacteria use light to move toward optimal light intensities (Clayton 1965; Hildebrand 1978), but most bacteria must use the strategy of klinokinesis (see Section 15-3), because they are too small to develop light receptors that have directional specificity (Foster and Smyth 1980). In contrast, single-celled eukaryotes are large enough that some have a photoreceptor area that is made directional by associating it with an absorbing or reflecting screen; these organisms frequently orient their swimming direction toward or away from the source of illumination.

Plants also obtain information about their environment from light, in addition to obtaining energy for photosynthesis. The higher plants seem to have at least two sensory photoreceptor pigments: **phytochrome** and one or more unidentified blue-light receptors (Pratt and Cordonnier 1989). Phytochrome exists in two interconvertible forms (Hoober 1984, 252–55). The red form absorbs maximally near 670 nm, whereas the far-red form absorbs maximally near 730 nm. The light absorbed by one form tends to convert that form into the other one so that in the presence of light the ratio of the two forms provides information about the ratio of red to far-red light in the environment. Since the absorption of photosynthetic pigments cuts off at about 700 nm and plants are relatively transparent to longer wavelengths (Yocum et al. 1964), this ratio provides a good indication of the presence of competing plants in the environment (Kasperbauer 1987; Bradburne et al. 1989; Ballaré et al. 1990). The phytochrome system is also capable of indicating the intensity of white light, because the red form is the first to be synthesized and the far-red form disappears in the dark; thus, the degree of concentration of the far-red form is indicative of the light intensity. The sensitivity of the phytochrome system is sufficiently high that it functions at light levels well below those required for photosynthesis. Many seeds use this sensitivity to determine whether they are close to the soil's surface and not too heavily shaded in order to make the decision to germinate (Pratt and Cordonnier 1989).

The **blue light receptors** mediate a variety of responses to low levels of blue light in some algae (Häder 1988) and fungi, as well as in plants (Pratt and Cordonnier 1989), but numerous attempts to identify them chemically have so far been inconclusive. The majority opinion suggests that these include a flavin molecule (Presti and Galland 1987; Pratt and Cordonnier 1989). In any case, the peak of activity is for a broad band of wavelengths centered near 450 nm. Neither of the photoreceptor pigments found in plants is associated with any specific structure.

Most animals—or at least the arthropods, mollusks, and vertebrates— seem to employ the same type of photopigment, called **rhodopsin**, for light detection. Rhodopsin consists of the twenty-carbon molecule 11-*cis*-retinal (or the 3,4-dehydro derivative, which absorbs at longer wavelengths) coupled to a membrane protein called opsin. Opsins vary in the extent to which they shift the absorption spectrum of retinal from the ultraviolet into the visible range. There is thus a wide variety of photopigments that absorb light most efficiently in different parts of the spectrum. There is also good evidence that some algae use rhodopsin (Foster et al. 1984; Martin et al. 1986), although other algae do not (Song 1983; Häder 1988), and an archaebacterium employs a receptor similar to rhodopsin (Schimz et al. 1982; Martin et al. 1986).

These photoreceptor pigments have been incorporated into specialized sensory structures with a variety of designs during the course of evolution (Figure 8-1). It is important to recognize that the photoreceptors of vertebrates are fundamentally different from those of arthropods. The design of the arthropod type makes them inherently sensitive to the polarization of light, because the membranes are in cylinders with axes perpendicular to the light path, whereas the vertebrate type is not sensitive to polarization, because its membranes are disk-shaped, with axes parallel to the light path.

The larger animals are able to determine the direction of many light sources simultaneously by employing an array of receptor cells in an eye that restricts the light impinging on different receptors to light that is coming in from corresponding directions. The amount of directional information thus obtained is generally proportional to the number of receptors in the array. There is a restriction on the minimum effective size of those receptor cells, to a radius approximately equal to the wavelength of light. The spacing cannot be less than twice this radius (spacing 1 μm) without photons "leaking" from one receptor to another (Land 1981). Consequently, the larger the eye, the more information it can obtain. In flying insects and birds (Pearson 1972, 279), the eyes may even be larger than the brain.

The simplest eyes are **pigment cup** and pinhole eyes in which directional specificity is provided by using an absorbing screen to admit only light passing through a small aperture to an array of receptor cells. Such eyes, which are found in a variety of simple invertebrates, are usually less than 0.1 mm in diameter. The best-known one has about 200 receptor cells (Land 1981). One disadvantage of this simple design is that to have high resolution requires an aperture much smaller than the array of receptors, so that most of the light that might fall on the receptors is instead screened out.

More sophisticated designs employ a **lens** that changes the direction of light by refraction (see Section 4-2), or use reflecting elements to direct light from a given direction onto a specific receptor cell. Since no photons need be discarded, the sensitivity is improved.

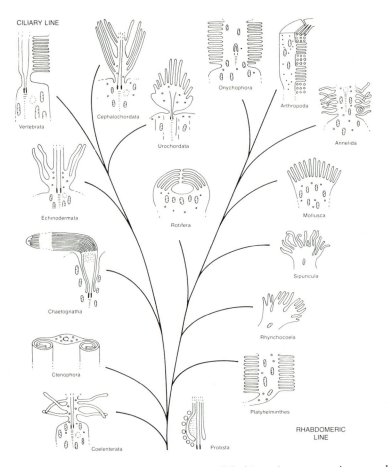

Figure 8-1　Photoreceptor cells have been modified in various ways to increase the amount of membrane area present in a small volume. There are two main lines of evolution. *Left:* The *ciliary line* led to the rods and cones of vertebrates, in which membrane disks are stacked along the direction of light propagation. This arrangement is symmetrical around the optical axis and is insensitive to the polarization of light. *Right:* The r*habdomeric* line led to the photoreceptor cells of arthropods, in which membrane cylinders are packed together, with the cylinder axes perpendicular to the direction of propagation of light. Such a cylinder composed of rhodopsin-containing membrane will absorb light polarized parallel to the cylinder axis more efficiently than light polarized perpendicular to the cylinder axis. Many photoreceptor cells contain groups of cylinders oriented perpendicular to one another, which eliminates the sensitivity of the receptor cell to the polarization of light. (From *Introduction to Nervous Systems*, by Bullock, Orkand, and Grinnell, W. H. Freeman and Company, 1977.)

Compound eyes consist of a hexagonal array of elements called ommatidia that each contain a small number of receptors (8) and optical elements that project light from a particular direction onto the appropriate receptor cells. The eyes of both crustaceans and insects are of this type, although there is a great deal of variation in the details of how the different systems work (Land 1981). This is an efficient design for small eyes (Kirschfeld 1976). In a compound eye of radius r, where adjacent ommatidia look in directions differing by ϕ radians and the diameter of the aperture of each ommatidium is d (= the distance on the surface between adjacent ommatidia), an eye that is optimized for resolution to the point where diffraction effects limit the resolution will have $\phi d = \lambda/2 \approx 0.2$ μm (Land 1981). Eyes optimized for light gathering will have a larger value for the product ϕd. In fact sensitivity (the photon catch) increases in proportion to $(\phi d)^2$ (Land 1981). These relations are useful because the two parameters ϕ and d can be determined by superficial observation of an eye. This calculation also indicates that to improve its resolution the eye must increase in size, specifically,

$$\lambda/2 = \phi d$$

$$= \phi r \sin\phi$$

$$\approx r \phi^2 \tag{8-1}$$

for small angles.

The resolution of a compound eye increases with the square root of the size of the eye (Kirschfeld 1976), which explains why some insects have eyes larger than the rest of their head.

In contrast, although a **single-lens eye** has a larger minimum size than a compound eye, its resolution increases in direct proportion to its size. Thus the advantages of larger eyes are best captured by single-lens eyes, and all large animals have eyes of this design (Kirschfeld 1976). Single-lens eyes probably cannot be made in small sizes and still have both high resolution and sensitivity over a wide field of view (Wehner 1981, 546). Spiders have single-lens eyes but employ several pairs that point in different directions to cover a wide field of view (Figure 8-2).

Vertebrates all have single-lens eyes of similar structure, except for their loss in a few cave-dwelling species. In these eyes a **lens** focuses light on a layer of millions of receptor cells and neurons called the **retina**. The

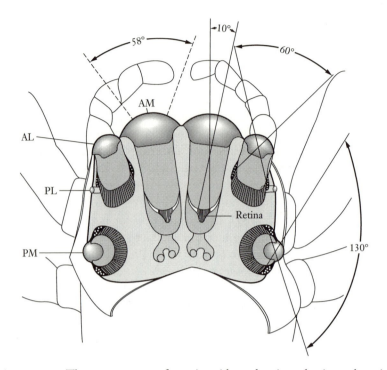

Figure 8-2   The arrangement of eyes in spiders, showing a horizontal section through a jumping spider. The two main eyes (anterior median, AM) have a small retina with only 10° field of view that can be moved with muscles to cover a total field of 58°. The other eyes have a fixed field of view, as indicated. The three main pairs of eyes cover all the horizon but a 30° sector to the rear. (After Foelix 1982, Figure 68.)

receptor cells are of two types: **rods,** specialized for dim light, and **cones,** specialized to function in bright light. In mammals the cones include several types, each containing a different photoreceptor pigment that confers sensitivity to different parts of the spectrum. Back of the retina is a layer of dense black pigment that prevents light from behind the eye reaching the receptors. In some benthic fish there are multiple layers of receptors, which probably increases the chances of the light being absorbed (Fernald 1988).

In some animals there is a reflective layer, the **tapetum,** in back of the retina to give light that was not absorbed in the first pass through the retina a second chance at it. This is why the eyes of some animals seem to glow when illuminated with a flashlight at night. The reflective eyes of spiders can often be seen at night by holding a flashlight close to one's head.

Across the retina there are often specialized areas for extracting different kinds of information from light coming from various parts of the environment. The separate parts of the retina may differ in their spectral sensitivity, with one or more areas being specialized for acute vision. Although such areas are usually circular (Lythgoe 1979, 149), there are horizontal bands present in fish that live near the air–water interface as well as in birds (Pearson 1972, 295–300), mammals, and perhaps arthropods that live in open, flat habitats (Wehner 1987). All these bands probably look at the horizon in the animals' normal posture. An even more specialized region, called a **fovea**, may be present within such an area. Many birds have two foveae, one that looks laterally along the optical axis, while the other looks forward in the binocular field of view.

In addition to there being a wide range in the number of receptor cells per eye, there is also a widely varying number of eyes per organism (Table 8-1). Arthropods usually have many eyes. Insects, for instance, often have six pairs of eyes, and other invertebrates have many more, with spiders having eight, scorpions, twelve, and Collembola, sixteen. Flying insects usually have one pair of large compound eyes and several pairs of simple **ocelli,** which are probably more sensitive and can provide information on light intensity over broad areas more rapidly than can the compound eyes. The ocelli probably provide information on the orientation of the horizon and help stabilize flight (Wehner 1987; Möhl 1989). In these circumstances, the best way of comparing the total amount of information an organism obtains from light is probably to count the total number of receptor cells.

A vertebrate eye images light coming from a range of directions (the **field of view**), including about 180°, or halfway around. Nearly all vertebrates have a bilateral pair of eyes, but there is much variation in where they are placed on the head. If they are on the side of the head, as in a rabbit, the total field of view can be 360° and the animal can see in all directions without moving its head. If the eyes are on the front of the head, as in primates, there is a large **binocular** field of view seen by both eyes, which improves sensitivity as well as depth perception (Lythgoe 1979, 148) (see pages 187–190).

In addition to these well-described, specialized light receptors there exists a variety of evidence that many organisms possess poorly charac-

Table 8-1 Number of Eyes and Receptor Cells in
              Various Organisms

No. of Eyes	No. of Receptor Cells per Eye	Organism	Reference
1	1	Unicellular algae	Foster and Smyth 1980
10^3	1	Volvox	Foster and Smyth 1980
2	200	Flatworm	Taliaferro 1920
1	1	Rotifer	Eakin and Westfall 1965
2	1	Certain nematodes	Burr 1984a
10^3	10	Chiton	Land 1984b
80	10,000	Scallop	Land 1984b
1	176	*Daphnia*	Macgno 1973
2+	8,500	Horseshoe crab	Land 1984a
8	10 to 10,000	Spiders	Foelix 1982, 85, 88, 92
12	~50	Ant lion	Land 1981
12	6,300	Tiger beetle	Land 1981
2+	80,000	Housefly	Shepherd 1988,,333
2	130,000,000	Humans	Levi 1980, 349

* 2+ signifies that the organism has two principal eyes, with additional much smaller eyes.

terized responses to light that are not mediated by any known receptors. In animals this is sometimes referred to as a **dermal light sense** or **extraretinal photosensitivity** (Wales 1975; Lythgoe 1984; Menaker 1989). In some cases, light reception may be mediated by heating or photochemical damage such as sunburn, and it might be considered a physical interaction rather than an informational one. On the other hand, some such interactions are probably caused by the presence of low concentrations of photoreceptive pigments in unspecialized cells. It has been demonstrated recently that an unpigmented nematode with no eyes exhibits a subtle response to light, having an action spectrum consistent with a rhodopsin photoreceptor, and carries genes for rhodopsin (Burr 1985 and personal communication). In some mollusks, defensive responses to shadows falling on the animal are mediated by such dermal light senses (Land 1984).

8-2 FREQUENCY

Light is an electromagnetic wave that oscillates at very high frequencies, about 10^{15} Hz. The frequency of oscillation (f) is what determines the perception of different colors, technically called **hues.** Frequency is inversely related to wavelength (λ), because the speed (v) is usually constant and $v = \lambda f$. Given this relation and the fact that it is usually more practical to measure wavelength than the very high frequencies, the distribution of light intensity over its frequency (a **spectrum**) is usually presented in terms of wavelength. Nonetheless, frequency is the more fundamental quantity. When light passes between different media its wavelength changes when its speed changes, but its frequency remains constant. In addition, frequency directly affects absorption, and absorption spectra are more symmetrical when displayed on a scale that is linear in frequency. Researchers often compromise between the two parameters by presenting data in units of **wave number,** the reciprocal of wavelength, which is proportional to frequency.

AVAILABLE FREQUENCIES

Almost all light in the natural environment originates from the sun. Its spectral distribution is similar to that of an efficiently radiating surface, known as a blackbody (described in Section 6-4) at a temperature of 6,000 °C (Figure 8-3). In passing through the atmosphere, a small proportion of this light is absorbed, and some is scattered. Short wavelengths are strongly scattered, and ozone absorption effectively eliminates wavelengths less than 300 nm. At long wavelengths, water vapor, carbon dioxide, and oxygen absorb light significantly at particular wavelengths, producing sharp dips in the spectrum, as shown in Figure 8-3. At still-longer wavelengths all objects in the environment become significant sources of radiation, depending on their temperature (see Section 6-4) and surpass sunlight in intensity beyond λ = 5 μm. These characteristics of the environment restrict the range of electromagnetic radiation that is most useful for obtaining information. There is little radiation below λ = 300 nm, and above a few micrometers, everything radiates from thermal motion at normal temperatures.

Light scattered from the atmosphere, which is what creates the sky, is often an important source of illumination that can also provide information about the direction of the sun (see pages 186–187). Because the

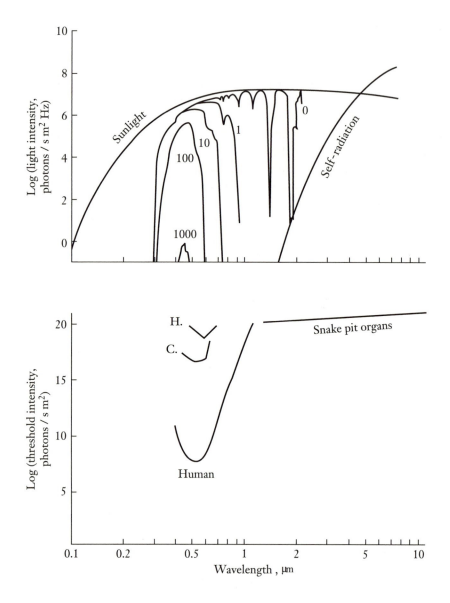

scattering by molecules decreases with the fourth power of the wavelength, short wavelengths make up a greater proportion of the light from a "clear blue sky" (Figure 8-4). Particles and droplets scatter light with less dependence on wavelength and, as can be seen in Figure 8-4, haze is closer to sunlight in color.

Figure 8-3 (*Opposite page*) Light intensity and threshold distributions. *Top:* The spectral light intensity (irradiance) distribution for different environments. "Sunlight" is the intensity from a blackbody (Davidson 1962, 218) of the temperature (6000 °C), size (radius = 7×10^8 m), and distance (1.5×10^{11} m) of the sun. "Self radiation" is the blackbody radiation of a source at 20 °C; all objects in the environment will radiate at close to this level. The curve labeled 0 is the estimated intensity of sunlight at sea level (Moon 1940, with estimates of the depth of the two largest dips from water vapor absorption data by Jeske 1988, 138). The loss of the short wavelengths is due to scattering and ozone absorption. Most of the sharp dips are the result of absorption by water, which is variable but is assumed to be equivalent to 2 cm in depth. The curves labeled 1, 10, 100, and 1,000 are the intensities of sunlight under water at a depth of the label in meters. The curves are calculated from the data for the light intensity at the surface (0), for the absorption of the clearest natural ocean waters (Højerslev 1986, 398), and for pure water above 800 nm (Curcio and Petty 1951). This calculation gives the irradiance averaged over all directions, in contrast to the often given downward irradiance.

    Bottom: The threshold intensity for several organisms. The curve labeled *H* is for the light attraction of the bacterium *Halobacterium* (Hildebrand and Dencher 1975). The curve labeled *C* is phototaxis of the flagellated alga *Chlamydomonas* (Foster et al. 1984). The curve marked "Human" is the dark-adapted threshold of human vision for light from a point source (C.I.E. standard; Sliney et al. 1976). "Snake pit organs" is the estimated sensitivity for pit viper detection of infrared radiation, which is based on heating the receptor (see Section 6-4).

    In terrestrial environments, vegetation often limits the light available. Chlorophyll captures blue and red photons quite efficiently. Consequently, plants have a minimum absorption of green light (Figure 8-5). They also absorb relatively little near-infrared (0.7–2.5 µm) light, which reduces their heating, and because they absorb strongly at longer wavelengths they can lose heat efficiently by radiation (Gates et al. 1965).

    As light passes through a few meters of water, absorption causes a narrowing of the spectral distribution. Pure water has a minimum absorption of blue light of 430 nm. Deep in completely clear water, light intensity averaged over all directions would be maximum at this wavelength. The minimum attenuation coefficient for downward propagation is shifted to longer wavelengths of 450 nm, because of the greater scattering of shorter wavelengths (Højerslev 1986, 398). This would be the wavelength of maximum intensity for illumination from above and is the parameter that is usually quoted in the literature. In the clearest natural waters the maximum intensity occurs at about 475 nm (Lythgoe 1979, 19), because of

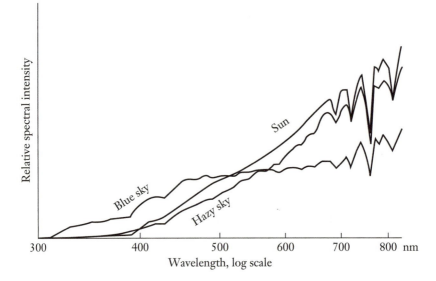

Figure 8-4   The spectrum of skylight. Intensity is proportional to the number of photons per unit of time per unit of surface area per unit of frequency interval (data for sunlight from Moon 1940; for skylight, Judd et al. 1964.) Note that a hazy sky has nearly the same proportion of different wavelengths as does sunlight, but clear sky contains relatively more short wavelengths.

various absorbing components other than water. Waters that contain enough nutrients and sunlight develop high levels of phytoplankton, and their chlorophyll may dominate light absorption, causing a minimum absorption at the green wavelength of 550 nm. Decomposing of organisms frequently produce a complex residue of humuslike material that absorbs increasingly strongly at shorter wavelengths and thus has a yellow hue (Lythgoe 1988). Such material is an important contributor to light absorption in some waters and, when highly concentrated, produces brown or even "black" waters. Soil also has a nearly linear reflectance spectrum, which corresponds to increasing absorption at shorter wavelengths (Kennedy et al. 1961).

Another restriction on the range of useful wavelengths is the sensitivity with which radiation of different frequencies can be detected. The most sensitive detection mechanism involves exciting a molecule to a higher energy electronic state upon its absorption of a photon. The excited molecule then triggers a biochemical reaction (*photochemistry*). For

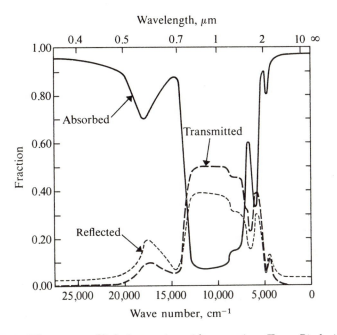

Figure 8-5 The spectra of light interaction with vegetation. (From *Biophysical Plant Physiology* from Nobel, 1974, 351)

excitation to be possible the molecule must have an available electronic state differing in energy from the regular "ground" state by the energy of the photon. If this energy gap is small, either thermal collisions between molecules or thermal radiation may excite the molecule, thus producing noise in the system. The long-wavelength sensitivity of human vision can in fact be explained by an energy gap of 1.4×10^5 J/mole in rhodopsin (Ben-Yosef and Rose 1978). Figure 8-6 shows the potential stability of molecules with various energy gaps between the ground and excited states. It can be seen that at terrestrial temperatures stability increases rapidly above one second for energies greater than about 1×10^5 J/mole. Energies this large are found only in photons with wavelengths of less than about one micrometer. Consequently, at normal temperatures radiation with wavelengths less than that can be detected much more efficiently than those of longer wavelength.

The only organisms known to make use of longer wavelengths are certain snakes having a detector that simply converts the radiation to heat and measures the subsequent temperature rise. This mechanism is much

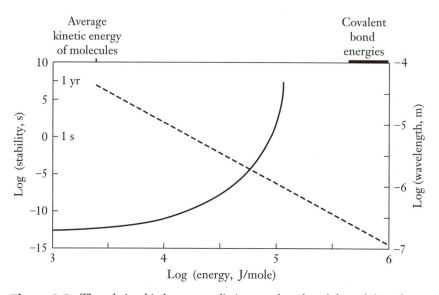

Figure 8-6 The relationship between radiation wavelength and the stability of a chromophore capable of absorbing it. The energy of a photon is inversely proportional to its wavelength (*dotted line*). The stability of a photoreceptor molecule (*solid curve*) is an estimate of the lifetime of a ground state before thermal excitation to an excited state, with the energy indicated. The calculation of the lifetime is the product of the collision frequency (10^{13} Hz) and the proportion of collisions having energy greater than the indicated energy E at ordinary temperatures (= $\exp[-E/k_B T]$). For an excited state lifetime greater than one second, an energy gap greater than 10^5 J/mole is required, which means that the wavelength of an exciting photon must be less than one micrometer. Molecules that absorb longer wavelengths will frequently become excited by thermal agitation and produce receptor noise. Radiation with wavelengths less than 300 nm has enough energy to break some covalent bonds.

less sensitive than photochemistry, as seen in Figure 8-3. Since the mechanism is based on heating, this strategy was discussed along with that of temperature stimuli (Section 6-4).

Another restriction on the useful frequency range is that shorter wavelengths are absorbed more and more strongly by more and more materials. Even air absorbs so strongly below 200 nm that laboratory instruments must operate in a vacuum. Proteins absorb strongly below 300 nm. Consequently, the absorption of shorter wavelengths restricts not only the available illumination but the distance that light can be transmitted through the environment as well as the design of light receptor organs.

The factors just discussed limit the wavelengths that are useful for obtaining information to the range of 300 to 1,000 nm. Human vision spans a visible range from about 400 nm for violet to 800 nm for red light. Shorter wavelengths are called **ultraviolet**, longer ones **infrared**. Although some vertebrates can see a degree of ultraviolet light, only insects are known to have receptor cells with a maximum sensitivity to ultraviolet light. The peak absorption of the receptor molecule with the shortest-known wavelength sensitivity is 345 nm, in insects; the longest known is 625 nm, in fish (Lythgoe 1979, 63).

The rhodopsin pigment itself absorbs in the ultraviolet as well as in the visible, but sensitivity to the ultraviolet is reduced by absorption in the eye before the light reaches the retina. Indeed, people who have had a lens removed because of a cataract can see well in ultraviolet light (Wald 1945). The lens acts as a short-wavelength cutoff filter, transmitting in young adults 15 percent at 405 nm and 0.1 percent at 365 nm on average. There is a great deal of individual variability, however, and the lens absorbs more with increasing age. This is probably one reason that different individuals sometimes perceive colors differently. Another probable factor is genetic variation between individuals in genes that code for opsins (Neitz et al. 1991).

The photoreceptor pigments of closely related species often differ in their wavelengths of maximum absorption. Are the pigments in a particular species adapted to the spectral distribution of its environment? The simplest case is that of fish living deep in clear ocean water where the available light is restricted to the blue wavelengths. Their rod pigments do indeed have their maximum absorption at the expected wavelength (Figure 8-7). Fish living in greenish coastal waters or yellowish fresh water have rod pigments that are shifted to longer wavelengths, but the shift is not large enough that the maximum absorption coincides with the maximum intensity. Deep-diving marine mammals and deep-sea crustaceans also seem to have pigments shifted appropriately toward the blue. Animals that migrate between habitats may also appropriately change the nature of their photoreceptor pigments (Lythgoe 1979, 87–89).

However, sensitivity is not the only consideration where the purpose is to discriminate between objects based on their light reflectance. If light levels are sufficiently high that receptor noise is not limiting, the contrast between an object and its background will also be important and the contrast may be higher at some wavelengths that differ from the wavelength of maximum intensity. Because of the square-root relationship

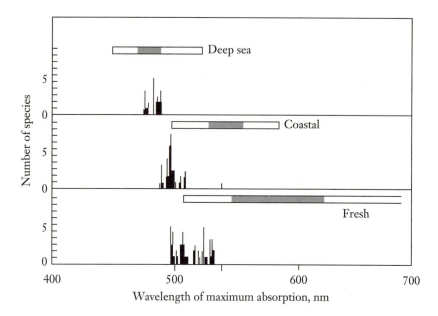

Figure 8-7 Receptor pigments in fish. *Vertical axis:* The number of fish species containing single-rod pigments of a known absorption maximum, grouped by environment. *Horizontal bars:* The range of wavelengths that would give maximum sensitivity in each type of environment. There is good correspondence to the maximum in the available light in the deep ocean. In shallow-water species there is a shift toward the red, in accordance with the fact that the available light is more red. However, the shift in pigments is clearly less than the shift in the available light. (After Lythgoe 1988, Figure 3.7.)

between relative statistical noise and sample size (see equation 5-2), a twofold increase in contrast can compensate for a fourfold decrease in intensity. Consequently, photoreceptor pigments may be chosen to absorb maximally either at wavelengths where the important contrasts are greatest or at a point of optimal compromise between contrast and intensity (Lythgoe 1984).

To summarize, the range of wavelengths exploited by organisms is exceedingly narrow compared to the range of electromagnetic radiation that is present in nature, which extends from cosmic rays ($\lambda = 10^{-14}$ m) to radio waves (10^4 m). Humans have learned to exploit some of these other types of radiation to acquire information, but the techniques involve nonbiological processes.

FREQUENCY DISCRIMINATION—COLORS

A molecule can be excited to a given electronic state by photons having a range of wavelengths near the one that is most efficiently absorbed. For example, the absorption spectrum of rhodopsin has a half-maximum width of about 1.2×10^{14} Hz, or 100 nm. Nonetheless, once a photon has been absorbed, the excited molecule retains no information about the nature of the photon that produced the excitation. For instance, a given level of stimulation of a green-sensitive photoreceptor could be produced by a green light or a brighter yellow light. Consequently, the frequency distribution (color) of light can be determined only by comparing the degree of stimulation of photopigments that absorb different frequencies of light. For example, human color vision is based on three pigments, which have absorption spectra centered on 420 nm (violet), 534 nm (green), and 564 nm (yellow-green) (Lythgoe 1979, 46). The relative degree of stimulation of these three pigments provides the information for the thousands of hues we sense. Pure spectral colors only 1 to 2 nm apart can be discriminated. Human color discrimination is sharpest between 500 and 600 nm but degrades rapidly below 440 and above 650 nm (Levi 1980, 407). In contrast, the bacterium *Halobacterium* (Armitage 1988) and plants detect hues by using a single molecule that exists in interconvertible forms with different absorption spectra, as explained in Section 8-1. Many animals, especially birds, use color filters to modify the color specificity of certain receptor cells (Bowmaker 1977).

Most insect receptors that have been investigated have their maximum sensitivity near 350 nm (ultraviolet) or 520 nm (green) (Lythgoe 1979, 188). The honeybee, which has been among the most carefully studied, has receptors with maxima at 345, 440, and 550 nm, with its best hue discrimination at 400 nm (Lythgoe 1979, 189). Insights into the visual perception of many insects can be gained by studying flowers, which have evolved to attract insect pollinators. As seen in Figure 8-8, a major feature of many flowers is that they look uniformly yellow to humans but have dramatic patterns in the ultraviolet (Eisner et al. 1969), a clue that they are insect pollinated. In contrast, many butterflies (Swihart 1970; Bernard 1979) and birds have more sensitivity to red light, and red flowers are usually pollinated by birds (Hinton 1973).

Since the sensation of hue is fundamentally based on patterns of stimulation across receptor cells having different frequency specificities, the other aspects of color sensation are discussed in Section 11-5, under stimulus patterns.

Figure 8-8 The ultraviolet light pattern in flowers. *Above,
top and bottom:* Flowers as seen by the human visual system.
Facing page, top and bottom: Flowers as seen in ultraviolet
light. *Top:* Marsh marigolds. *Bottom:* five species of yellow-
petaled Compositae from central Florida. (From Eisner et
al. 1969.)

COLOR GENERATION

The distribution of light intensities over frequency or wavelength, which
produces the perception of color, contains useful information primarily
because different objects in the environment vary in the relative effectiveness with which they reflect ambient light of different wavelengths. Thus,

vegetation is green because chlorophyll strongly absorbs blue and red light and can be distinguished from brown dirt, which absorbs shorter wavelengths more strongly than longer ones. Subtle changes such as the ripening of fruit are readily apparent to an animal with good color vision.

In the natural environment, yellow, orange, and brown colors are generally produced by pigments that absorb light having wavelengths below a certain value. They consequently act as long-pass (long wavelength) or low-pass (low frequency) filters. The cutoff frequency determines the color perceived (Lythgoe 1979, 31). Mixtures of organic materials tend to absorb more strongly at the shorter wavelengths within the visible and ultraviolet range and thus to act as long-pass filters. The cutoff wavelength moves to longer wavelengths with more and larger conjugated double-bond systems in the molecules of the material, and it shifts from acting as an ultraviolet filter to a yellow filter, then to brown and finally to black.

Structural colors are produced by separating photons of different wavelengths rather than from differential absorption. **Scattering** by small particles or molecules is more effective for short wavelengths than long. Therefore, scattered light is bluer than the incident light, and unscattered light is redder. This is the mechanism that produces blue oceans, blue sky, and red sunsets. The blue colors found on animal surfaces are often produced by scattering.

Another type of structural color is **iridescence**, which is produced by a stack of parallel, transparent layers of different refractive index. Reflections from different layers interfere constructively or destructively with one another, depending on wavelength and orientation of the light. In direct sunlight, which provides a parallel beam of light, the appearance of an iridescent surface may vary dramatically in color and intensity as its orientation is altered.

Some organisms actually produce light themselves, by chemical reactions. Such **bioluminescence** is another source of light that is important for acquiring information. In this case, organisms are not constrained to employ the light frequencies that exist in the environment, and there are advantages to using different frequencies. Although no organism is known to produce light outside the visible range, some deep-water fish with reddish bioluminescence may have a relatively private channel, because of the preponderance of photosensitivity in the blue-green region of maximum light intensity in their environment (Lythgoe 1979, 194). There is also some variation in the colors of the bioluminescence produced by fireflies—it has been argued that they maximize the contrast for the typical lighting conditions existing when they are active (Lee 1989).

8-3 SENSITIVITY

Organisms have developed photoreceptor cells that operate close to the maximum possible sensitivity of detecting single photons. Within this limit there is a wide scope for compromise, between minimizing eye size and maximizing sensitivity, spatial resolution, temporal resolution, and color perception. Figure 8-9 illustrates the effects of photon noise on sensitivity as well as spatial and temporal resolution.

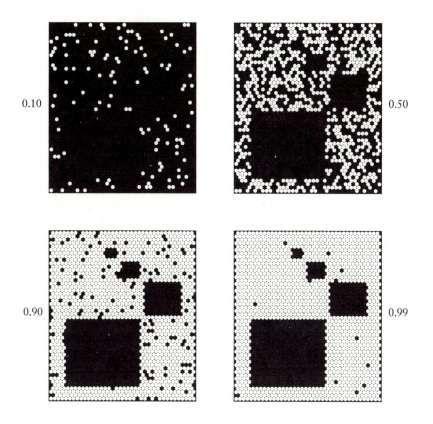

Figure 8-9 Photon noise. This is a model of a receptor array stimulated with an image of five dark rectangles that are arranged in order of size along diagonals and against a bright background. Each white circle represents the activation of one receptor cell in the array during a certain time interval. The activation patterns are illustrated for light intensities and time intervals that activate each receptor with probabilities of 0.1, 0.5, 0.9, and 0.99, as indicated. The larger rectangles are apparent under lower stimulation. Even with a probability of activation of 0.99, the one-receptor-sized rectangle is not clearly defined.

DETECTION SENSITIVITY

All known photoreception mechanisms in animals are based on having rhodopsin molecules embedded in a membrane. The simplest probably consist of a patch of rhodopsin molecules in the plasma membrane (Burr 1984b). In the cases examined, the density of packing of rhodopsin appears to be about 20,000 molecules/μm^2 of membrane (Krebs and

Kühn 1977). Given the absorption of rhodopsin at the wavelength of maximum absorption ($\varepsilon \cong 40,000$ M^{-1} cm^{-1} (Stavenga and Schwemer 1984)), the fraction of light absorbed in passing perpendicularly through a stack of N_m such membranes is given by

$$\frac{I_0 - I}{I_0} = 1 - 10^{-0.00013\,N_m}$$

(8-2)

The fraction absorbed is only 0.0003 for a single membrane, which greatly limits its sensitivity.

Single-cell organisms that use rhodopsin probably have only a small patch of a single membrane containing thousands to millions of rhodopsin molecules (Foster and Smyth 1980), which would not be very sensitive. However, many single-celled eukaryotes seem to increase their sensitivity and directional specificity by associating the membrane with a stack of reflecting layers that establish standing waves of light. If the receptor membrane is located at the position of maximum amplitude in the standing waves, the sensitivity and directional specificity of the detector can be greatly enhanced. The whole mechanism is similar to elaborate television or FM radio antennas except scaled down a millionfold. Some examples of such sensitivity in microorganisms are presented in Table 8-2.

Table 8-2 Light Sensitivity in Single-celled Organisms

Organism	λ_{max} (nm)	Threshold (Photons m^{-2} s^{-1})	Integration Time (s)	Reference
Bacteria				
Halobacterium	565	10^{18}	3	Hildebrand 1978; Schimz et al. 1982
Rhodospirillum	870	10^{16}	?	Hildebrand 1978
Flagellated algae				
Euglena	480	10^{15}	1	Song 1983
Chlamydomonas	503	10^{15}	1	Foster and Smyth 1980
Fungi				
Phycomyces	450	10^9	10^3	Galland and Lipson 1987

Many animals increase their sensitivity to light by stacking up many rhodopsin-containing membranes in the direction of the light path. In this regard vertebrates use modified cilia, arthropods modified microvilli. However, there is a trade-off between sensitivity and color specificity in that maximizing sensitivity precludes using color filters. Furthermore, in a receptor cell that absorbs a significant fraction of the incident light self-screening will tend to flatten out the absorption spectrum. As seen in Figure 8-10, light-catching efficiency increases in proportion to the number of membranes in a stack, up to about 1,000 membranes, at which point 25 percent of the photons are absorbed at the absorption maximum. A further tenfold increase in the number of membranes increases the efficiency at the absorption maximum to only 0.95, which is less than a fourfold increase. In contrast, photons that are of a wavelength away from the absorption maximum are still absorbed in proportion to the number of membranes. Thus, the spectral sensitivity of a receptor becomes flattened out and less discriminating with regard to color. Consequently, maximum sensitivity is incompatible with color discrimination. To have maximum sensitivity and still good color discrimination, receptors should have about 1,000 membranes per stack, which

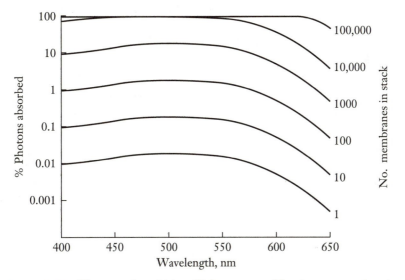

Figure 8-10 The spectral sensitivity of light receptors. The photoreceptor here is assumed to consist of a stack of membranes containing rhodopsin at a density of 20,000 molecules per square micrometer. Notice that as the absorption maximum approaches 100 percent of photons absorbed at the peak wavelength, the spectrum flattens out.

is often found (Schmidt 1978, 134). In the mud puppy *Necturus*, the rods average some 2,200 membranes, whereas the cones have 1,500 (Wald et al. 1963). Greater sensitivity can be obtained in retinas with more membranes, but at the expense of color resolution. They will absorb with increasing probability the photons of a wavelength away from the absorption maximum. This is probably the reason some mesopelagic fish are equipped with retinas having rods with extremely long outer segments or multiple layers of rod cells (Locket 1975). Another limitation on increasing the number of membranes is that the long receptor cell cannot all be in the focal plane of the lens. Consequently, eyes designed for high resolution, such as those of hawks are probability limited to short receptor cells (Snyder et al. 1990).

The overall sensitivity of an eye cannot be described by a simple number without specifying the information the eye must obtain. However, a useful summary is provided by the relation (Land 1981):

$$N = I_R \, (\pi/4)^2 \, (f\text{–number})^{-2} \, d^2 \, (1 - e^{-kx}) \qquad (8\text{-}3)$$

where N is the number of photons absorbed per receptor per second, I_R is the radiance in photons s^{-1} m^{-2} sr^{-1}, f-number is the ratio of the focal length to the aperture diameter, as in photography, d is the diameter of a receptor ($\cong 1$ μm minimum), k is the absorption coefficient per unit of path length of the pigment-containing region of the receptor cell (0.007 μm^{-1} in arthropods and 0.015 μm^{-1} in vertebrates), and x is the optical path length of the receptor cell (10–1,000 μm). By varying the f-number, receptor diameter (d), and receptor length (x) between species of arthropods, eyes can be seen to vary over a range of more than a thousandfold in sensitivity. The absorption coefficient (k) seems to be relatively fixed, presumably maximized within the relevant constraints. In general, sensitivity can increase (with the resolution held constant) in proportion to the square root of the size of the eye.

Most vertebrate eyes are nearly spherical allowing them to be rotated in their sockets. However, certain animals that depend on having eyes of great sensitivity like owls and mesopelagic fish have given up this advantage in order to get a large aperture without requiring them to have a large eye. They have thus developed **tubular eyes**, which are usually immobile. Owls compensate for their immobile eyes by having unusually mobile necks.

IMAGING SENSITIVITY

Imaging sensitivity depends on the contrast between the objects that are of interest and the background light. For an object to be detected it must differ in radiance (see Table 8-4 below) from the background radiance in at least one appropriate range of wavelengths. **Contrast** (C) is defined quantitatively as the relative difference between target ($I_t r$) and background (I_b^r) radiance, as follows:

$$C \equiv \frac{I_t^r - I_b^r}{I_b^r}$$

$$(8\text{-}4)$$

Defined thus, contrast is negative for a dark object against a light background and ranges from −1 to infinity. Contrast thresholds for humans vary from 1,000 for low light and small targets down to 0.001, for targets > 10° wide and high intensities (Jeske 1988, 314–18). For small targets < 10° wide, the contrast threshold is inversely proportional to the target area.

The emphasis on contrast by the human visual system produces many optical illusions like the two shown in Figure 8-11. This feature of information processing in the nervous system enhances the visibility of the boundaries between objects.

The ability to detect an object is usually limited by noise. Consider your own visual sensations when light stops entering your eyes—does vision stop, and do you see only blackness? You probably "see" patterns and scintillations of light. These sensations are a consequence of noise in the visual system. Some sensations might be due to thermal excitation of rhodopsin, but many probably arise from noise in the rest of the visual system, which could produce sensations of patterns. The state of adaptation clearly influences such noise, as in the production of afterimages. This type of noise is receptor noise in the sense described in Section 5-2. As a first approximation, this type of noise in the visual system is considered to be constant, independent of light level, and is usually called **dark noise**.

Another basic type of noise is that referred to as **channel noise** (Table 17-3), which is associated with the stimulus. In the case of light a fundamental contribution to channel noise derives from its photon nature and the resulting statistical fluctuations in the capture of photons. Such **photon noise** can be accurately estimated as the square root of the number of photons captured (see Section 5-3). Thus, its level should be proportional to the square root of the light intensity. The limitations on contrast threshold and spatial frequency at different light levels as a result of photon noise in human vision are depicted in Figure 8-12. Table 8-3 shows examples of trade-offs in eye size, spatial resolution, and threshold intensity in eyes of various species.

Figure 8-11   Optical illusions. *Top:* In this series of grey steps each step has uniform density, but most of the steps look lighter along the edge next to a darker step and look darker along the edge next to a lighter one. *Bottom:* In this pair of concentric squares the central squares are both the same density, but the one in the darker surround looks lighter, and vice versa.

These concepts can be used to estimate the contrast threshold (Levi 1980, 403ff). Let us assume that detection occurs when the contrast threshold (C_{Tb}) is some factor times the ratio of root-mean-square luminance noise to background luminance. Noise from several sources is combined by adding their squares, as follows:

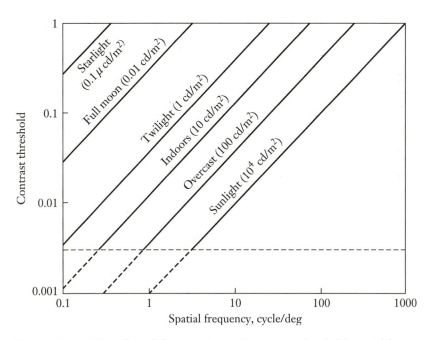

Figure 8-12 The effect of photon noise on the contrast threshold, spatial frequency, and luminance. The contrast threshold is a measure of the differences in intensity that can be detected. Spatial frequency is a measure of the sizes of objects of interest. For contrast thresholds below about 0.003, other factors limit detection. (From Levi 1980, 405.)

$$C_{Tb}^2 = \frac{k_d^2}{L_b^2} + \frac{k_p^2}{L_b} + k_n^2 \, (1 + K \, L_b^{-\beta} \,)^2$$

$$(8\text{-}5)$$

where L_b is the background luminance, the k's are constants for the appropriate signal-to-noise ratios for dark noise (k_d), photon noise (k_p), and neural noise (k_n), and β ($0 \leftrightarrow 1$) describes the effect of adaptation. It is clear from this relation that at sufficiently low light levels the first term dominates, the contrast is limited by the dark noise, and the contrast threshold varies inversely with intensity. At sufficiently high light intensities, the last term dominates, the contrast sensitivity is limited by neural noise, and the contrast threshold approaches a constant value that is independent of intensity. At intermediate levels the photon noise dominates the other two types, at least for human vision.

Table 8-3 Some Examples of the Visual Capabilities of Eyes

Organism	Aperture (mm)	f-number	Field (Degrees)	Resolution (Degrees)	Threshold* s^{-1} m^{-2} sr^{-1}
Alga[†]	0.001	——	60	60	?
Flatworm[‡]	0.03	——	?	35	10^{11}
Scallop[‡]	0.4	0.6	?	2	10^{11}
Nautilus[§]	0.4	25	?	2	10^{14}
Octopus[‡]	10	1.2	?	0.01	10^{11}
Horseshoe crab	?	?	?	8[‖]	10^{9} [§]
Daphnia[‖]	?	?	?	38	10^{11}
Deep-sea shrimp[‡]	0.6	0.4	?	8	10^{8}
Sawfly larvae[‡]	0.2	1.0	180	4	10^{10}
Net-casting spider	1.3[#]	0.6[#]	?	15[‡]	10^{10} [‡]
Jumping spider	0.4[#, ***]	1.7[#] 2.1[***]	?	0.1[‡]	10^{13} [‡]
Housefly	0.02[#]	2.0[#]	?	2.5[***]	?
Bee	0.02[#, ***]	2.4[#, ***]	160[***]	2[**], 1.4[***]	10^{12} [‡]
Moth[‡]	0.3	0.5	?	3	10^{10}
Fish	?	1.2[‡‡]	180[‡‡]	?	(10^{9})[§§]
Pigeon	2.0[#]	4.0[#]	162	0.03[‖‖]	?
Eagle[##]	6	2.6	?	0.004	?
Owl[#]	13	1.3	?	?	?
Cat	14[#]	0.9[‡‡]	143	0.1[**]	?
Human	8[#]	2.1[‡‡]	169	0.01[**]	10^{10}[‡]

* Radiance (photons s^{-1} μ^{-2} sr^{-1}). Unless otherwise indicated, this is the calculated (Land 1981) radiance required to excite receptors with 1 photon per second.
[†] Foster and Smyth 1980.
[‡] Land 1981.
[‖] Land 1984a.
[§] Land 1984b.
[#] Lythgoe 1979, 39.
[**] Ibid., 45
[††] Ibid., 68
[‡‡] Fernald 1988.
[§§] Northmore 1977; this is the behavioral threshold of goldfish for a large field.
[‖‖] Pearson 1972, 314.
[##] Shlaer 1972.
[***] Wehner 1981.

For human vision the constants for object diameters < 0.6° may be estimated as

$$C_{Th}^2 = \left(\frac{4 \times 10^{-8}}{I_b \, \Omega}\right)^2 + \left(\frac{6 \times 10^{-4}}{I_b \, \sqrt{\Omega}}\right)^2 + \left(\frac{0.014}{\sqrt{\Omega}}\right)^2 (1 + 19 \, I_b^{-0.32})^2 \tag{8-6}$$

For object diameters > 0.6°, the constants are

$$C_{Th}^2 = \left(\frac{4 \times 10^{-6}}{I_b \, \Omega}\right)^2 + \left(\frac{6 \times 10^{-4}}{I_b \, \sqrt{\Omega}}\right)^2 + (0.0012)^2 (1 + 19 \, I_b^{-0.32})^2 \tag{8-7}$$

I_b is the background flux per unit solid angle entering the eye ($I_b = L_b \times$ area of pupil) and Ω is the apparent object size in steradians (Levi 1980, 406).

There is also a trade-off between spatial resolution and temporal resolution, since long sample times can reduce noise. This becomes clear from Figure 8-13, which summarizes a great deal that has been learned about human vision.

In many species of flying insects the males use visual information to chase females on the wing for mating (Wehner 1981). In flies these chases occur at high speed, which puts a premium on both high spatial and high temporal resolution, as shown in Figure 1-1. The males of these species have unusually large eyes, as might be expected. Some species specialize even further by expanding the dorsal part of the eye (which is used in pursuing females viewed from below as they are silhouetted against the sky) to be much larger than the ventral part. Having a larger eye allows for both high spatial and high temporal resolution.

Contrast is diminished by light scattering in the environment (see pages 184–186) and also by scattering in the eye (Lythgoe 1979, 116). Since scattering often increases with a decrease in wavelength, contrast is often greater at longer wavelengths. The lenses of most animals act as a type of short-wavelength cutoff filter. In many diurnal animals the cutoff wavelength moves into the visible range, thus producing **yellow filters,** which probably function to improve contrast (Lythgoe 1979, 118–24) in the way that yellow sunglasses do. Primates have a yellow filter that covers only the fovea, and also the blue-sensitive cones are absent, which explains why it is so difficult to read yellow print on white paper. The human lens gets progressively more yellow with age, but whether this is a nonadaptive result of aging or instead an adaptive response to increased scattering

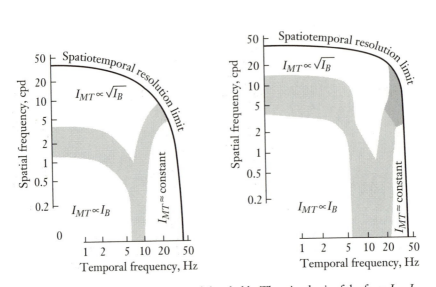

Figure 8-13 Human spatiotemporal thresholds. The stimulus is of the form $I_B + I_M$ $\sin(2\pi f_s x) \sin(2\pi f_t t)$, where I_B is the background intensity, I_M the modulated intensity, f_s the spatial frequency, and f_t the temporal frequency. The experiment is designed to determine the minimum modulated intensity (I_{MT}) that can just be detected (the threshold) at various values of the other parameters. The limit for complete modulation is indicated by the solid curve. Within this area the threshold varies with the background intensity in three ways: (1) At high spatial frequencies the threshold varies with the square root of the background intensity; (2) at high temporal frequencies the threshold is nearly independent of the background intensity, which is a property of any linear system; and (3) when both the spatial and the temporal frequencies are low, the thresholds are nearly proportional to the background intensity. The plot on the left is for a lower light intensity (50 trolands) than the one on the right (200 td). (From Kelly 1972.)

within the older eye is not known. Another probable function of yellow filters is to protect the eye from photochemical damage by ultraviolet light. These considerations may be why more animals do not make use of the ultraviolet region of the spectrum.

SENSORY ADAPTATION

Because organisms that live where sunlight is present experience tremendous changes in ambient light intensity during the course of a day (Figure 8-14), many have developed mechanisms to adjust their sensitivity. The

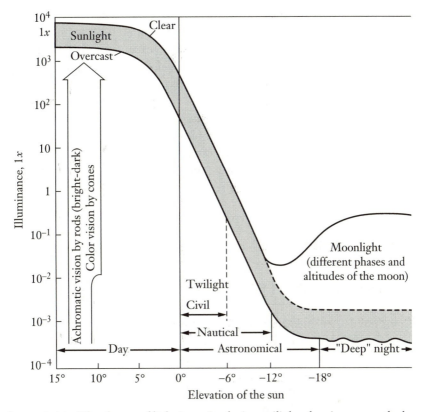

Figure 8-14 The change of light intensity during twilight, showing a smoothed illuminance on a horizontal surface for different elevations of the sun. (From Jeske 1988, 308.)

most common mechanism is the movement of screening pigments. For vertebrate eyes a change in pupil aperture is the most rapid mechanism. Animals like humans with circular pupils have mechanical limits on how small the aperture can be made, so that this mechanism provides only about a tenfold change in intensity. Many nocturnal animals like cats have slit pupils that can probably be closed down much further (Lythgoe 1979, 79). Full adaptation often takes close to an hour, which is probably rapid enough to keep up with changes at dawn and dusk. However, diving animals face severe problems as rapid changes occur in ambient light intensity (Lythgoe 1979, 93).

The fact that physiological adaptation is not 100 percent effective is part of the reason most animals are active during daylight (diurnal) or at night (nocturnal) but not both. In terrestrial environments some daily activity patterns may arise from changes in temperature or humidity, but coral reefs are exceptionally stable habitats, except for daily changes in light intensity. On coral reefs most fish species are active only in daylight or at night, with there being a quiet period at dawn and dusk when most fish take refuge. However, predation is at a peak during the transitional periods, and Lythgoe (1979, 95–97) suggests that this is because the prey are at a disadvantage from the rapidly changing light levels, which predators are adapted to take advantage of. In the open ocean, pelagic animals often stay within a narrow range of light intensities by migrating up and down during the course of a day (see pages 197–198).

8-4 INTENSITY

UNITS OF INTENSITY MEASUREMENT

Light intensity is measured in many different units, which may be confusing. This book usually measures intensity as a flux density of energy per unit time per unit area, and light can also be measured this way (radiant flux density). However, since light is quantized into photons and can be detected only one photon at a time, it usually makes more sense to measure intensity as the rate at which quanta (photons) impinge on a given unit area. It is important to remember that since the energy of a photon is proportional to its frequency ($E = h f = h c /\lambda$), a spectrum may look quite different depending on whether one uses energy units or the number of photons. A second factor is that most researchers measuring light intensity are interested primarily in how it affects human vision, so they often measure light in terms of how effective it is in stimulating the human visual system, a characteristic known as **illuminance.** A third factor is that it is often important to distinguish between light coming from a particular direction and light from all directions. In order to measure light coming from only a particular direction, the flux density is normalized to the solid angle, measured in steradians, within which light is collected. The relationships between these units are summarized in Table 8-4. It can be seen that the terms *radiant* or *radiance* are generally applied to quantities measured in terms of energy, with *luminous* and *luminance* being used for measurements based on the sensitivity of human vision.

Table 8-4 Measures of Light Intensity

Quantity	Measures (units)		
	Energy	Quanta	Human Vision
Flux	Radiant flux (J/s = Watt, W)	? (Photons/s)	Luminous flux (lumen, lm)
Flux density at a surface	Irradiance (W/m^2)	Fluence (Photons s^{-1} m^{-2})	Illuminance (lm/m^2 = lux)
Flux per unit solid angle	Radiant intensity (W/sr)	? (Photons s^{-1} sr^{-1})	Luminous intensity (lm/sr = candela, cd)
Flux per unit solid angle per unit area	Radiance (W sr^{-1} m^{-2})	? (Photons s^{-1} sr^{-1} m^{-2})	Luminance (lm sr^{-1} m^{-2} = cd/m^2)

* sr = steradian, there are 4π steradians in the complete solid angle of a sphere.

INTENSITY IN NATURAL ENVIRONMENTS

Table 8-5 describes typical light intensities under various environmental conditions. The units are roughly equivalent to what could be captured by a typical amount of rhodopsin in an eye having a 1 mm aperture. In full sunlight there is a strong enough stream of photons to provide detailed information. On a clear, moonless night, however, or at a depth of 1,000 m in clear water the intensity is reduced by 100 million times, making the usefulness of vision marginal.

Underwater, because the environment is relatively homogeneous it becomes possible to make some general statements about the intensity and color of background light from different directions. These generalizations are sufficiently valid that the eyes, body reflectance patterns, and behavior of pelagic animals seem to be adapted to them (Lythgoe 1979, 14) (see Section 8-9). Near a calm surface, the intensity is strongest in the direction of the sun's image, whereas surface waves refract sunlight into complex patterns. If there were absorption without scattering, a depth would occur at which only light from above was greater than threshold. If there were scattering without absorption, a depth would occur at which light of an equal intensity came from all directions. Real environments have some combination of these conditions. For a given ratio of absorption to scattering and, below a few hundred meters, the directional distribution of light is independent of the depth.

Table 8-5 Light Intensities in Various Environments

Situation	Intensity * (Photons cm^{-2} s^{-1} nm^{-1})
Full sunlight	10^{14}
Overcast daylight	10^{13}
Twilight	10^{11}
Moonlight	10^{8}
Clear moonless night (starlight + airglow)	10^{6}
Overcast moonless night	10^{5}
Full sunlight at 1000 m in clear ocean waters	10^{6}

* Spectral density near 500 nm (Lythgoe 1979, 4–6; Brines and Gould 1982).

As indicated in Figure 8-5, in the clearest water sunlight can provide an effective source, down to depths of roughly 1,000 m. At the other extreme, sunlight can effectively penetrate in blackwater rivers and swamps to only a few centimeters. At night, of course, such extraterrestrial sources as the moon and stars provide light to much shallower depths.

In marine environments, significant amounts of light are generated at depths below the reach of sunlight by the bioluminescence of a wide variety of organisms (Herring 1987). Bioluminescence in clear ocean water becomes significant below about 500 m in the daytime and 100 m on moonlight nights (Lythgoe 1979, 29). In turbid, in-shore waters bioluminescence can be significant at less than 30 m, even at midday. This kind of light is produced by many marine species of bacteria, protozoa, fungi, and some animals. These organisms produce light with spectra peaking variously between 420 nm (violet) and 550 nm (green) (Lee 1989). Thus, they produce frequencies in the same range as the most useful sunlight, colors that are minimally absorbed by clear or chlorophyll-containing water.

A few freshwater and terrestrial organisms also produce light, notably the fireflies, which employ flashes (λ = 552–582 nm) for communication (see Chapter 13) between mates. Other animals probably employ bioluminescence as a lure or warning. Bioluminescent bacteria typically produce 10^{3} photons/s and a single dinoflagellate about 10^{8} photons in a single flash lasting a fraction of a second (Lee 1989).

8-5 TRANSMISSION

Visible light can travel long distances through the atmosphere. Absorption, by O_3 and H_2O, is important only in considering how much light of various frequencies gets through the atmosphere from the sun. Nonetheless, attenuation between transmitters and receivers on the earth's surface can be severe because of scattering, especially by water drops in the atmosphere (fog). Underwater, light absorption is much greater than in the air, but scattering is also greater and usually limits visibility more than does absorption (Lythgoe 1979, 7). That there is a greater range of vision in air than in water is dramatically illustrated by the behavior of some aquatic animals that jump out of the water to see a predator. This technique has been called "spy hopping" by Hawkins and Myrberg (1983, 403).

The phenomenon of absorption significantly limits the transmission of information by light only in water and in vegetation. Absorption by clear water was illustrated in Figure 8-3. The absorption of light by vegetation often limits the range at which information can be obtained by vision in terrestrial environments.

Light scattering by particles much smaller than the wavelength (Rayleigh scattering) is simply proportional to the reciprocal of the fourth power of the wavelength (see Section 4-1). This is the prime reason the sky looks blue and the sun is red near the horizon. Scattering by particles of a size near that of the wavelength is very complex, whereas scattering by larger particles is largely independent of the wavelength, which makes clouds, snow, and foam look white or gray.

The reflection of light is of the utmost importance, because the major use of light to acquire information involves the reflection of sunlight off objects of interest. In such cases the reflecting object is considered the transmitter of the information, since the sun is such a predictable source. Reflection can also interfere with information transmission if the light carrying the information is reflected away from the receiver or the reflection adds noise to the transmission. This problem is particularly evident when trying to see from air down into water, because of reflections off the water's surface.

The refraction of light also affects the transmitting of information. This is again a problem when the light passes between air and water. In transmission from water to air the major problem is that the transmitter is located in a direction different from what it appears. Birds like kingfishers amd herons that hunt for fish from the air must correct for this effect in making their strikes. Aquatic animals wanting to see what is above the water find that the hemispherical world above the surface is compressed into a cone extending only 48.5° from the vertical in perfectly calm conditions (Lythgoe 1979, 13). At larger angles a phenomenon called *total internal reflection* occurs, making visible only light reflected off the surface from underwater. Under natural conditions the water surface is rough, and the images from above and below are intermixed.

Under certain atmospheric conditions **mirages** are produced, in which the bending of light rays causes images to appear that are misleading. A common example occurs when light from the sky appears to come from below the horizon, suggesting its reflection off a body of water (Levi 1980, 87). Many other peculiar distortions occur under specific circumstances, as described by Minnaert (1954).

8-6   THE RANGE OF ACTION

If nothing else impedes the propagation of light, the earth's curvature sets a limit on the distance at which objects can be seen. The **horizon** is the distance at which the curvature of the earth limits the propagation of light from a given source at the surface. If the receiver's height is h, the horizon extends to approximately the square root of the product of h and the earth's diameter (Levi 1980, 86). Under average atmospheric conditions, refraction increases the distance about 7 percent, making it possible to approximate the distance to the horizon as

$$r_H \approx 3.8 \sqrt{h} \qquad\qquad (8\text{-}8)$$

with h in meters and r_H in kilometers. Thus, the horizon for a human whose eyes are 2 m above the surface is about 5 km away. For an insect having eyes 1 mm above a smooth, horizontal surface the horizon would be about 100 m. For a source elevated above the surface, the maximum range of transmission becomes simply the sum of the horizons of the source and the receiver.

The other major limit on the **range of visibility** is the ability to distinguish an object from its background. The ease with which this can be accomplished depends on a combination of factors involving contrast, light intensity, and the apparent size of the object (see pages 168–174). As an object recedes from a receiver, the size of the receptive field it occupies shrinks and its contrast decreases as scattering intervenes until at some distance its image can no longer be detected. Thus, contrast is a major limitation on the range over which light can carry information. It is usually the predominant limitation in daylight, when intensity is high, and under water, where scattering is usually severe.

The range of visibility is controlled by the beam attenuation coefficient, α, which includes the effects of both absorption and scattering. Deep in a homogeneous medium illuminated from above, the contrast for horizontal viewing is calculated as

$$C(r) = C_0\, e^{-\alpha r} = C_0 e^{-r/L_A} \tag{8-9}$$

where C_0 is the inherent contrast at close range (Jerlov 1968, 139; Lythgoe 1988) and L_A is the attenuation length. In the atmosphere the range at which $C = 0.05\, C_0$ is called the meteorological optical range. It varies from ≈ 100 km in exceptionally clear air through 2 km in haze down to < 50 m in dense fog (Jeske 1988, 319). In exceptionally clear ocean water a large, dark object is visible out to only about 30 m, and turbid water may reduce the range to just a few centimeters (Lythgoe 1988). Some values for beam attenuation length in various environments are presented in Table 8-6.

The ecology of fish in many pelagic environments is controlled by visual predation. In this regard the range of detection has been measured for some freshwater gamefish (O'Brien 1979). The general result is that the range of detection is found to be proportional to the prey's size, which

Table 8-6 Light Beam Attenuation Length in Various Environments*

	Wavelength (nm)					
Environment	300	400	500	600	700	800
Pure air[†]	7.0 km	22 km	55 km	120 km	220 km	370 km
"Clear air"[†]	3.8 km	5.0 km	6.0 km	6.7 km	7.4 km	7.9 km
Moderate fog[‡]			— 50 m —			
Pure water[§]	?	23 m	28 m	5.4 m	2.0 m	0.49 m
Oceans[§]	?	1–10 m	1–15 m	1–5 m	1–2 m	?
Tissue[‖]	?	?	0.5 mm	?	1.5 mm	?

*The beam attenuation lengths given are the reciprocals of the beam attenuation coefficients.
[†]Levi 1980, 91, 988.
[‡]Jeske 1988, 327.
[§]Jerlov 1976, 52, 62.
[‖]Grossweiner 1989.

is what one would expect if the prey were required to occupy a certain visual receptive field size in order to be detected. The range is also sensitive to the light intensity at low levels, which is probably why most plankton migrate downward during the day.

8-7 INFORMATION EXTRACTION

Light is used to obtain a variety of kinds of information. Photosynthetic organisms are interested in light for its own sake, but others use it mostly to obtain information about objects in their environment. In this section the basic types of information that can be obtained are described, along with strategies for obtaining them. Color, because it depends on the pattern of spectral distribution, is discussed in the chapter on patterns in Section 11-5.

DIRECTION

In order to determine the direction of a light source, an organism must possess both a light detector and a light screen of some kind, to provide directional selectivity. Such a screen can be either an absorbing or a reflecting element that prevents light from certain directions from reaching the detector or it may be a refractile element that focuses light from only specific directions on the detector.

Most bacteria are too small compared to the wavelength of light to be able to create differential light intensity; thus they cannot determine the direction of a light source (Foster and Smyth 1980). Bacteria can still make use of light, but can only determine its intensity, so they must employ klinokinesis (see Sections 15-2 and 5-3) to move in a light-intensity gradient. However, gliding filamentous bacteria that move very slowly can compare the light's intensity at different positions and make directed turns a process called tropotaxis (see Table 17-1 and Section 17-3) (Häder 1987). In contrast, single-celled algae are large enough to determine the light's direction and can move along a beam of light by scanning their directionally sensitive receptor as they rotate while swimming (Foster and Smyth 1980). Amoeboid cells can detect differences in light intensity across the cell and move directly toward higher or lower intensities (Nultsch and Häder 1988). Some fungi and slime molds make use of refraction at their tissue–air interfaces to map light from specific directions to specific parts (see Figure 8-15) (Galland and Lipson 1987; Nultsch and Häder 1988). Plants respond to partial shadowing of their leaves by moving the leaves aside to avoid having them shaded.

The pinnacle of directional information extraction occurs in animals, particularly the vertebrates. Although visual resolution is difficult to determine experimentally, good estimates can be made from the anatomy of different kinds of eyes (Land 1981). In eyes with good lenses, spatial resolution is limited by the receptors' spacing and the size of the aperture.

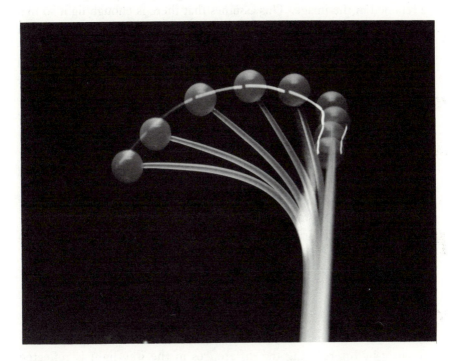

Figure 8-15   The phototropic response of a sporangiophore of the mold *Phycomyces*. This multiple-exposure photograph was made at four-minute intervals with flashes of orange light, which were not perceived by the cell. This sporangiophore was exposed to balanced intensities of blue light from its right and left sides for the first thirty-five minutes, then the source on the right was extinguished and the one on the left simultaneously doubled in intensity to maintain a constant total intensity. The bright streaks are glints of blue light from having the camera's shutter left open. With unequal illumination the cell starts bending toward the light source, moving horizontally much faster than its upward rate of growth. The spherical sporangium, which contain the spores, is about 0.6 mm in diameter. (Photo courtesy David S. Dennison, Dartmouth College.)

The maximum spatial frequency in cycles per radian that can be detected is limited by receptor spacing to the focal length of the lens divided by twice the receptor spacing, and by diffraction to the aperture diameter divided by the wavelength (Land 1981). There is no benefit in providing resolution in one component, at the expense of sensitivity, that is only to be lost in another component. Consequently, the receptor spacing in most eyes matches the aperture in terms of the resolution limits. Remembering that the minimum useful receptor spacing is about 2λ, the eye should have an f-number in the range of 1 to 4, depending on the degree of contrast to be retained in the image. This assumes that there is enough light so that noise is not limiting; otherwise, smaller f-numbers are advantageous. Most eyes in diurnal animals fall within this narrow range of f-numbers, although their resolution varies (with a subsequent trade-off in size and sensitivity) over a thousandfold range (see Table 8-3 above). Generally the resolution can increase, with the sensitivity held constant, in proportion to the increased size of the eye (Land 1981). As shown in Figure 8-16, there is a fairly good correlation between the size of an animal and the resolution of its eyes.

Humans have close to the best visual resolution of any known animal. However, this high resolution is only for an image falling on the fovea. Birds also have very good visual resolution, with some of the large birds of prey probably surpassing humans in resolution. In addition, birds have good resolution over wide areas of the retina and can probably see much more detail at a single glance than can humans (Pearson 1972, 314).

CONTRAST

Light is used to obtain information about objects in the environment primarily by interpreting the differences in the way light is reflected from different objects. Thus, at least in vertebrates the visual system emphasizes contrasts, which may involve comparing intensities that differ in time, direction, or frequency. Contrasts in time are important in all the sensory modalities, most generally for detecting changed environmental conditions. Contrasts in wavelength, which are what produce color sensations, are most generally useful for distinguishing one object from another (see Section 11-3). Light is unique in the accuracy with which directional information can be obtained. Spatial contrast allows different objects to be distinguished and their relative positions in space estimated.

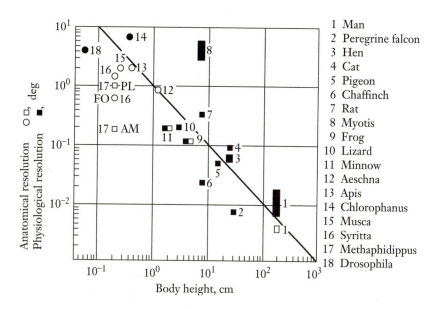

Figure 8-16 The visual resolution of eyes in various-sized animals. The body height is measured at the center of the eyes above the ground. The diagonal line represents a resolution of one degree per centimeter of body height. (After Kirschfeld 1976, 356.)

To a great degree visual systems alter the spatial and temporal integration of signals to fit the current level of luminance, so that the temporal and spatial resolution are maximized within a fixed level of photon noise (Levi 1980, 380). Spatial integration can generally be approximated as the difference between two Gaussians (see page 104) with a ratio of widths of 1.75 (Marr 1982, 62). Even though this property of the visual system results in several types of optical illusions (Blakemore 1973; Lythgoe 1979, 197; Levi 1980, 372; Marr 1982, 251; Land 1986), it has the benefit of emphasizing the contours of objects seen against backgrounds with a different luminance (see Marr 1982, 58–59 for examples).

Snow and the whitest of clays reflect about 93 percent of the light at all visible wavelengths (Weast 1985, E390). At the other end of the range, soot reflects only about 5 percent as does an air–water interface. (The reflectivity of other natural materials has seen collected in Levi (1980, 1,000).) Thus, the contrast of one material as seen against another is limited to the range −0.95 to 18 (Equation 8-4) and is usually much closer to zero. A contrast of 30 or more is perceived as a light source (Marr 1982,

86). Shadows often have high contrast, depending on the amount of scattered light in the environment, so that camouflaged animals avoid producing them (see pages 199–201). Differences in shape, color, and texture are also important in discriminating objects.

POLARIZATION

In light, the electric and magnetic fields oscillate in directions perpendicular to each other and to the direction of propagation. A particular photon is described as being polarized in the direction in which its electric field is vibrating. (This direction is two ended, since the field alternately points in opposite directions; that is, the full range of possible polarizations is 180° rather than 360°.) Most light sources, including the sun and bioluminescence, produce photons having randomly oriented polarizations and their light is said to be unpolarized. However, certain processes can separate photons with different polarizations, to produce polarized light. One such process is reflection off of water. The reflected light is dominated by photons polarized parallel to the water's surface, with complete polarization occurring where the light intercepts the surface at 53° (Cox 1974). Polarized sunglasses that absorb light polarized in the horizontal plane more strongly than in the vertical plane thus absorb a greater proportion of the light reflected off the surface than off of underwater objects.

The scattering process that makes the sky blue also polarizes the scattered light. Light scattered in a particular direction tends to have more photons polarized in some directions than in others. The predominant direction of polarization is at right angles to the plane containing the receiver, the sun and the sky that is scattering the light (Lythgoe 1979, 10). The degree of polarization can be as high as 90 percent but drops to less than 10 percent in directions near the sun. In addition, scattering from water droplets in the air tends to depolarize light.

Given the pattern of light scattered over the sky, an organism can in principle obtain useful information about the direction of the sun even if it can see only a small patch of blue sky. Knowledge of the sun's position is used by many animals in navigation (see Chapter 18). Recent experiments have demonstrated that the direction of polarization is a more reliable indication of the sun's position than are either the degree of polarization or the light's hue or intensity distribution (Brines and Gould 1982). The sun is located on the great circle through the observed point in the sky and perpendicular to the direction of polarization. Although the mechanisms and strategies for using this information are obscure (Waterman 1984; Wehner 1987), there is evidence that honeybees use an approximate rule for relating the direction of polarization in a small patch of blue sky to solar position (Rossel and Wehner 1982; Wehner and Rossel 1985). This ability could be of great usefulness in a forest, where the sun is often not directly visible. A different use for the polarization of light is made by certain flying insects that appear to recognize water surfaces by the horizontally polarized reflections they throw off (Waterman 1984). Another potential use, by squid, is to break the mirror camouflage of fish by making use of the fact that reflected light is often polarized (Land 1981; Land 1984).

It is not difficult to design polarization-sensitive receptor cells—in fact, the structure of most invertebrate receptor cells makes them inherently sensitive to polarization. Under certain circumstances, even humans can detect polarization. The polarization of ultraviolet light seems to be most used by insects (Duelli and Wehner 1973), a finding consistent with the high proportion of ultraviolet in sky light. Detailed experiments with the desert ant suggest that only the part of its eye that looks 50° above the horizon and 45° to the side from a point directly ahead is used to determine skylight polarization (Duelli 1975).

DISTANCE

Humans judge distance in part by the slight differences in direction of an object viewed by their two eyes. This strategy, based on binocular disparities, is called stereoscopy or binocular parallax. Humans' ability to detect differences in direction of only a few seconds of arc is sufficient to let them distinguish between a one-kilometer distance and a much larger one (Levi 1980, 420).

For stereoscopy to work the *field of view* of the two eyes must overlap. Since the field of view of a single eye is limited to about 180°, a large area of binocularity can be achieved only by having a large blind area behind. The extent of this trade-off varies a great deal among different animals, particularly between prey and predators. For instance, rabbits have a full 360° field of view but only a 24° binocular field; a cat has only a 187° field of view but a 99° binocular field. Humans have a 130° binocular field. Consider the difference in the locations of the eyes of a chipmunk and a cat in Figure 8-17. In spite of these differences, the extent to which other

Figure 8-17   The difference in the positions of eyes in prey and predators. *Top:* The predatory cat has its eyes on the front of its head, thus providing much overlap in the visual fields of the two eyes and a large field of binocular vision that has good depth perception. *Bottom:* In contrast, the preyed-upon chipmunk has its eyes on the sides of its head, giving it a wide field of view to detect danger in almost any direction.

animals use stereoscopy to increase their depth perception is not clear. Predatory insects probably use binocular cues when grasping their prey, but the small separation of their eyes limits their useful range to just a few millimeters (Wehner 1981, 438).

There are also monocular cues to distance. One, called **motion parallax**, is related to stereoscopy except that instead of comparing images between two eyes one eye is moved and images from its two different positions are compared. This technique is particularly effective in determining the relative distances of an object and its background, as the closer the object is as compared to its background, the more it will appear to move across the background. The reason owls often bob and wave their heads around in a peculiar fashion is probably that it affords them greater depth perception, which is important because their eyes cannot be rotated to make their sight converge on an object. Also, locusts sway from side to side before jumping to a new perch. If their target is moved in the direction opposite of the sway, they jump short of the target, an observation consistent with their use of motion parallax (Wehner 1981, 442).

Another monocular clue used by vertebrates is derived from the lens adjustments called **accommodation** that are required to bring an object into focus. One detailed study of a chameleon indicated that it judged depth primarily by the accommodation required for focus on the target (Lythgoe 1979, 148). This is a strategy that cannot be employed by invertebrates whose compound eyes have a fixed focus (Wehner 1981, 435).

Some of the other monocular cues that are also used, at least by humans, include

1. The object's position in the field of view. Since the eyes are located above the ground surface, more-distant positions on the surface correspond to higher locations in the field of view.
2. Perspective, based upon the fact that parallel lines appear to get closer together as they recede from the observer.
3. Texture. The surface texture of any object appears smoother the farther away it is. The limit for resolving texture is about four cycles of contrast per degree of angle (Marr 1982, 43).
4. The apparent size of familiar objects. Male flies pursuing females probably use this strategy (Wehner 1981, 439).
5. Overlap of further objects by nearer objects.
6. Contrast and color. Light scattering reduces the contrast of distant objects and shifts hue toward the blue.

All these cues are often employed by visual artists wanting to create an appearance of depth in works that are in reality flat (Blakemore 1973).

THE THREE-DIMENSIONAL ENVIRONMENT

In higher animals the primary function of vision is to provide information about the three-dimensional spatial relationships of objects in their environment. The first step is to identify separate objects in the visual field. The principal means of establishing this separation is to identify the edges surrounding individual objects and separate the objects from one another. This requirement accounts for the emphasis on spatial contrast that is such an important factor in processing visual information (Marr and Hildreth 1980).

Psychologists, particularly those of the Gestalt school, have identified many of the principles used by the nervous system to extract a four-dimensional (three of space and one of time) interpretation of the environment based on information from light and sound (Bergman 1990). Among these principles are that perceptual qualities like a color, a boundary, and a pitch are associated with a mental construct that normally corresponds with an object or event in the environment. A quality can belong to only one construct at a time, as illustrated by the "vase vs. faces" illusion shown in Figure 8-18. Another principle is that related qualities tend to belong to the same construct, and relatedness is based on the likely properties of physical objects. For example, a change of position or pitch is likely to continue even if the sensory information concerning it is interrupted. A common case of such interruption occurs in vision when a closer object eclipses a more distant one. One's visual system tends to relate the apparent boundary between two perceived areas to the nearer object, as illustrated in Figures 8-19 and 8-20.

The characteristic of *shape* is obtained from a variety of cues (Marr 1982, 215–48). In the absence of other cues, the shape of a silhouette is assumed to come from generalized cylinder shapes, which are often good approximations of images of organisms. Collections of visual features appropriately distributed may be interpreted as arranged uniformly on the actual surface (the texture) or in parallel lines on the surface, and the apparent distribution of the features of an image is used to infer the three-dimensional shape of its surface.

Illumination in most natural environments comes predominantly from above. Thus, the surfaces of three-dimensional objects that face

Figure 8-18 The vase vs. faces illusion. Most humans interpret this drawing alternately either as a vase or two faces, depending on whether black or white is taken as the background, but the visual system does not provide for accepting both interpretations simultaneously.

upward are more strongly illuminated and look brighter than those facing downward. Such variation in *shading* is another cue used to infer the three-dimensional shapes of objects. This interpretation is so strong that, in pictures lacking other cues, projections look like indentations when the

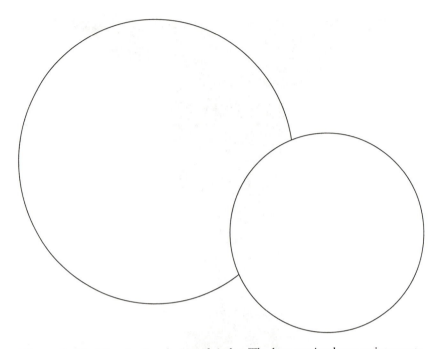

Figure 8-19   The visual occlusion of circles. The human visual system interprets this pattern as being formed by two circles at different distances, with the nearer one occluding the farther one. Logically, this pattern could also represent two objects in contact, one of which is a circle with a piece missing, but the first interpretation is more likely to be correct for our environment.

pictures are turned upside down, and vice versa (Figure 8-21). Another clue to the importance of this mechanism is the prevalence of counter-shading as a strategy for camouflage (Cott 1941, 35ff) (see pages 198–199). However, Figure 8-21 also demonstrates that familiar shapes like a face are perceived in normal relief at the price of assuming that the light is coming from below, even when this is erroneous.

Another important clue to an object's three-dimensional shape comes from the *shadows* it casts. When an object is illuminated directly by the sun so that its shadow falls on a smooth surface, its shadow has high contrast that reveals information about the shape of the object (Figure 8-21). Humans, at least, make use of this information.

A different kind of information about three-dimensional environments concerns the receiving animal's motion through them. Visual information provides useful feedback, even for maintaining one's position, as

Figure 8-20 The visual occlusion of complex objects. The gray areas are much easier to connect into a pattern of the letter *R* when the occluding area is distinguished from the background (*below*) than when it matches the background (*above*).

can be appreciated by trying to stand on one leg with one's eyes closed. Such information is even more important during locomotion, particularly for the flying and swimming animals that do not maintain mechanical contact with a substrate. This may be why many moths have eyes with a

Figure 8-21 (*Opposite page*)   The effect of shading on perceived relief. *Top:*
Look at this photo briefly, then turn the book upside down. You are likely to see that
the apparent depths have reversed—the light–dark boundary through the center may
look like the bottom of a canyon or the top of a ridge. This unfamiliar pattern, which
is on the bottom of a sand dollar skeleton, is interpreted by the visual system as
having the illumination coming from above, which is the usual case. Consequently,
when the book is rotated, features that first appeared to be projections now look like
depressions, and vice versa. *Middle:* In contrast, familiar shapes like faces are assumed
to have their usual shape, and the illumination is deduced as coming from the direction indicated by the shadows. This photo is of a pair of identical thin masks, the one
on the right being viewed from the front and the other from behind. Both masks
were illuminated by the same light from above, but our visual system interprets the
view of the back side of the mask (*left*) as that of a face viewed from the familiar side,
but with the illumination coming from below. *Bottom:* This photo differs from the
middle one only in the position of the camera, to illustrate the dramatic fact that as
the position of the observer changes, from left to right, the two masks appear to
rotate in opposite directions, because of incorrectly interpreting the depth of the
mask on the left. In the middle photo the two faces appear to be facing toward a common point, whereas in the bottom one they are looking away from each other.

large field of view that is directed downward, as can be dramatically seen
by looking through a window at night-flying moths resting on the other
side of the glass and illuminating them with a light source close to your
head. The resulting retroreflection makes their eyes appear to glow, so
that the extent of their visual field can be easily determined.

When one travels with one's eyes in a fixed orientation, the images of
stationary objects in the environment move across one's retina in a simple
pattern called **visual flow**. This flow of images is away from a point that
corresponds to the direction of travel at a rate that increases from zero at
the point straight ahead to a maximum for points out to the side. If an
object's position on the retina is measured by angle θ of the object from the
direction straight ahead, the rate of visual flow is described quantitatively
by

$$\frac{d\theta}{dt} = \frac{v \sin \theta}{d} \tag{8-10}$$

where v is the animal's speed of travel and d is the distance from the animal
to the object (Land 1983, 23). Thus the pattern of visual flow provides

information about the distances of objects as well as the animal's direction of travel. In particular, the time to collision with an object is available simply from the visual flow:

$$t = \frac{d}{v} \cong \frac{\sin \theta}{d\theta/dt} \tag{8-11}$$

where t is the time to collision with a small object ahead, θ the apparent position of its edge, d its distance, and v the velocity. A short time to collision stimulates eye blink in mammals and landing in flies (Collett 1983).

Many flying insects tend to maintain a constant rate of image flow (Baker and Haynes 1987). This simple strategy works out so that, flying into a headwind (causing a smaller v), maintaining a constant rate of image flow ($d\theta/dt$) in each direction (θ) with a constant effort requires flying closer to the ground (smaller d), where the wind velocity is lower. This process is reversed for tailwinds. An even more important function of visual flow is probably that it provides information about the true direction of travel. Consequently, it can be used to detect and correct for crosswinds.

8-8 INFORMATION CAPACITY

As previously pointed out, the rate of information transmission depends on the level at which it is analyzed, because information is selected as it passes through the nervous system. In natural daylight, the human eye captures something like 10^{10} photons in an integration time of 0.1 s. This many photons could, in principle, carry 10^{11} bits/s of information. At the receptor-cell level there are 10^8 receptors, meaning that if each acts as though it had only "on" and "off" states, the receptors could transmit 10^9 bits/s. This estimate is close to one made in a somewhat different argument (Ackerman et al. 1979, 473). At the level of the optic nerve there are about 10^6 axons. If their maximum frequency of action potentials is 10^2 Hz, the optic nerve could transmit 10^8 bits/s to the brain. These numbers are summarized in Table 8-7.

At the level of perception, the information rate can be estimated roughly as follows. Since there are thirty to forty distinguishable steps of intensity (Grüsser and Grüsser-Cornehls 1978), the contrast information represents about 5 bits ($2^5 = 32$). There are roughly 17,000 distinguishable hues (Le Grand 1968, 300). As a result, color vision represents 14 bits ($2^{14} = 16,384$). This total of nineteen bits of intensity and hue are available for

Table 8-7 An Estimate of the Maximal Rates of Information Transmission in the Human Visual System

No. of Elements	Time Interval	Information Rate (Bits/s)
10^{10} photons	0.1 s	10^{11}
10^8 receptors (binary)	0.1 s	10^9
10^6 optic nerve axons	0.01 s	10^8

each of many directions, a figure estimated as 5×10^5 (Kirschfeld 1976). This suggests that a total of 1×10^7 bits are acquired in 0.1 second, or roughly 10^8 bits/s. These estimates are only very rough ones, because of the interaction of the parameters (e.g., the contrast threshold depends on size). A more sophisticated analysis (Snyder et al. 1977) has suggested that the maximum information that can be obtained from an eye having a given aperture increases with the mean contrast in the scene as well as with its effective intensity (Figure 8-22). In a scene with a mean contrast equal to 1 that is illuminated by daylight, the human eye could acquire a maximum of about 10^5 bits in one sample, corresponding to a rate of 10^6 bits/s. From even these rough estimates it is clear that vision is capable of providing enormous amounts of information, more than any other type of stimulus, primarily because its high spatial resolution effectively provides a large number of parallel channels.

8-9 THE IMPORTANCE OF INFORMATION FROM LIGHT

SPECIES SPECIALIZATIONS

The dominant importance of light as a carrier of information is demonstrated by the fact that most animals confine specific activities to given times of the day. For many animals, food is easier to find in daylight, but this may also make them vulnerable to predators. Most animals therefore specialize in seeking food under particular lighting conditions and have become adapted to perform most successfully at those times. This specialization occurs even in aquatic environments and the tropics where the daily temperature changes are insignificant (see pages 174–176).

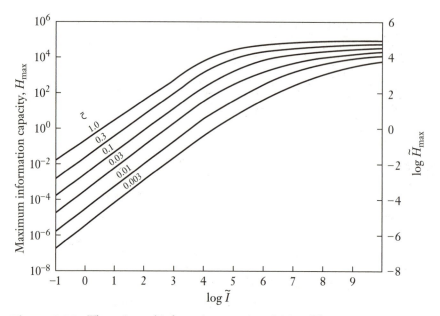

Figure 8-22 The estimated information capacity of vision. The maximum information capacity (H_{max}) as defined here is based on the natural logarithm instead of base two, as in Section 3-2. Thus, it should be multiplied by a factor of $\log_2 e = 1.44$ to convert it to bits. $\tilde{C}$ is the mean contrast in a scene of random objects. The intensity parameter, $\tilde{I}$, is the mean number of effective photons entering the eye in a sample time per square degree. (From Snyder et al. 1977, 1169.)

The importance of vision in the interactions of organisms is further indicated by the fact that, in the relatively homogeneous environment of open water, the presence or absence of enough light for vision determines a boundary between major ecological zones. Marine ecologists identify three fundamental depth zones, each containing its own species. The shallowest, the *euphotic* zone, extends to a depth of ≤ 140 m, to which sufficient light penetrates to support photosynthesis. The next zone, the *mesopelagic*, appears to be determined by the depth to which sufficient sunlight penetrates to support vision, which is $\leq 1,000$ m. The third (*bathypelagic*) zone receives no useful light from above, and the organisms within it do not make daily vertical migrations (Lythgoe 1979, 100). Mesopelagic animals generally have large, well-developed eyes, are bioluminescent, and have bodies that are either transparent, reflective, or countershaded. In contrast, bathypelagic animals have poorly developed eyes, little bioluminescence, and nonreflecting bodies that strongly absorb blue light and thus look black, brown, or red when illuminated with white light (Lythgoe 1979, 104).

DAILY MIGRATIONS

Another indication of the importance of vision to predatory interactions comes from analyzing the vertical migrations that can commonly be observed in both oceanic and fresh-water environments. These migrations often involve movement of several hundred meters by copepods, hydrozoans, shrimp, squid, and fish (Gauthreaux 1980). The existence of these migrations has long been known and there have been more than a dozen explanations suggested (Mangel and Clark 1988, 149). However, recent research suggests that the dominant factor is visually directed predation, which is strongly affected by changes in the level of illumination.

An animal intermediate in the food chain is likely to find its prey at a much higher density than the density it presents to its own predators. At appropriate light levels such an animal may have a detection range as large as the spacing typical between its prey, making it able to feed efficiently. But at the same light intensity its own predators may have a detection range that, although larger, remains smaller than the typical distance between their prey, thus forcing them to spend most of their time searching for sight of their prey. This situation suggests that there is a window of light intensity within which an organism can maximize the ratio of its feeding rate to that of its mortality (Clark and Levy 1988).

Many organisms in fact migrate toward the surface, where the food chain starts, at night but then move away from it during daylight. Others make two round trips per day to take advantage of the special conditions of twilight. In a long food chain these interactions can be quite complex. The fact that fish are often attracted to artificial light can now be understood as providing an unusually favorable opportunity to feed.

CAMOUFLAGE

The extremes to which animals go to camouflage themselves provide further testimony to the extent that predators employ vision to locate and identify their prey. In his extensive study of camouflage, Cott (1941) identified four fundamental steps in effective camouflage:

1. **Color resemblance** to the background.
2. **Countershading**, which uses absorptive pigmentation and reflection to destroy the natural patterns of brightness that would otherwise reveal the shape of the organism.
3. **Disruptive coloration**, which uses high-contrast color patterns to interfere with perception of the true outline of the organism or its motion.
4. **Shadow elimination**, by modifying the form or orientation of the organism.

Disruptive coloration is optimized by creating camouflaging patterns in which the brightest and darkest colors are adjacent, which produces boundaries with the highest contrast (Cott 1941, 51–61). Furthermore, subtle shading within color patches often generates impressions of false surface relief, contributing to an impression that the boundaries between the patches are ones between three-dimensional objects (Cott 1941, 62–65). The actual contour of the animal is thus concealed by patterns in which contrasting boundaries intercept its contour at nearly perpendicular angles, as in a zebra, which is surprisingly well concealed in its natural habitat (Cott 1941, 93; Lythgoe 1979, 203).

Motion can be camouflaged by having stripes run parallel to the direction of motion. Snakes that avoid predators by escaping them generally have longitudinal "racing stripes," but those that escape by hiding have disruptive cross banding (Jackson et al. 1976). This correlation exists even within species such as the garter snake, which includes individuals that differ in their behavior and body color patterns (Brodie 1989). Some fish actually change from displaying longitudinal stripes when in motion to transverse bars when stationary (Cott 1941, 30).

In the uniform midwater environment, several special camouflage mechanisms can be employed (Lythgoe 1988), the simplest being **mirror camouflage**. In such a homogeneous environment, which is equally illuminated from all directions at a given angle to the vertical, a vertical mirror will be invisible, because when it is viewed from any angle except directly above or below it reflects light that is propagating at the same angle from the vertical as the background light against which it is viewed (Figure 8-23). Pelagic fish often have mirrorlike silvery flanks with reflecting guanine crystals in the scales that are oriented vertically (Lythgoe 1979, 173).

Another problem of camouflaging in such an environment is that downwelling light is typically fifty times more intense than upwelling. Any nontransparent object is readily detectable from below as a dark silhouette. Predators often take advantage of such situations by striking upward out of

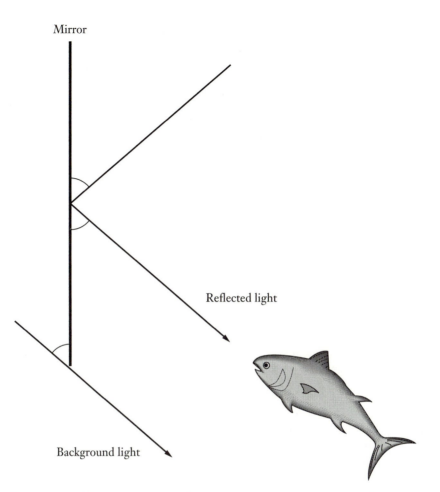

Mirror

Reflected light

Background light

Figure 8-23 Mirror camouflage. To the receiver (*lower right*), the vertical mirror reflects light coming from the same angle to the vertical as is the background light. Both usually have about the same intensity and color for a given depth underwater.

the darkness toward prey silhouetted against the bright downwelling light (Lythgoe 1979, 162). Many mesopelagic fish have tubular eyes, for greater sensitivity and in most of these fish the immobile eyes are aimed upward (Lythgoe 1979, 169). The only effective camouflage in this situation is to become transparent, which many pelagic animals indeed are (Lythgoe 1979, 174), or else to emit light to balance out the downwelling backlighting. Many mesopelagic fish, crustaceans, and cephalopods do produce bioluminescence on their bottom surfaces, adjusting the intensity to match the radiance of the downwelling light (Buck 1978; Lythgoe 1979, 175). It is probably difficult for an organism to match the spectral distribution of its bioluminescence to that of the prevailing light, suggesting why some predatory fish may use a yellow filter to change their spectral sensitivity from that of the majority of predators, to let them detect a mismatch (Lythgoe 1979, 195).

CHAPTER 9

Sound

■ ■

> On days of marvellous optical clearness
> the atmosphere may be filled with impervious acoustic clouds,
> while days optically turbid may be acoustically clear.
> —John Tyndall, 1874

Sound is a mechanical vibration that propagates through an elastic medium such as air or water. It travels as waves of alternating high and low pressure accompanied by back-and-forth movement of the medium in the direction the wave is propagated. Close to the sound's source, the movement of the medium is more like the flow of an incompressible fluid with no elastic properties. It is useful to distinguish between this **near field behavior** and the **far field behavior** of true sound, although there is often a significant transition region in which both properties are important. The boundary between the two is usually at a distance from the source on the order of the wavelength of the sound. In familiar terms, this distinction is similar to the one between vibration and sound. Although the distinction normally seems clear, it is difficult to pin down a clear boundary between the two; both humans and other animals automatically generalize between vibration and low-frequency sound (Schwartzkopff 1976 and discussion following his paper). Near-field effects, which are also a source of information, are discussed in the next chapter, along with other types of mechanical vibration.

9-1 MEASURING SOUND

A **free field** is a space in which sound propagates away from a small source without interference from features in the environment that might change the direction of propagation. When it is sufficiently far from the source, the wavefront approximates a plane, and there is a simple relationship between the speed of sound propagation (v_s), the density (ρ) of the medium, the particle velocity (v_p), and the sound pressure (p_s):

$$p_s = v_p \, v_s \, \rho \qquad\qquad (9\text{-}1)$$

Note that p_s is the small fluctuation from the average ambient pressure. Since water is a thousand times more dense than air, the pressure change in water is much larger than that in air for an equal intensity (see Table 9-2).

Sound intensity (I) is the sound energy per unit time that passes a unit area perpendicular to the direction of propagation of the sound wave. In a free field with a plane wave, the intensity is simply the sound pressure times the particle velocity (Michelsen 1983):

$$I = p_s \, v_p = \frac{p_s{}^2}{v_s \, \rho} \qquad\qquad (9\text{-}2)$$

Since v_s and ρ are properties of the medium and independent of intensity, the intensity increases as the square of the sound pressure, which is the parameter most easily measured by electronic equipment.

Sound levels are normally reported in unusual units called **decibels** (*dB*), which have proven to be a source of much confusion to those not familiar with the field of acoustics. A decibel is a logarithmic measure of a ratio. *Sound pressure levels* (SPLs) are defined by

$$SPL \ (\mathrm{dB}) = 20 \log \frac{p_s}{p_0} \qquad\qquad (9\text{-}3)$$

where p_0 is a reference sound pressure. The reference pressure for sound in air is usually 2×10^{-5} *N* m^{-2}, which is approximately the human threshold at 1 kHz, but other reference levels are also used. For sound underwater, the previous reference of 1 µbar (1×10^{-1} *N* m^{-2}) has now been replaced by 1 µPa (1×10^{-6} *N* m^{-2}) (Urick 1983, 14). Only decibel values using the same reference can be directly compared. *Intensity levels* (*ILs*) are similarly defined by

$$IL \ (\mathrm{dB}) = 10 \log \frac{I}{I_0} \qquad\qquad (9\text{-}4)$$

where I_0 is a reference intensity. The reference intensity is often taken as 10^{-12} W m^{-2}, which under standard conditions corresponds to the commonly used reference sound pressure level (2×10^{-5} N m^{-2}). Some correspondences between decibels and various ratios are presented in Table 9-1.

It should be emphasized that strictly speaking the decibel is a measure of intensity equivalent to a plane wave having the appropriate sound pressure (Urick 1983, 15), understanding that distorted waves may have a different relationship between pressure and intensity. Usually pressure is the parameter that is actually measured.

Decibels are a convenient measure to use, because the range of human hearing is approximately from zero to 100 dB, and a one-dB difference is barely discernible by the human ear under ordinary circumstances. Furthermore, the effects of different factors on sound intensity are calculated simply by adding or subtracting decibels rather than multiplying very large or very small numbers that are directly related to intensity.

The acoustic properties of air and water are quite different, and it is important to understand these differences so that we do not make false assumptions about sound stimuli in aquatic environments, based on our

Table 9-1 Relation Between Decibels and Sound Intensity

			Plane-Wave Intensity, W/m^2		
			In Air	In Water	
dB	I/I_0	p_s/p_0	$p_0 = 20$ μPa $= 2 \times 10^{-5}$ N/m^2	$p_0 = 1$ μbar $= 1 \times 10^{-1}$ N/m^2	$p_0 = 1$ μPa $= 1 \times 10^{-6}$ N/m^2
−20	0.01	0.1	1×10^{-14}	7×10^{-11}	7×10^{-21}
0	1	1	1×10^{-12}	7×10^{-9}	7×10^{-19}
1	≈ 1.26	≈ 1.12	1.26×10^{-12}	9×10^{-9}	9×10^{-19}
3	≈ 2.00	≈ 1.41	2×10^{-12}	1.4×10^{-8}	1.4×10^{-18}
6	≈ 3.98	≈ 2.00	4×10^{-12}	3×10^{-8}	3×10^{-18}
10	10	≈ 3.16	1×10^{-11}	7×10^{-8}	7×10^{-18}
20	100	10	1×10^{-10}	7×10^{-7}	7×10^{-17}
30	10^3	≈ 31.6	1×10^{-9}	7×10^{-6}	7×10^{-16}
40	10^4	100	1×10^{-8}	7×10^{-5}	7×10^{-15}
60	10^6	10^3	1×10^{-6}	7×10^{-3}	7×10^{-13}
120	10^{12}	10^6	1	7×10^3	7×10^{-7}

greater experience with terrestrial ones. Some of the important differences are summarized in Table 9-2.

Note the large difference between the acoustic impedances of air and water in Table 9-2. Using equation 4-4, one can calculate the proportion of a sound wave propagating perpendicular to a flat boundary between air and water that will be reflected at the boundary as 0.9989, meaning that only 0.1 percent is transmitted. An air–water interface is, then, an almost perfect reflector of sound.

Because of the complexity of sound, which varies independently in frequency as well as intensity, a specialized instrument called a **sound**

Table 9-2   Comparisons of Atmospheric and
Aquatic Acoustics

Property	Terrestrial Environments	Aquatic Environments
Speed v_s	340 m/s	1,500 m/s
Particle velocity at 1 W/cm^2	50 cm/s	8 cm/s
Sound pressure at 1 W/cm^2	0.02 μPa	12 μPa
Impedance* $= \dfrac{p_s}{v_p} = v_s\,\rho$	410 kg m^{-2} s^{-1}	1,500,000 kg m^{-2} s^{-1}
Speed profile	Large changes with altitude (up to 300%), due to winds and large temperature changes	Changes with depth are small (less than 6%)
Profile variability	Extreme; diurnal	Surface only; seasonal
Reflection	Ground may help or hinder	Surface interferes (phase inversion)
Scattering	Temperature and wind inhomogeneities; vegetation	Bubbles and swim bladders
Attenuation	Moderate	Low
Transmission range	Long-range is seldom possible	Long-range transmission is often possible
Reciprocity	Often not reciprocal, due to winds	Always reciprocal, since currents are negligible

Adapted from Rogers and Cox (1988, Table 5.1.) and Hawkins and Myrberg (1983, Table 1). The data are for typical open environments and do not apply to burrows, streams, etc.
* Free-field impedance. Solids are usually 100 to 1,000 times higher than water.

spectrograph is often employed to describe it. This device produces a plot called a sound spectrogram or sonogram of frequency versus time, with all the frequencies above a certain intensity being indicated visually for each instant of time. This mechanism provides a convenient visual and printable representation of the pattern of a complex sound. An example can be seen in Figure 9-14.

9-2 FREQUENCY

Frequency is the measure of how rapidly a sound wave oscillates, it is usually measured in Hertz (Hz), which is one complete cycle per second. As illustrated in Table 9-3, sound speed does not vary much with frequency, but wavelength and attenuation vary a lot, which means that sound propagation is influenced by frequency. In addition, as will be shown next, a sound's frequency is closely related to the size of the structure generating the sound. As a result, the frequency distribution of sound carries information about the source as well as the environment through which the sound has traveled. Frequencies above the range of human hearing, or about 12 kHz, are called **ultrasonic,** while those below the range at about 20 Hz are called **infrasonic.**

Sound is usually generated by a vibrating structure acting on the surrounding medium. Because muscles are unable to contract faster than 1,000 times per second, the frequency of biological sounds would be limited to less than 1 kHz if animals had to rely only on direct muscle contraction to generate sound (Michelsen 1983). However, animals use a variety of mechanisms to produce vibrations at higher frequencies than muscle contraction. These mechanisms include **stridulatory organs** that move a sharp-edged "scraper" over a rough "file" as in crickets, or force air over **vocal cords** as in mammals, or that have **click-producing** mechanisms in certain insects and in some human languages, or that create **aerodynamic turbulence,** as in the hiss of a cockroach or the spoken *s* sound in languages.

Where sound is produced by vibrating air in a cavity or stretching a string that is in contact with air, simple physical laws relate the size of these structures to the frequencies they produce. In a cylindrical cavity of length L that is open at one end, the frequencies produced (the harmonics) are evenly spaced, according to

$$f_n = \frac{(2n - 1)\, v_s}{4L} = (2n - 1) f_1$$

(9-5)

Table 9-3 A Comparison of Frequency Effects

Property	Frequency (Hz)		
	10	1,000	100,000
Wavelength			
Air	34 m	34 cm	3.4 mm
Water	150 m	1.5 m	1.5 cm
Wave velocity (m/s)			
Air*	344.057	344.078	344.17
Attenuation length ($1/\alpha_e$)			
Air*	> 200 km	1,000 m	1 m
Fog[†]	?	200 m	?
Grass[‡]	900 m	40 m	?
Deciduous trees[‡]	200 m	50 m	?
Evergreen trees[‡]	60 m	26 m	?
Fresh water[§]	≈ 1,000 km	≈ 1,000 km	2,000 m
Ocean[‖]	≈ 1,000 km	70 km	140 m
Mud[#]	20 km	5 km	1,400 m
Ambient intensity (W/m^2)			
Terrestrial	?	$10^{-11} \leftrightarrow 10^{-9}$	?
Ocean	$10^{-11} \leftrightarrow 10^{-10}$	$10^{-13} \leftrightarrow 10^{-11}$	$10^{-16} \leftrightarrow 10^{-14}$

* Air at 20° and 50% relative humidity. (From Weast 1985.)
[†] White 1975, p. 70.
[‡] Ford 1970, p. 18.
[§] From Equation 9-13 at 20°C.
[‖] From Equation 9-14.
[#] Sündermann 1986, p. 340.

where $n = 1, 2, 3,...$ and v_s is the velocity of sound. The lowest frequency (or the **fundamental**) is $f_1 = v_s/4L$. Similarly, the natural modes for a stretched fiber are

$$f_n = \frac{n}{2L} \sqrt{\frac{F}{\mu}} = (2n - 1)f_1 \qquad (9\text{-}6)$$

where F is the tension in the fiber and μ is its mass per unit length. For both types of sound generators, the lowest frequency is inversely related to the length of the device. This basic relationship is why within a family of

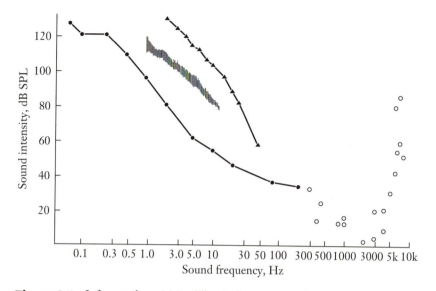

Figure 9-1 Infrasound sensitivity. The circles represent the hearing thresholds of homing pigeons. The triangles are the thresholds for humans. The vertical bars in the center indicate the sound levels in the environment. (After Kreithen 1978.)

musical instruments the larger ones produce the lower frequencies. The same is true for organisms in that larger animals can efficiently produce lower sound frequencies. Elephants can in fact produce some sounds that are too low in frequency for humans to hear (Payne 1990). Thunderstorms and winds blowing over mountain ranges produce infrasound (Figure 9-1) that may be audible to birds thousands of kilometers away (Kreithen 1978).

For sounds of a single frequency, which is what pure tones are, humans can determine their relative frequency quite precisely and distinguish between frequencies that differ by only about 0.1 percent (Fay 1974), but above 5 kHz this ability degrades (Moore 1988, 121). Bottle-nosed dolphins have a similar capability (Hawkins and Myrberg 1983; Ehret 1989), but fish seem to be less discriminating, with limits of 5 to 10 percent (Hawkins 1981). The frequency dependence for many vertebrates has been presented by Fay (1974) and Dooling (1982).

The abilities of humans to analyze sound spectra can be thought of as employing a bank of one-third octave band-pass filters to determine the intensity of sound in each frequency band (Searle 1982). Again, dolphins seem to have similar abilities (Ralston and Herman 1989). Such a signal-processing analysis, in the frequency domain, requires that several cycles of all the relevant frequencies be received, which makes the process relatively slow for low frequencies. For instance, at 50 Hz a 60-ms duration is required to make it possible to generate an accurate perception of frequency (White 1975, 144). An additional limitation of this signal-processing mechanism is that it discards information about the relative phase of the different frequencies (Figure 9-2). For example, humans are insensitive to whether a sound wave starts with a compression or a rarification of the medium.

For sounds containing more than one frequency, the human perception of frequency is usually broken down into two categories. **Pitch** is the perception of a fundamental frequency of vibration that produces harmonics in the sound. It is perceived even when the fundamental frequency is absent from the sound and only a few harmonics are present (Moore 1988, Ch. 4). This perception allows sounds to be ordered on a musical scale. The actual mixture of frequencies generates a multidimensional perception of **timbre** or **quality**. In this case the relative phase of different frequencies is sometimes perceptible as a change in timbre or the clarity of the pitch (Moore 1988, 243).

It is interesting to note that our auditory system retains more information about sound frequency than our visual system does about color (White 1975, 116). In vision, all the relative frequency components of light are merged the two stimulus dimensions of hue and saturation. All the sensations of color can be generated with the three primary colors. However, sensations of sound frequency cannot be similarly generated from a small number of primary frequencies. Human hearing involves about twenty-four distinguishable frequency bands (White 1975, 119), which are not merged into a single sensation.

The ability to discriminate frequencies is very important in human hearing. Many cases of hearing impairment involve a decrease in frequency discrimination, which cannot be alleviated by using a hearing aid. This loss contributes to the common complaint of not being able to understand conversation in a noisy environment (Moore 1988, 107).

Remarkably, human sound perception often ignores some components, synthesizes others, and makes mistakes (Moore 1988, Ch. 4). For example, single tones are often mistaken for those an octave above or below, and mixtures of tones such as the sound of a bell are normally perceived as containing tones that are not in fact present (Richardson 1953, 479; Meyer and Neumann 1972, 248). Furthermore, a sound and its delayed reflection can generate a sensation of a pitch that is not present in

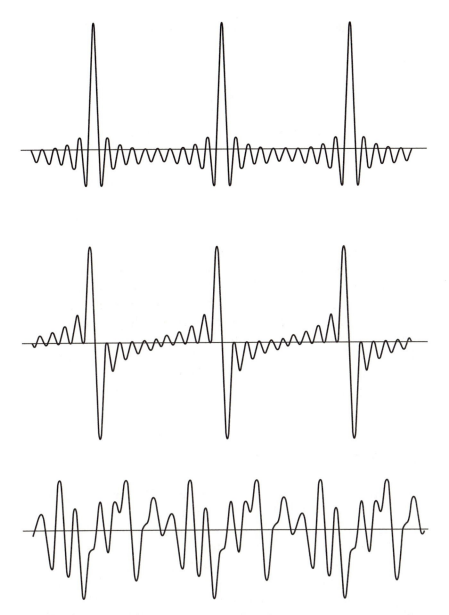

Figure 9-2 A demonstration of phase information. Each of the three samples is composed of ten harmonic frequencies of equal amplitude. The only difference between the three samples is in the relative phase of the different frequencies. *Top:* All the frequencies are in phase. *Bottom:* The phases were selected randomly, an example of white noise. To the human ear, all three samples would sound nearly alike.

either the stimulus or the source (Bilsen and Ritsma 1969). Although the telephone does not reproduce the fundamental frequency of the human voice, we hardly notice its absence (Truax 1984, 134). In contrast, songbirds seem to be much more aware of absolute pitch than are humans (Hulse 1989), although their ability to discriminate between different frequencies is not as good (Dooling 1982).

Fish seem to use a quite different strategy for analyzing sound, one that is more in the time domain. The frequency range of fish is limited (Hawkins and Myrberg 1983), but they can determine whether a sound wave starts with a rarification or a compression and can respond to a single cycle of a sound wave (Piddington 1972; Blaxter et al. 1981; and following discussion). This type of analysis may be appropriate for fish to adopt in order to trigger escape responses with less time delay and to coordinate their movements in fast-swimming schools.

In many animals the frequency of their most sensitive hearing matches their specific needs. For example, frogs exhibit a good match between the most sensitive frequency of their hearing and the fundamental tone in their mating calls. The correspondences even include geographical variation within a species and sexual dimorphism (Hopkins 1988; Ryan and Wilczynski 1988). A similar correlation has been observed in damselfish (Myrberg and Spires 1980).

The frequency range of hearing for a variety of animals is shown in Figure 9-3.

9-3 SENSITIVITY

Sound sensitivity varies widely among animals. Human hearing is most sensitive at about 3 kHz. The threshold is a sound pressure of about 2×10^{-5} N/m² (20 μPa), which corresponds to a plane-wave intensity of about 1×10^{-12} W/m². In comparison, the thermal noise of air molecules corresponds to a sound pressure 20 times lower (Meyer and Neumann 1972, 242). At higher frequencies this difference is reduced; at lower frequencies, the threshold sensitivity is close to the estimated noise level generated by blood flow and muscle contractions (Shaw 1982). In mammals the frequency of maximum sensitivity often coincides with the resonance frequencies of the ear canal (Ehret 1989). The most sensitive of the mammals that have been studied thus far is the cat (Ehret 1989). As we shall see, most environments have sound intensity levels that are well above mammals' thresholds at the most sensitive frequencies.

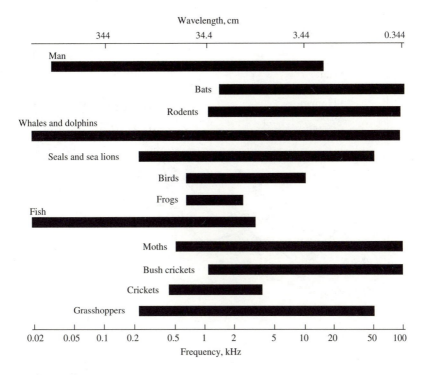

Figure 9-3 The hearing-frequency ranges of different animals. (After Lewis and Gower 1980.)

The ability to hear a pure tone mixed with noise is usually described by the **critical ratio**, defined as the ratio of the power in the pure tone to the power per Hz of the background noise (Dooling 1982). As seen in Figure 9-4, the critical ratio for most vertebrates increases about twofold for each doubling of frequency, which is due to the mechanics of the ear. Certain species of bats, and probably some birds, have specialized ears that dramatically improve their performance at specific frequencies.

In fish, sensitivity to sound has been demonstrated scientifically only within the last fifty years (von Frisch 1938). In them particle displacements in the range of 1 to 100 Å may be detected (Fay 1988). Fish with swim bladders can also detect pressure changes, with a sound pressure threshold that is typically about –40 dB relative to 1 µbar, which corresponds to about the same intensity as that of the human threshold. Fishes' sound

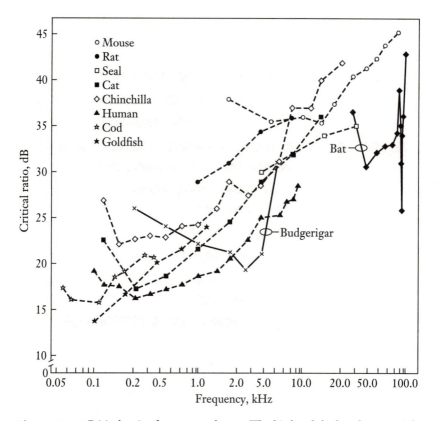

Figure 9-4 Critical ratios for ten vertebrates. The bird and the bat show specializations for high performance at certain frequencies, but the others all follow the same pattern. (After Dooling 1982.)

sensitivity may extend from 50 to 1,000 Hz (Hawkins 1981). In many cases, pressure changes can be detected even when the particle motions are below threshold (Fay 1988).

A summary of the acoustic sensitivities of a variety of animals is presented in Table 9-4.

9-4 INTENSITY

In natural environments sound intensity usually includes sounds from a variety of sources. Functionally these sources are separable into sources of

Table 9-4 Some Sound Sensitivities

| Organism | Best Frequency (kHz) | Free-field Threshold | | Reference |
		SPL	Intensity	
Cricket	4	44 dB(20 µPa)	10^{-8} W/m^2	Michelsen and Larsen 1983
Moth	30	40 dB(20 µPa)	10^{-8}	Fullard and Thomas 1981
Bullfrog	0.3 ↔ 1.8	20 dB(20 µPa)	10^{-10}	Gerhardt 1989
Snake	0.25	35 dB(20 µPa)	10^{-9}	Buning 1983
Birds	2 ↔ 4	4 dB(20 µPa)	10^{-12}	Saunders and Henry 1989
Bats	10 ↔ 200	0–20 dB(20 µPa)	10^{-12}	Surlykke 1988
Listening bat	20 ↔ 30	−10 dB(20 µPa)	10^{-13}	Surlykke 1988
Mouse	15	− 8 dB(20 µPa)	10^{-13}	Ehret 1989
Cat	?	−18 dB(20 µPa)	10^{-14}	Ehret 1989
Human	3	−5 dB(20 µPa)	10^{-12}	Richardson 1953, 250
Flatfish	0.1	−10 dB(1 µbar)	10^{-9}	Chapman and Sand 1974
Salmon	0.1 ↔ 0.2	−30 dB(1 µbar)	10^{-11}	Hawkins and Johnstone 1978
Cod	0.02 ↔ 0.3	−40 dB(1 µbar)	10^{-12}	Hawkins 1981
Goldfish	0.1 ↔ 1	−40 dB(1 µbar)	10^{-12}	Hawkins 1981
Catfish	0.2 ↔ 2	−40 dB(1 µbar)	10^{-12}	Hawkins 1981
Ringed seal	1 ↔ 40	−20 dB(1 µbar)	10^{-10}	Hawkins and Myrberg 1983
Dolphin	20 ↔ 100	−60 dB(1 µbar)	10^{-14}	Ralston and Herman 1989
Killer whale	16	−72 dB(1 µbar)	10^{-15}	Hawkins and Myrberg 1983

background noise that contain no information of interest but may interfere with the reception of sound from other sources, and sources that provide sound that is of interest. Background noise is important because it often limits the range at which sounds that are of interest can be detected. This section discusses the level of background sound intensity in different environments and the question of how precisely the intensity of sound can be determined, which provides information about the range of a source.

BACKGROUND SOUNDS

Background noise often limits the ability to perceive important sound signals. For example, atmospheric noise is produced naturally by wind and various biological sources. A sample of background noise spectra for three different habitats is given in Figure 9-5. The spectral distribution varies moderately from one habitat to another and according to the time of day. In open habitats near midday, wind is likely to dominate other noise sources, while biological sources dominate in a rain forest (Waser and Brown 1986) and in the morning and evening (Brenowitz 1982). In all these cases as well as others (Brown and Schwagmeyer 1984), noise levels of about 40 dB (re 20 μ Pa) are common. In these noise distributions, monkey calls and human speech were detectable at signal-to-noise ratios of around −24 and −14 dB, respectively (Brown 1989).

A useful strategy for separating a specific signal from the background noise is to separate sounds according to the direction to the source. Humans can do this effectively, an ability referred to as the **cocktail party effect** (White 1975, 125), because of the experience of being able to tune into one conversation at a party while blocking out many other simultaneous conversations. There is some evidence that fish can also make use of this strategy with an improvement of 6 to 8 dB in their ability to separate sounds of interest from background noise (Fay 1988).

In water, background noise is significant because of the long distances sound propagates in that medium. Noise is caused naturally by wind, waves, surf, ice movement, and molecular motion as well as by organisms. At ordinary temperatures in the ocean, thermal noise is given by

$$NL = -15 + 20 \log f \qquad\qquad \text{dB re 1 μ Pa} \qquad (9\text{-}7)$$

where NL is the noise level and f is the sound frequency in kHz (Urick 1983, 208). This source of noise dominates at frequencies above about 100 kHz. From this frequency down to about 100 Hz, noise is caused primarily by wind interacting with the surface of the water in the vicinity of the receiver, and the noise level depends on the wind speed or sea state (Urick 1983, 207–208). Even a moderate amount of wave action can generate noise sufficient to raise the hearing threshold for fish (Figure 9-6). From 100 Hz down to 10 Hz, noise at sea is presently dominated by ships (Urick 1983, 207). And from 10 Hz down to 1, noise is thought to be caused primarily by the ocean's turbulence (Urick 1983, 205). Some typical levels of noise in deep water at various frequencies are shown in Figure 9-7. In

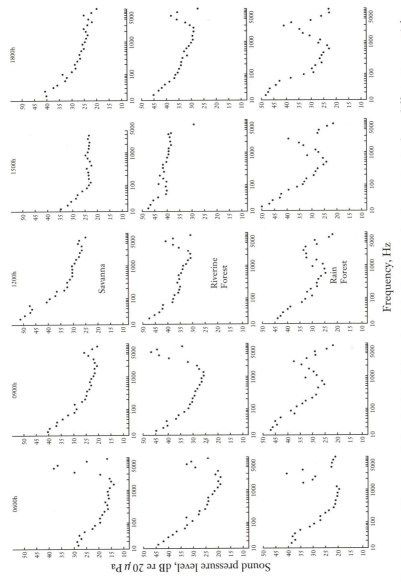
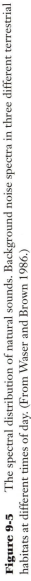

Figure 9-5 The spectral distribution of natural sounds. Background noise spectra in three different terrestrial habitats at different times of day. (From Waser and Brown 1986.)

217

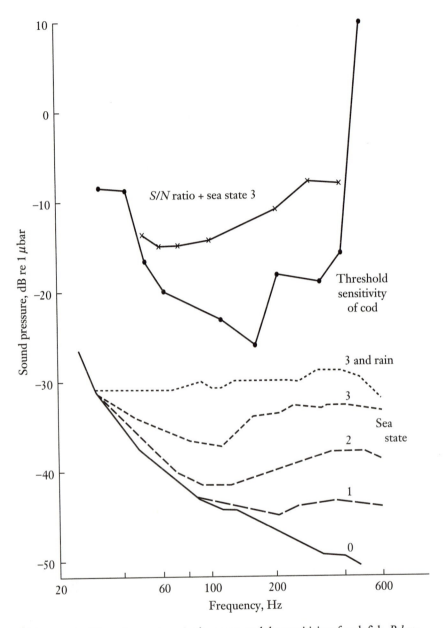

Figure 9-6   The noise spectrum in the ocean, and the sensitivity of each fish. *Below:* Ambient noise spectra in different weather conditions in a sheltered area. The circles represent the threshold sensitivity for cod in the absence of noise. The crosses show cod sensitivity in the presence of noise equivalent to sea state 3, as defined by nautical almanacs. (After Blaxter 1988.)

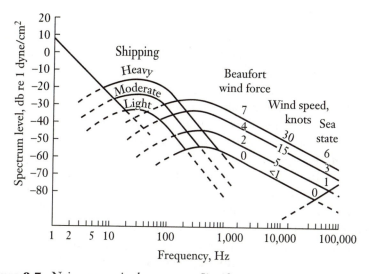

Figure 9-7 Noise spectra in the open sea. Significant variations are associated with weather and ships. (From Urick 1967.)

the open ocean, noise generally decreases with depth. If a "mixed layer channel" exists, much surface-generated noise will be confined to this channel and the noise level will be lower below it (Urick 1983, 222–223).

In shallow water, noise is much more variable with the main sources being ships, wind, and organisms (Urick 1983, 212). When local shipping and biological sources of sound are absent, the noise levels are similar to those in the open ocean and correlated strongly with wind speed. At low frequencies shallow water may be quieter than the open ocean, because of the poor propagation of low frequencies in shallow water.

Rivers are noisy because of current-generated sound. The few noise spectra that are available (Hawkins and Johnstone 1978; Evans 1988) have levels about 35 dB below 1 μ bar for 50 to 500 Hz frequencies. These levels will raise the already relatively high thresholds for salmon (see Table 9-4).

SOURCES OF SOUND

The production of sound is generally a complicated process. For high efficiency, the vibrating source must be as large across as the wavelength of the sound (Michelsen 1983). A theoretical pulsating sphere is the simplest sound source and very efficient because all its parts are moving in the direction of propagation and are producing compressions or rarifications at the same time.

In other vibrating structures, by comparison one part is acting to compress the medium in one place, while another part, such as the back side of a vibrating fiber, is acting to rarify the medium in another place. The flow of the medium between these two places, as around a vibrating fiber, acts to short circuit the changes in pressure in the medium, greatly reducing sound output. Acoustic short circuiting is particularly severe if the interfering parts are much less than one wavelength apart. For this reason sound engineers often mount a speaker in the side of a box so that the sound produced by the back side of the speaker's membrane is isolated inside the box from that produced by the front side. Certain crickets have been found to use a similar strategy by cutting holes in the center of a large leaf and placing themselves in the hole while singing to reduce the short circuiting, because the short-circuiting flow now has to go around the leaf, instead of around the much smaller cricket. This arrangement provides the crickets with about a 10 dB increase in sound output. Another strategy, exploited by mole crickets, is to use a specially shaped burrow that acts much like a horn musical instrument to increase the effective size of the vibrating structure and thus increase the sound output (Michelsen 1983).

The actual sound energy radiated from some biological sources is presented in Table 9-5. In conversation the human voice radiates an average sound intensity of about 25 μW (Meyer and Neumann 1972, 269). In a noisy river a salmon could hear the voice of a human on the bank only if that person were shouting, but his footsteps would be readily heard (Hawkins and Johnstone 1978). Swimming fish generate sounds themselves, and schools of fish generate sounds of sufficient intensity that native fishermen can determine their direction by using only a paddle to transmit the underwater sound to the ear of the fisherman in a boat (Moulton 1960).

INTENSITY DISCRIMINATION

Humans can discriminate sound intensities that differ by about 1 dB, which is a 25 percent difference (Coleman 1963; Moore 1988, 54). Small mammals and birds require two or three times larger differences for intensity discrimination (Dooling 1982). Although few data are available for fish, it appears that they can discriminate intensity changes of a few dB, with higher limits at higher frequencies (Hawkins 1981; Blaxter 1988).

Table 9-5 Biological Sound Sources

Animal	Frequency (Hz)	Power *	Reference
Terrestrial			
Fruit fly	166	10^{-16}	Bennet–Clark 1971
Ant	500–30,000	10^{-10}	Markl and Hölldobler 1978
Cricket	4,000	10^{-4}	Michelsen and Larsen 1983
Mole cricket	3,500	10^{-3}	Bennet–Clark 1971
Blackbird	3,000	10^{-2}	Brenowitz 1982
Bat clicks	30,000	10^{-3}	Miller 1983
Human voice	$100 \leftrightarrow 1,000$	$10^{-9} \leftrightarrow 10^{-3}$	White 1975, 367, 369
Marine			
Toadfish	20	?	Urick 1983, 217
Bowhead whale	$100 \leftrightarrow 2,000$	$0.01 \leftrightarrow 10$	Cummings and Holliday 1987
Finback whale	20	$0.1 \leftrightarrow 30$	Watkins et al. 1987
Dolphin whistle	$5,000 \leftrightarrow 10,000$	10^{-4}	Ralston and Herman 1989
Dolphin click	25,000	10^5 max	Ralston and Herman 1989
Croaker	500	?	Urick 1983, 217

* Using $I = p_s^2 / v_s \, \rho$, in water 1 μPa at 1 m corresponds to 6.7×10^{-19} (W/m^2) $\times 4\pi$(m^2) = 8×10^{-18} W. In air, 20 μPa at 1 m corresponds to 9.8×10^{-13}(W/m^2) $\times 4\pi$(m^2) = 1.2×10^{-11} W.

One of the most remarkable aspects of human hearing is that it can discriminate intensities over a range of 120 dB, which is twelve orders of magnitude. The intensity appears to be integrated (that is, averaged) over a period of about 0.1 s (Moore 1988, 52).

9-5 TRANSMISSION

As will be seen, the transmission of sound through the natural environment can be complex and is often not amenable to accurate prediction. This section considers the factors influencing the speed, direction of propagation, and attenuation of sound.

SPEED

The speed of propagation of sound through a homogeneous medium is relatively simple to analyze. As previously described (Box 4-1), the speed at which a mechanical wave propagates is proportional to the square root of the ratio of a restoring force to the inertial forces. The relationships for sound waves are presented more specifically in Box 9-1. From these relations, it can be seen that the speed of sound in air varies with the square root of the absolute temperature and with the composition of the medium; water content is most important. Under natural conditions the speed of sound is about 340 m/s.

In water, which is relatively incompressible, the speed of sound is increased to about 1,500 m/s. This speed varies linearly with temperature, salinity, and water depth. The difference in speed between fresh water and seawater is only about 2 percent (see Table 4-1). The speed of sound in the ocean can be summarized by

$$v_s = 1449.2 + 4.6\,T - 0.055\,T^2 + 2.9 \times 10^{-4}\,T^3$$
$$+ (1.34 - 0.01\,T)\,(S - 35) + 1.6 \times 10^{-2}\,d \tag{9-8}$$

where v_s is the speed in m/s, T the temperature in degrees centigrade, S the salinity in parts per thousand, and d the depth in meters (Urick 1983, 113).

REFLECTION

Reflection is the redirection of sound back into its original medium after encountering a boundary between two different media. Reflection occurs when the media differ in their acoustic impedance as explained (in Section 4-1). In the simplest situation, water has an acoustic impedance that is 3,000 times that of air (see Table 9-2). Consequently, sound impinging on an air–water interface is almost totally reflected. At normal incidence only 0.1 percent of it is transmitted (from Equation 4-4), and sound reflection is total for angles of incidence from air to water that are greater than about 13° (Ford 1970, 75).

When it impinges on an air–water interface from the water side, sound is reflected back with an inversion of its phase, meaning that a pressure maximum occurs when a pressure minimum would have occurred

---

Box 9-1 Speed of Sound Waves

The speed at which a mechanical wave propagates is proportional to the square root of the ratio of a restoring force to the inertial forces, as described in Box 4-1. For sound this can be expressed as

$$v_s = \left(\frac{\delta p}{\delta \rho} \right)_s^{1/2} \tag{a}$$

where δ indicates a partial derivative, p is pressure, ρ is mass density, and the subscript s indicates that the derivative is taken with constant entropy. For a gas, this becomes

$$v_s = \sqrt{\frac{\gamma p}{\rho}} \tag{b}$$

where γ is the ratio of the specific heat at constant pressure to the specific heat at constant volume and depends on the number of degrees of freedom of individual gas molecules. For diatomic molecules, which make up the bulk of air, $\gamma = 1.40$. If we assume that air acts like an ideal gas ($PV = nRT$), then

$$v_s = \sqrt{\frac{\gamma RT}{M}} \tag{c}$$

where R is the gas constant, T the absolute temperature, and M the molecular mass of the gas (White 1975, 28). For condensed materials, meaning liquids or solids, it is more convenient to use the following relations:

$$v_s = \sqrt{\frac{B}{\rho}} = (\kappa \rho)^{-1/2} \tag{d}$$

where B is the bulk modulus of the medium or κ is its reciprocal, the compressibility of the medium.

---

in the original wave. As a result, if the reflected wave travels a path similar to a direct wave, as in the case of a source and a receiver just below the surface of a body of water, the two waves tend to cancel one another out so that the transmitted sound intensity is low (Figure 9-8). Away from the surface, positions of weak and strong intensity will be intermixed.

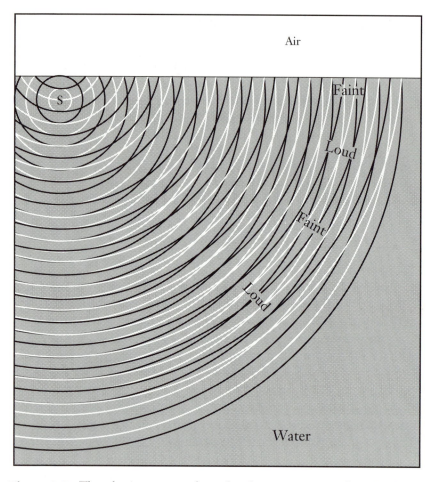

Figure 9-8 The relection pattern of sound under an air–water surface. The black and white curves represent the wavefronts of maximum and minimum pressure. Where the direct and reflected waves are in phase (where black intersects black and white intersects white) they sum to produce a loud sound. Where they are out of phase (where black and white intersect), they tend to cancel each other out, so that the sound will be faint. With the source (S) near the surface, the sound will always be faint for a receiver near the surface, and positions of alternating loud and faint sound will vary with the depth and distance.

If the surface is rough, with surface waves that are large compared to the wavelengths of the sound waves, the sound will be reflected in many directions (scattered) rather than reflected in a coherent path. Quantitatively, the amplitude ratio of a coherently reflected wave to an incident wave can be predicted by exp($-2 \pi h \sin \theta/\lambda$), where h is the height of the surface waves, θ the grazing angle, and λ the wavelength of the sound wave (Urick 1983, 129).

At the bottom of a body of water similar interactions occur, but the bottom varies much more in its acoustic properties than does air and may even consist of layers with varying properties. For example, the speed of sound in solid rock (3,000–7,000 m/s) is higher than in water, but certain sediments have a speed of sound below that of water (Urick 1983, 139). Most typically, the reflection losses at the bottom are about 10 dB, but they may be much lower at grazing angles and can be as high as 30 dB (Urick 1983, 142–143).

REFRACTION

Refraction is the bending of a propagating wave as a result of changes in the speed of propagation (see Section 4-1). The change can result from a sharp interface between two media, like the air–glass interface in a lens, or from a gradual speed gradient with no interface.

In air, refraction occurs as a result of temperature gradients, which near the ground can be quite large. On a sunny day the air is warmer close to open ground, which increases the speed of sound as the ground is approached, causing the sound waves to bend upward. This bending upward creates a sound shadow beyond a certain range (Figure 9-9). On cool nights, the opposite will occur (Ford 1970, 17). When such a temperature inversion occurs, sound waves are bent toward the ground and sounds may be heard for much greater than normal distances. A sound channel may then be created by repeated refraction in the atmosphere toward the earth's surface and reflection from the surface upward into the atmosphere. This channeling is particularly effective if the surface is smooth enough to reflect sound coherently instead of scattering it in all directions. A lake under windless conditions is ideal for this; under such circumstances it is sometimes possible to hear a conversation a kilometer away (White 1975, 65–66). Similarly, the early morning heating of the canopy in a forest can create a sound channel between the canopy and the ground (Wiley and Richards 1978).

Sound waves in air are also bent by wind gradients. Wind speed always decreases as a stationary surface is approached. Because wind "carries" sound waves, they will be bent upward in the upwind direction and downward in the downwind direction (Figure 9-10). Beyond a certain distance upwind there may be a sound shadow in which sound intensity is

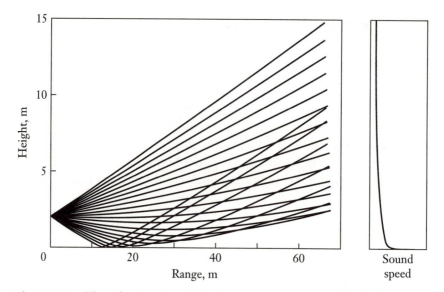

Figure 9-9 The refraction of sound by daytime temperature gradients. Shown are the paths of rays emitted from a source 2 m above the ground at 1° intervals within 10° of horizontal for 0.2 s travel time. The temperature, in centigrade, was assumed to be described by $30 - 1.5 \ln (1 + z/0.001)$, with z in meters. Under these conditions a receiver that is also 2 m above the ground will not receive any rays if it is more than 70 m from the source.

much less than would normally be expected (Ford 1970, 16). Sound thus becomes detectable farther downwind. Moreover, sound propagation between two parties is not reciprocal near the earth's surface in a wind. Quantitative estimates of the magnitude of these effects are difficult to make, but as a rule of thumb one can expect them to increase the sound by no more than 5 dB (less than a doubling of the range) or decrease it by more than 20 dB (a tenfold reduction in range) (Ford 1970, 17). If the wind velocity changes sharply with altitude, sound shadows may be followed by areas reached by sound traveling high in the atmosphere, thus creating a zone at ground level that the sound appears to skip (Ingård 1953).

More complex effects occur when sound propagates near and parallel to the ground (Michelsen and Larsen 1983). A so-called **ground wave** can carry low-frequency (<1 kHz) sounds relatively well, a situation that many tropical birds as well as grouse may make use of (Wiley and Richards 1982).

In aquatic environments, horizontal stratifications of temperature, salinity, and pressure cause changes in sound speed that refract sound

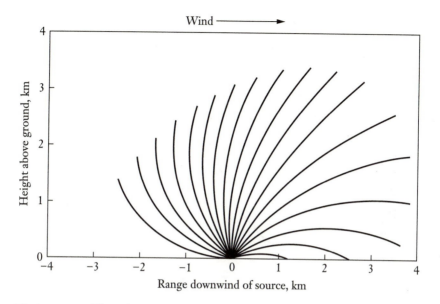

Figure 9-10 The refraction of sound by wind. The paths of rays emitted at 10° intervals from a source 2 m above an absorbing ground, for a time interval of 10 s. The wind was assumed to increase by 10 m/s for every 100 m above the ground, an assumption that is reasonable near the ground but that exaggerates normal conditions at higher levels.

waves in the direction of lower speed. Deep water is usually isothermal near 4° C, the temperature at which water is at its maximum density. Thus, sound speeds up with the increasing pressure at increasing depth and is refracted toward the surface. Water is usually warmer near the surface, and often enough so to cause an increase in sound speed. As a result, there is often a minimum level in sound velocity at some depth that causes sound waves to be refracted back into this layer and retained there over long distances without their being absorbed or scattered at the surface or bottom (Figure 9-11). Thus, a channel for favorable sound propagation is formed that spans depths in which the sound speed is less than that at the surface (Urick 1983, 163–164) analogous to a light pipe. This **deep sound channel** is also called the **SOFAR channel**. As might be expected from the considerations above, the depth of the SOFAR channel is greatest near the equator and disappears in the arctic. Because sound within this channel never encounters the absorptive bottom or scattering from surface waves, it can propagate for thousands of kilometers.

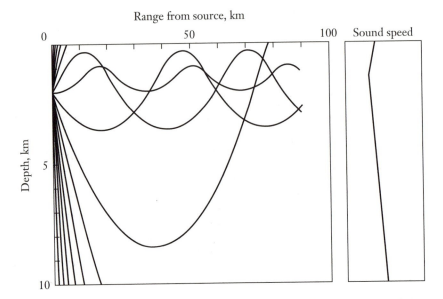

Figure 9-11 The deep sound channel in the ocean, also called the SOFAR channel. The paths of rays separated by 10° intervals are shown here for a travel period of 60 s after leaving the source. The sound speed (km/s) profile was assumed to be 1.538 km/s at the surface, declining to a minimum of 1.490 km/s at a depth of 1.2 km and increasing by 0.015 km/s for each km of increased depth. Complete absorption of rays reaching the surfaces is assumed, though in reality waves that are scattered and reflected from the surfaces would complicate the picture. Observe that rays leaving the source over a range of at least 30° are trapped in the channel and that rays emitted at angles closer to the vertical will sooner or later reach one of the surfaces, where they are likely to be scattered or absorbed.

The sound field in the SOFAR channel is composed of a complex pattern of convergence and shadow zones (Urick 1983, 167). In addition, the sound is distorted by having different arrival times for sound traveling via different paths. Surprisingly, the first sound to arrive at a distant location in the channel is the one following the longest path, because it has the greatest excursion from the axis and is often at depths where the speed of sound is at its greatest (Figure 4-2). Sound traveling along the axis of the channel has the longest travel time but also carries the most energy. Consequently, a sound pulse from a source on the axis of the channel is received at a distant location on the axis as a roar of increasing intensity, followed by a sharp cutoff as the sound traveling along the axis is received.

Near the surface of the ocean the sound-speed profile will vary with the seasons. In winter, mixing of the surface water by storms often creates an isothermal layer near the surface so that sound speed decreases with pressure as the surface is approached. Sounds therefore tend to be refracted toward the surface and away from the boundary between this **mixed layer** and the thermocline, which forms the upper part of the SOFAR channel. Since as we have seen sound is reflected from the surface back into the water, any sound originating within the mixed layer tends to be trapped in it (see Figure 9-12). Data for the depth of the mixed layer in the North Atlantic (Urick 1983, 158) indicate that it has a median depth of

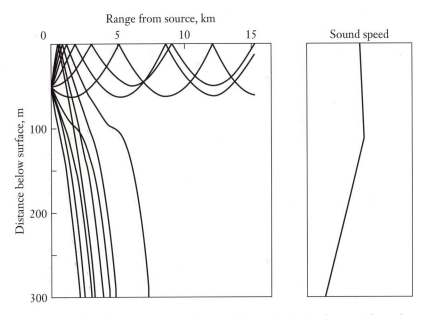

Figure 9-12 Sound propagation in the mixed layer of a body of water. The paths are shown for rays emitted within 10° of horizontal for a travel period of 10 s. The depth is exaggerated fiftyfold with respect to the horizontal distance. The temperature was assumed to be uniform within the 100-meter deep mixed layer and to decline by 4° C per 100 m in the thermocline below. The sound speed was calculated from Equation 9-8. Speed increases with depth in the mixed layer because of increasing pressure and it decreases with depth in the thermocline. This produces a depth that gives the maximum speed of propagation and waves are refracted away from this layer. If the surface is smooth enough that scattering is minimal, sound can propagate long distances, trapped in the isothermal layer.

30 m in summer and 80 m in winter, but this layer can be as deep as 400 m (Rogers and Cox 1988). Any sound duct of a given thickness has a maximum wavelength that can be transmitted, as well as a corresponding cutoff frequency, discussed below. Sound is lost from the mixed layer by scattering at the surface, the amount of which increases with frequency, and by transverse diffusion from the diffusely defined bottom, the amount of which decreases with frequency. Thus, there is a frequency that is best retained within the layer. Under natural conditions this optimum lies between a few hundred and a few thousand Hz (Urick 1983, 153).

In summer, the water's surface is heated and the thermocline extends all the way to the surface from the deep channel. Sound near the surface is therefore refracted toward the deep channel axis, creating a sound shadow near the surface beyond a certain range in a manner analogous to that of refraction in the atmosphere on a hot day, as shown earlier in Figure 9-9 above. In natural situations, intensities within the sound shadow are reduced by 40 to 60 dB below what would occur in a free field (Urick 1983, 135).

Shallow water forms **sound ducts** in which sound waves interact with both the top and bottom surfaces, making analysis of them more complex (Rogers and Cox 1988). In this case there is a lower frequency limit on the sound that can propagate. With given depth d of water with sound velocity v_w and a bottom with velocity of sound v_b, this limit or cutoff frequency f_c is given by

$$f_c = \frac{v_w}{4d}\left(1 - \left(\frac{v_w}{v_b}\right)^2\right)^{-1/2} \tag{9-9}$$

If $v_b \gg v_w$, then $f_c = v_w/4d$ and the limiting wavelength is four times the depth. In less than 10 cm of water, very little sound can propagate that is within the hearing range of fish, and the propagation of a 20 Hz sound is limited to depths of over 100 m. In addition, near a coastline, complex patterns of sound refraction are produced by variations in salinity as well as temperature (Urick 1983, 120). Similar considerations apply to the propagation of sound in tubes. For instance, terrestrial burrows often carry 440 Hz sound most efficiently, and many burrowing animals seem in fact to specialize in these low frequencies (Burda et al. 1990).

These considerations emphasize that refraction and reflection can produce a complex pattern of positions in which sound transmission is either favorable or unfavorable, making it difficult to determine the extent to which aquatic animals make use of sounds from distant sources.

DIFFRACTION AND SCATTERING

Diffraction is the redirection of a wave by an obstacle that interferes with part of the wavefront. A sound wave passing the edge of a barrier that absorbs or reflects sound will undergo diffraction that will cause some of the sound to propagate into the area behind the barrier that would be in a shadow for a light wave. This phenomenon is much more common with sound waves than light waves, because their much longer wavelengths are frequently on the same size scale as impeding objects and also on the same scale as the range of transmission. Estimates of the sound intensity expected in different locations behind a barrier are presented in Figure 9-13.

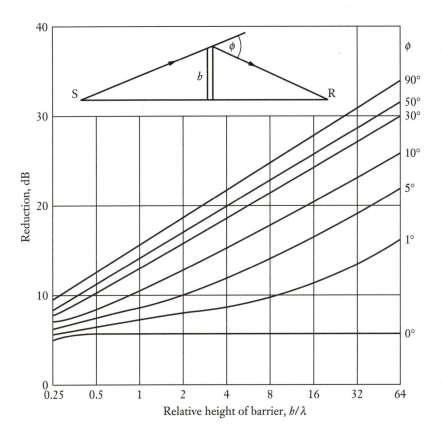

Figure 9-13   Refracted intensity behind a barrier. (From Ford 1970, 19.)

Many sound sources have transmission patterns that are uneven in different directions because of diffraction around the source, and vocal animals vary a great deal in the extent to which their transmitted sounds are confined to a beam (Brown 1989c).

Scattering is the altering of sound waves in ways that are neither simple reflection nor refraction. The altered wave propagates into new directions in complex patterns related to the size and properties of the scattering object. Scattering generally results from interaction with objects comparable in size to the wavelength of the sound. Although scattering can sometimes increase the intensity of sound in the shadow of a barrier, it usually has negative effects on information transmission (Wiley and Richards 1978). Scattering along the direct path of the sound diverts some of the sound energy into new directions, reducing the intensity received. In addition, multiple scattering may result in the arrival of sound that has traveled farther and thus been delayed, called **reverberation**. Reverberation tends to "smear out" any temporal pattern and make detection more difficult (Figure 9-14).

For a small homogeneous sphere, Lord Rayleigh (1945, 284) calculated that intensity (I_r) reflected by a sphere (radius $<< \lambda$) from an incident plane wave of intensity I_i is

$$\frac{I_r}{I_i} = \frac{\pi^2 \, V^2}{r^2 \, \lambda^4} \left(1 - \kappa_r + \frac{3(\rho_r - 1)}{1 + 2 \, \rho_r} \cos \theta \right)^2 \qquad \lambda >> \text{radius} \qquad (9\text{-}10)$$

where

V is the volume of the sphere

r is the distance at which I_r is measured

κ_r is the ratio of compressibilities of the sphere and the surrounding medium, with $\kappa_r = 0$ being equivalent to a rigid target

ρ_r is the ratio of densities of the sphere and the surrounding medium, with $\rho_r \to \infty$ equivalent to a fixed target

θ is the angle between the reverse direction of the incident wave and the direction to where I_r is measured, with $\theta = 0$ for backscattering.

When the differences in compressibility and density between the target and the surrounding medium are small ($\kappa_r \approx 1$; $\rho_r \approx 1$), equation (9-10) applies to a target of any shape (Rayleigh 1945, 284).

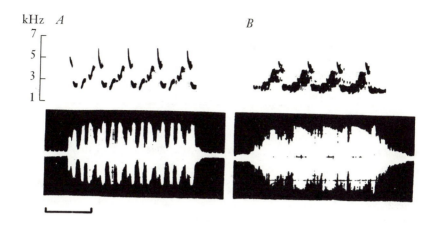

Figure 9-14   The smearing of sound by reverberation. *A*: Songs of Carolina wrens recorded at 10 m away. *B*: The same song recorded 50 m away. The upper figures are sound spectrographs showing the frequency distribution, and the lower markings are direct recordings of pressure amplitude. The horizontal line indicates a duration of 1.0 s. The basic pattern of the frequency sweeps (*above*) is preserved at the greater distance, but the amplitude pattern (*below*) is obscured. (From Wiley and Richards 1982.)

The effects of the composition of the sphere are entirely contained within the number determined by squaring the sum of the terms in the large brackets. For backscattering (θ = zero) this number is calculated for several materials in Table 9-6. It can be seen that scattering can be either stronger or weaker than for the ideal fixed, rigid sphere, but ordinary situations do not differ much in magnitude from the ideal, except that gas bubbles in water scatter 100 million times more strongly. However, the strength increases with the sixth power of the diameter, so that a bubble of about 4 percent the diameter of a rigid, fixed sphere would backscatter at the same intensity.

Sound in the atmosphere is scattered significantly both by density and by wind inhomogeneities (Ingård 1953), which are greater during periods of sunshine. Sounds are thus usually more distinct and carry further at night (Geier 1965, 141). Significant scattering, which also occurs in forests, can obscure sound patterns by reverberation (Wiley and Richards 1982; Waser and Brown 1986). Since atmospheric inhomogeneities are usually smaller within vegetation than out in the open, the effects of the vegetation itself on sound propagation are less than might be supposed (Wiley and Richards 1978). In a wind, the reverberation pattern from the vegetation changes as the vegetation moves, producing large intensity

Table 9-6 Effect of a Material's Composition on Sound
Scattering by Small Spheres

Nature of Components				Backscattering	
Sphere	Medium	κ_r	ρ_r	Strength*	Ratio[†]
Matching	Matching	1	1	0	0
Water at 0°C	Water at 20°C	1.1051	1.0016	2.9×10^{-5}	1.8×10^{-4}
Water at 20°C	Water at 0° C	0.9049	0.9984	1.5×10^{-3}	9.3×10^{-3}
Mineral	Water	0.1	2.6	2.8	0.45
Water drop	Air	5.3×10^{-5}	833	6.24	0.998
Mineral[‡]	Air	5×10^{-6}	2,000	6.24	0.9991
Fixed and rigid	——	0	∞	6.25	1.0000
Gas bubble[§]	Water	19,000	0.0012	5.8×10^7	3.6×10^8

Source: Based on equations in Box 9-1, equation 9-10, and data from Urick (1983, 300) and Weast (1985).
* The value of the term in large brackets in equation 9-10.
[†] The ratio of the backscattered intensity to that of the ideal fixed, rigid sphere.
[‡] Could be rock or bone.
[§] Atmospheric pressure is assumed for the gas.

fluctuations. For ultrasonic frequencies this pattern can be significant over a range of only a few meters (Michelsen and Larsen 1983).

In aquatic environments the most important scattering objects are gas bubbles, including the swim bladders of fish (Rogers and Cox 1988). At very low frequencies scattering by inhomogeneities dominates attenuation (see Figure 9-16).

ATTENUATION

Attenuation is a reduction in intensity that occurs as a stimulus travels away from its source. The attenuation of sound is usually measured in decibels, with the attenuation coefficient α_{dB} defined so that

$$\frac{I_1}{I_2} = 10^{-(r_1 - r_2)\,\alpha_{dB}/10} \tag{9-11}$$

These values are then converted to the attenuation coefficient α_e that is used throughout this work (see Section 4-1) by

$$\alpha_e = \frac{\alpha_{dB}}{10 \log e} \cong \frac{\alpha_{dB}}{4.343} \tag{9-12}$$

Attenuation of sound may be due to (1) geometrical spreading, discussed in Chapter 4, on transmission; (2) absorption, which is the conversion of sound energy to some other form of energy; and (3) scattering. Attenuation as the result of absorption generally increases with the square of the frequency. The absorption of sound is vastly greater in air than in water, as seen in Table 9-3. Sound absorption in air is influenced by humidity in a complex fashion; maximum absorption occurs at a particular water content, which increases with frequency (Griffin 1971). Multiple scattering is likely to lead to an attenuation of about 6 dB for doubling the range (Michelsen 1983). Wind fluctuations can cause an attenuation of 4 to 6 dB per 100 m and create large (20 dB) fluctuations in transmitted intensity (Ingård 1953).

In forests, absorption and scattering both contribute to the attenuation of sound. The data in Figure 9-15 demonstrate that the frequency dependence varies with the habitat. An apparent "sound window" at about 200 Hz results from refraction caused by a heating up of the canopy (Waser and Brown 1986). Sound that travels high in the atmosphere and is returned to the surface by wind gradients, at a point that may be many kilometers from its source, may have a surprisingly high intensity, because of a lack of scattering in the relatively homogeneous higher atmosphere (Ingård 1953).

Absorption in pure water which is due to viscosity's effects, is proportional to the square of the frequency and is dependent on temperature in a way that is approximated by:

$$\alpha_e \cong f^2 (1.137 \times 10^{-4} - 5.96 \times 10^{-6} \, T$$
$$+ 2.10 \times 10^{-7} \, T^2 - 3.5 \times 10^{-9} \, T^3) \, \text{km}^{-1} \tag{9-13}$$

where T is the temperature in °C and f the frequency in kHz (Sündermann 1986). At 20 °C this reduces to $\alpha_e \cong 5.1 \times 10^{-5} f^2$.

Sound propagation in the ocean has been studied intensively because of its importance to submarine warfare. Over a broad range of frequencies, a number of attenuation mechanisms come into play, not all of which are well understood. The following equation (Urick 1983, 108) is a useful summary:

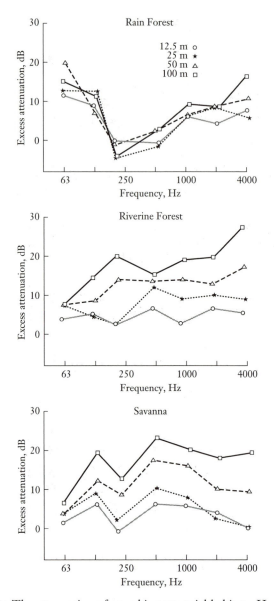

Figure 9-15 The attenuation of sound in terrestrial habitats. Here measurements of excess attenuation over that which is due to geometrical spreading are presented for three environments over a range of frequencies and distances. Excess attenuation, which generally increases with distance, can be highly dependent on the sound's frequency and the type of environment. (After Waser and Brown 1986.)

$$\alpha_e = 6.3 \times 10^{-5} f^2 + \frac{9.2 f^2}{4,100 + f^2} + \frac{0.023 f^2}{1 + f^2} + 7 \times 10^{-4} \text{ km}^{-1} \quad (9\text{-}14)$$

where f is the frequency in kHz. This relation is for a temperature of 4 °C and a depth of 1 km. At the ocean's surface the attenuation would typically be reduced about 10 percent. The first term relates to the viscosity of water, the second is for $MgSO_4$ dissociation, the third is probably caused by ionization of boric acid, and the fourth is due to scattering caused by inhomogeneities in water masses. The latter term varies from a low of 5×10^{-5} near the equator to 1×10^{-3} near the poles (Sündermann 1986, 333); more-precise relations incorporating temperature and salinity variations are also given in Sündermann (1986, 324–337). These relations are plotted in Figure 9-16. A rough calculation for figuring transmission loss (TL) in the ocean between distances r_1 and r_2 combines spherical spreading with attenuation: $TL (r_1, r_2) = 20 \log (r_2/r_1) + \alpha_e 4.343 (r_2 - r_1)$ dB. This rela-

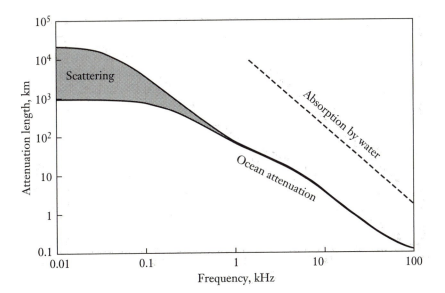

Figure 9-16 The attenuation length ($1/\alpha_e$) of sound in water. The curve for absorption by pure water is from equation 9-13, that for attenuation in the ocean from equation 9-14.

tion often provides a useful approximation, even when spherical spreading is not really appropriate (Urick 1983, 111).

9-6 TRANSMISSION RANGE

Given the parameters previously presented in this chapter, it is possible to make some estimates of the range at which some natural sounds can be heard. The primary considerations are the power of the source, the geometry of spreading, the attenuation and noise in the environment, and the threshold intensity of the receiver. In the absence of conditions suggesting otherwise, spherical spreading is the best assumption; even if there are boundaries containing the sound, absorption at the boundaries may produce attenuation approximating spherical spreading. The intensity at different ranges is then given by

$$I(r) = \frac{P}{4\pi r^2} \exp\left(-\alpha_e\, r\right) \tag{9-15}$$

where P is the power of the source. Some estimates of the detection range, given typical noise at the appropriate frequency in the environment, are displayed in Table 9-7. An examination of this table will reveal that the range of detection is much more dependent on attenuation in the environment than on the power of the source.

Table 9-7 The Calculated Range for Some Sound Sources

Source	Frequency (Hz)	Power (W)	Attenuation $\alpha_e(\mathrm{km}^{-1})$	Noise Intensity (W/m^2)	Threshold Intensity (W/m^2)	Range (km)
Human speech	1,000	10^{-5}	30	10^{-9}	10^{-11}	0.4
Human yell	1,000	10^{-3}	30	10^{-9}	10^{-11}	0.6
Dolphin click	25,000	10^{5}	1.3	10^{-15}	10^{-14}	30
Dolphin whistle	10,000	10^{-4}	0.25	10^{-14}	10^{-14}	70
Finback whale	20	10	0.0007	10^{-11}	10^{-13}	10,000

Spherical spreading is assumed and that the threshold is 20 dB below the noise level (i.e., that the signal-to-noise ratio is −20 dB at detection [Brown 89b]), except that the absolute threshold for dolphins is 10^{-14} W/m^2.

To analyze this relationship further, Figure 9-17 plots intensity as a function of range for these cases. At ranges much less than the attenuation length, spreading dominates and intensity falls as $1/r^2$. However, for ranges much larger than the attenuation length, attenuation dominates and intensity falls exponentially, which is much more rapid. In all these examples, the sound source is strong enough as compared to the threshold that the limiting range is in the zone dominated by attenuation and any increase in the strength of the source would increase the range relatively little. This is a clear demonstration of the importance of making an appropriate choice of frequency for long-range communication by sound.

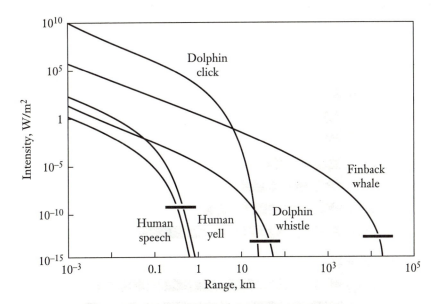

Figure 9-17 Examples of sound intensity as a function of range. Here spherical spreading and appropriate attenuation are assumed for vocalizations by several mammals. At short ranges the decaying of intensity with distance is dominated by geometrical spreading, which produces a relatively slow decay. At distances beyond the attenuation length attenuation dominates and the decay becomes much more rapid. The short horizontal bars indicate the limits of hearing in the appropriate environment with the dolphin being limited by its sensitivity and the others by noise. In all these cases, the strength of the source and the sensitivity of the receiver are sufficient to extend the range of communication to distances at which attenuation becomes significant. To increase the range of communication further would require greatly increased energy output.

There have been a few actual observations made of the range of sound detection in natural environments. Brenowitz (1982) has estimated that the maximum range for detecting the red-winged blackbird's song was about 200 m, and Brown (1989a, 1989b) has provided figures for the effective range of primate vocalizations in a forest habitat (Figure 9-18). The audible range of the calls Brown studied varied from about 80 m to 2 km, with a cluster of short-range calls near 100 m and one of long-range calls over 1,000 m. Similar estimates of a human call "Hey," closer to the ground, produced a range near 200 m. Although it was the loudest call, attenuation was greater near the ground (Brown 1989). Less than 30 percent of the variance in audible range of monkey vocalizations came from the intensity of the sound's production. Half the long-range calls utilized a low attenuation "sound window" near 200 Hz, while the others were more intense at the source.

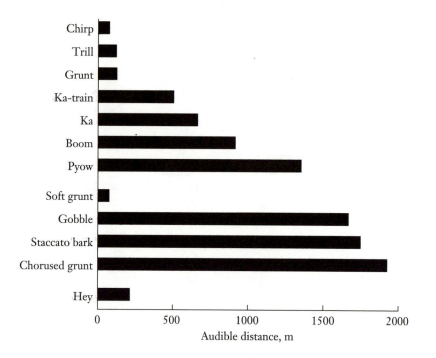

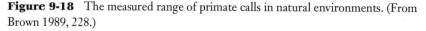

Figure 9-18 The measured range of primate calls in natural environments. (From Brown 1989, 228.)

There is also evidence that some moths can detect the sounds produced by echolocating bats at about 35 m and thereby take effective evasive action (Miller 1983).

There is convincing evidence that sharks can locate natural sounds made by struggling fish from a distance of at least 30 m (Nelson and Johnson 1976); reports of their being attracted from hundreds of meters away have employed loud artificial sound sources (Myrberg et al. 1972). In shallow water 1 m deep, Banner (1972) found that small sharks responded to natural sounds within only a meter or two, with a maximum of 4 m.

9-7 GENERAL INFORMATION EXTRACTABLE FROM SOUND

Albert Bergman (1990) has brought to auditory analysis much the same kind of thinking David Marr brought to vision. In particular, Bergman points out that many of the information-processing principles originally identified by Gestalt psychologists for vision (see page 190) have close analogs in auditory sensation, and the principles are easily understood as mechanisms for extracting information about external objects and events from the sensory information provided by light and vision. For example, the synchronous onset of two sound frequency components is interpreted as indicating that they belong to the same event. This strategy helps allocate different sound components appropriately to different sound sources, which often exist simultaneously.

Besides any particular signal that is generated by a sound transmitter, there is also some general information that can be obtained from a given sound stimulus. The most important kinds of general information concern the direction and distance to the source, and the nature of the acoustic environment between transmitter and receiver.

DIRECTION TO SOURCE

The direction of a sound's source can be determined by various strategies. In air, the speed of propagation is sufficiently low that differences in the sound's arrival time at each ear can be used. This is the most important cue for humans at frequencies below 3 kHz (Richardson 1953, 267). If d is the distance separating the ears, and α the angle of the sound source from the plane perpendicular to the line between the two ears (i.e., the sagittal plane, for bilaterally placed ears), the extra travel distance to the farther ear is $d \sin \alpha$. If v_s is the speed of sound, the time delay will be $d \sin \alpha / v_s$. Delays of as little as 10 μs can be detected by humans (Moore 1988, 155). With $d = 0.17$ m, departures as small as 1° from the midline can be detected. Various other mammals, including dolphins, have been found to have a directional resolution of 1° to 6° (Hawkins and Myrberg 1983; Lewis 1983). Most birds have a resolution of only 10° to 20°, probably because of the small separation of their ears, although owls are able to orient to sound with an accuracy of 1° to 2° (Saunders and Henry 1989). Walking crickets have been shown to have an accuracy of orientation toward a sound source of about 30° (Kennedy 1986). A grasshopper can orient to sound within 10° (Helversen and Helversen 1983).

Another clue to sound direction results from shadowing by the head, which reduces intensity at the more distant ear. The difference in intensity depends on the sound's direction and frequency in complicated patterns (Sivian and White 1933). Consequently, each direction probably has its own unique spectrum of intensity differences, and comparing the intensity at the two ears provides information as to direction. Humans use this strategy for frequencies above 3 kHz (Stevens and Newman 1936, 36; Richardson 1953, 267; Moore 1988, 152). For speech sounds, the maximum difference is 7 dB and changes linearly between 0° and 50° from the midline, but very little at larger angles (Richardson 1953, 428). Similarly, moths can distinguish right-to-left ultrasonic sound sources by means of diffraction caused by their bodies (Miller 1983).

Both strategies indicate which side a sound source is on and supply its angle from the midline but they provide no clue as to whether it is ahead of or behind or what its elevation is with respect to that of the receiver. More precisely, in three dimensions the sound cues define a cone with its axis through the ears, upon which the source might lie (Wallach 1939). Information to resolve what is referred to as this **cone of confusion** is obtained both from head rotations (Wallach 1939; de Boer and van Urk 1941) and from distortions in the sound spectrum produced by frequency-dependent diffraction of the head and the external ear, or the pinna in humans (Butler and Belendiuk 1977) and the facial ruff in barn owls (Konishi 1983). The complex shape of the pinna causes sounds of different directions and frequencies above 4 kHz to be coupled with varying efficiency to the eardrum (Shaw 1982a; Shaw 1982b). Therefore, comparing the frequency distribution either between the two ears or against the expected characteristics of the frequency distribution, provides a mechanism for locating the sound's source, even in the plane of symmetry

(Lewis 1983). This strategy requires having a source containing multiple frequencies or else movement of the ears or head. The necessary frequency characteristics can apparently be learned within a matter of a few seconds of exposure to new situations (Moore 1988, 162). Elevation can be estimated by the way the sound intensity changes as the head rotates around a vertical axis (Wallach 1939). Frequency ranges and some estimates for the accuracy of some of the different cues used by humans are presented in Table 9-8.

Many mammals have pinnas of a simpler structure but more mobile than those of humans; prey usually have particularly mobile pinnas (Knudsen 1983). Barn owls, which hunt on the wing at night, using sound to locate prey, need precise localization in the vertical plane, as well as horizontal plane and have ears aimed toward different elevations, which is thought to enhance localization in the vertical plane through the ability to make interaural comparisons (Konishi 1983). Crickets also have structures that permit them to determine frequency and direction specificity (Lewis 1983). Dolphins apparently have a narrow "beam" of hearing (Ralston and Herman 1989).

Small animals may not be large enough for diffraction to produce significant differences in sound intensity at their two ears except at very high frequencies. In addition, the difference in time of arrival is reduced for closely spaced ears and may not be a practical cue for small animals. Many small vertebrates solve this problem by channeling sound to the back side of the tympanum in a way that converts the ear from being a pressure detector to a pressure *difference* detector sensitive to the direction of propagation (Michelsen 1983). Some insects also have ears with tympanal membranes, which may be connected to act either as pressure

Table 9-8 Directional Cues Used by Humans in Sound Localization

Cue	Frequency Range (kHz)	Accuracy in Azimuth (S.D.)
Time delay + head shadow	Broad	5.6°
Interaural pinna	4–12	18°
Monaural pinna	4–12	23°
Shoulder bounce	2–3	67°

Source: From Searle 1982.

detectors or pressure difference detectors. In general, it appears that animals large enough, compared to the wavelength of the sound that is of interest (e.g. a moth's detecting the ultrasonic sounds of bats), use the technique of pressure detection, which is inherently more sensitive. In fact, the same ear may act as a pressure detector at high frequencies and pressure difference detector at low ones, where the intensity differences caused by diffraction are small (Michelsen 1983).

In water, sound travels so fast that the differences in its time of arrival at different receptors are thought to be too small to be useful. In addition, sound's wavelengths in water are so large compared to the receiver's body size, and their tissues are so similar in their acoustic properties to those of the external environment, that in water there will be little shadowing to use as a directional cue. By itself, sound pressure is a scalar parameter (that is, a single number, in contrast to a vector) that provides no directional information. However, sound waves do set water particles in motion, and these motions are vectors that have direction as well as magnitude, and both properties can be detected.

In principle, fish having swim bladders can determine the direction of a distant source from the components of acoustic particle velocity and the pressure (Popper et al. 1988). A swim bladder provides maximum sensitivity in detecting sound. Directional cues are obtained from less-sensitive displacement detectors that have their axes oriented at an angle to one another (Fay 1988). Comparing the relative levels of activation (in the members of a pair of such detectors will localize the source to a plane, and employing three or more detectors could pinpoint the source to a line. However, ambiguity remains as to which direction along this line the source lies. This ambiguity can be eliminated by using the pressure variation detected through the swim bladder as a timing reference to determine the direction of propagation, since particle velocity is in the same direction as propagation during the high-pressure phase and in the opposite direction during the low pressure phase (Schuijf 1976).

Experiments have indicated that cod can discriminate sound sources that are about 20° apart in the horizontal plane and somewhat closer in the vertical one (Blaxter 1988) and that the phase relations between pressure and particle motion are important (Buwalda et al. 1983). In addition, it has been determined that sharks can orient themselves accurately toward sounds (Myrberg et al. 1972; Myrberg et al. 1976; Nelson and Johnson 1976), although they lack a swim bladder and any other known pressure transducer. It has been proposed (Schuijf 1981) that they have a mechanism for determining direction without a pressure reference using phase differences between direct sounds and those reflected off the surface.

A final point is that nearly all mechanisms for determining direction will be hindered in environments having substantial scattering or reverberation (Wiley and Richards 1978).

RANGE TO SOURCE

There is no obvious mechanism used for determining distance to a sound source, but humans are surprisingly accurate at it (Richardson 1953, 265). There are various possibilities to acount for this ability.

The intensity of sound would be a cue if the receiver had information about the likely intensity at the source (Coleman 1963), but this cue is subject to serious distortion by the acoustic environment, and experiments with humans indicate that intensity provides no absolute perception of distance (Mershon and King 1975).

The time of arrival of the sound could be used if there were information available about when the signal left the source. In fact, humans frequently estimate the distance to lightning by the difference between the time of the light's arrival and the sound of the thunder. However, this technique is useful only in very special circumstances, and if a source is visible then vision usually provides better evidence of its distance.

A more useful strategy would be to compare sound to itself. For example, in solids the mechanical waves of different types travel at different speeds, so that the range to a source can be determined by comparing the time of arrival of different types of waves generated at the same time. Geophysicists use this technique to estimate the distance to an earthquake. However, as seen in Table 9-3, the variation in sound speed with frequency in air is extremely small, the relative change being only about 0.0003 over the full range of frequencies relevant to animals.

In contrast, the same table also shows that attenuation varies by a factor of 100,000, which is a change larger by eight orders of magnitude than sound speed. As a result, comparing the relative intensities of various frequencies is more likely to provide useful clues about distance (Coleman 1963). However, the receiver also needs to know something about the relative intensities originally transmitted. This information might derive from familiarity with the specific sound at various distances or perhaps from knowledge of the general relations between harmonics generated by common types of physical vibrations. The case for the former familiarity is supported by evidence that humans judge the distance to a source of speech sounds more accurately than they do other sounds (Mershon and King 1975). Another limitation on this strategy is that attenuation is relatively low, as can be seen from Table 9-3, so that this strategy might be practical only over large distances. In any case, there are data indicating that sounds with an artificially reduced high-frequency content are perceived as originating from further away (Mershon and King 1975).

At close range, near-field effects may be used, such as the ratio of pressure to flow velocity, or distorted ratios of binaural intensities. There is in fact some evidence that humans use such cues within a meter of the source (Coleman 1963). Although our knowledge of the capabilities of fish is limited (Fay 1988), there is some evidence of an ability in them to estimate at least relative distances (Schuijf and Hawkins 1983), along with suggestions of a mechanism that would make range estimation possible, based on the relative magnitudes of pressure and particle displacement (Popper et al. 1988).

Another strategy for determining distance to a sound source would be to analyze the amount of reverberation in the signal. Reflective or scattering objects in the environment would increase the amount of reverberation with increased distance between the source and the receiver (see Figure 9-14). This degree of reverberation would provide a cue to distance, especially if the receiver were familiar with the acoustic environment (Wiley and Richards 1978). In fact, humans interpret sound with more reverberation as originating from farther away (Mershon and King 1975), as do birds (Wiley and Richards 1982).

THE ACOUSTIC ENVIRONMENT

Listening to a sound source provides information about the environment as well as about the source, because the sound may be altered by interaction with objects in the environment (Fay 1988). Although we are not often aware of this information, blind people often become quite adept at making use of it (Supa et al. 1944). A more familiar example is that of movies in which the acoustic environment frequently does not match the visual one because the sound track was recorded by microphones placed close to the actors rather than at the position of the camera.

There is some evidence that fish also may obtain information about their environment from the propagation of sounds (Fay 1988), a strategy that may be of particular importance to fish that live in the darkness of the deep ocean. It has been suggested by Peter Rogers of Georgia Tech that fish may be able to sense the presence of other individuals with swim bladders by the efficient scattering of ambient sound from the swim bladders of their neighbors.

9-8 INFORMATION-CARRYING CAPACITY

To conclude this discussion of sound, let us estimate the amount of information that it can provide. As discussed in Section 3-6, such an estimate depends on the level of processing at which it is determined, because information is selected as it progresses first from the physical stimulus to a signal in the sense organs, then to a signal in the central nervous system, and then to a stored memory.

The frequency range of human hearing encompasses a bandwidth of about 10 kHz. Using Equation 3-6, the rate at which information could be transmitted within this frequency range could be calculated if we knew the signal-to-noise ratio. For this let us use the data from Table 9-7 for speech in a typical environment (s/n = 10^{-5} W/10^{-9} W = 10^4). Under these conditions, sound is capable of transmitting about 10^5 bits/s.

Let us now consider information available at the level of perception. Intensity differences of 1 dB can be perceived over a range of about 120 dB. These 120 discriminable levels could be encoded by about 7 bits (2^7 = 128). Another consideration is that there are about 24 nonoverlapping frequency bands within which tones "mask" one another (White 1975, 119). If the intensity within each of these is independently determined, a total of 170 bits is acquired in one sample.

To estimate a rate of information transmission, then, we also need to estimate the rate at which independent sound samples are acquired. In humans, sound pulses separated by 0.05 s are perceived as being distinct (White 1975, 71). Intensity is integrated for about 0.1 s (Moore 1988, 52). Furthermore, sound sensations take about 0.14 s after the sound's termination to decay (White 1975, 120). And the time required to determine pitch varies from about 0.01 to 0.06 s (White 1975, 144). Considering all these measurements, it seems appropriate to estimate that, typically, about ten independent samples are acquired per second. Significantly, about ten phonemes per second typically occur in speech (White 1975, 373).

We can thus conclude that the rate of information transmission at the perceptual level can be estimated as on the order of 10^3 bits/s. This rate is in fact similar to another estimate of 200 bits/s, though made by a somewhat different argument (Ackerman et al. 1979). Although there is much uncertainty in this estimate, it is clearly orders of magnitude lower than the rates carried by the sound stimulus, partly because most of the phase information has been lost.

CHAPTER 10

Other Stimuli

■ ■ ■

The spider's touch, how exquisitely fine!
Feels at each thread, and lives along the line.
Alexander Pope, *An Essay on Man*, 1733

This chapter discusses the types of stimuli that have not previously been covered. These stimuli are relatively minor ones, at least in terms of our present knowledge.

10-1 MECHANICAL STIMULI

Arthropods usually detect mechanical stimuli through the movement of innervated hairs formed of and protruding outward from the cuticle. These cuticular hairs may be coupled to various types of mechanical stimuli either by their mechanical properties or through contact with other structures. Spiders also have slit sensilla in their cuticles that provide sensitivity to deformations (Barth 1982). Displacements of the substrate of only one nanometer may be detectable (Masters et al. 1986).

Vertebrates are most sensitive to the mechanical stimuli that are detected by **hair cells,** which are the receptors in the lateral-line system in fishes and amphibians and in vertebrate organs of hearing and equilibrium. Hair cells' name derives from the cilia projecting from the distal end of each cell. Bending the cilia in one direction leads to strong excitation of the cell, while bending in the opposite direction produces a weaker inhibi-

tion of the ongoing activity. A 1 mm displacement or a 10° bend is sufficient to produce a maximal response (Hudspeth and Corey 1977). In most
cases the cilia are in contact with a structure that moves with a specific type
of stimulation. In the simplest case this type of structure is a gelatinous
projection or cupula that sticks out into a flowing fluid to couple the flow
to the cilia of several hair cells. In a steady flow, the displacement will be
approximately proportional to the flow velocity and the hair cells will
become velocity detectors. However, in a changing flow the thickness of
the boundary layer will vary, changing the force acting on the cupula. The
cells then become sensitive to acceleration as well, an effect that has often
been overlooked (Kalmijn 1988). Hair cells can thus function to obtain
either information about the flow pattern over the body surface, as in the
lateral-line system, or information about the acceleration of the animal, as
in our own semicircular canals. If the cilia are attached to an object that
differs in density from the surrounding fluid, a gravity detector is produced, which must also be sensitive to acceleration of the sense organ.

NEAR-FIELD ACOUSTICS: UNSTEADY FLOWS

An object moving through any medium, particularly one of low compressibility such as water, generates a flow pattern or wake around itself. If the
object accelerates, and especially if it vibrates, sound waves propagate away
from the object. In the so-called near field, flow effects predominate over
effects resulting from the compressibility of the medium, which dominate the
properties of sound in the far field. The extent of the near field is not well
defined, but generally the near-field effects of flow dominate when the
distance to the source is small compared to the wavelength of the sound or the
size of the source. Kalmijn (1988) furnishes an extensive discussion of the
importance of the near field to fish. In fact, he proposes that the basic
functions of the lateral line and the inner ear originated to extract information
from the near-field flows. For a 10 Hz vibration in water, the near field effects
can extend from at least 20 to over 100 m, depending on orientation.

Fish and the aquatic forms of amphibians can readily detect the relative motion of any solid object within a few centimeters of their bodies
from the flow produced around the object acting on the lateral-line system
(Dijkgraaf 1962). The hair cells on the surface, called free neuromasts,
respond to a flow over the surface with sensitivities on the order of 1
mm/s. Most fast- and persistently swimming fish have some of their
lateral-line hair cells enclosed in canal organs that respond to local pres-

sure differences ($\propto$ acceleration of the fluid) (Denton and Gray 1988). The lateral-line system usually contains organs oriented in different directions so that the direction of stimulus flow can be resolved. The wide variety of lateral-line geometries (Coombs et al. 1988) can be understood as providing different speed-of-response and sensitivity trade-offs as well as directional and frequency specificity (Denton and Gray 1988). Lateral lines seem to be more extensively developed in fish that live in the darkness of caves or the deep ocean (Dijkgraaf 1962). Lateral-line organs function at frequencies below 100 Hz and match the frequencies of vibration of the animals during locomotion (Montgomery and Macdonald 1987; Kalmijn 1988).

The lateral-line system probably functions mainly to detect and localize obstacles in the environment along with prey, predators, and conspecifics. The chief mechanism for obstacle avoidance in fish is probably an increase in pressure around the head as water pushed ahead of a swimming fish is restricted by an obstacle. The array of lateral-line sense organs on the head probably provide some indication of the direction and size of the obstacle. For example, blind cave fish can choose between openings of different shape (von Campenhausen et al. 1981) probably by detecting distortions in the flow pattern around the swimming fish, which could be considered a type of sounding (see Chapter 12). Similarly, an object approaching a fish pushes water ahead of itself, a flow that is detectable by the lateral-line system of the fish. In this regard, many blind fish have been observed to move toward small approaching objects but away from large ones (Dijkgraaf 1962).

Such interactions with conspecifics as schooling and sexual display are probably also mediated, at least in part, by flow patterns. It has been observed, for instance, that transparent partitions alter schooling patterns (Cahn 1972); that blind fish can maintain their position in schools (Moulton 1960); and that eliminating of the lateral-line system leads to collisions during rapid escape movements by schools (Partridge and Pitcher 1980).

The lateral-line system is probably useful only for detecting flows from sources at a distance less than about the size of the recipient or the size of the source, if it is larger than the recipient. At much larger distances, the flow would be uniform over the volume occupied by the recipient and the recipient would simply be carried along by the flow without causing the flow relative to the body that is required for stimulation of the lateral line. However, motions of the whole animal might be detected by the inner-ear otolith organs that detect acceleration of the whole fish (Kalmijn 1988).

Near-field effects also occur in air (Bailey 1991, 45–47, 58–63). Under certain circumstances humans may detect near-field effects, which provide information on the distance to nearby sources (Verheijen 1976). It has been demonstrated that near-field acoustics serves as an essential component of the famous dance language of honeybees (Towne and Kirchner 1989). In it the signal is transmitted by air-particle oscillations in the near field that are generated by wing vibrations. The effective range of such signals is limited to a few millimeters, because bees' wings are small compared to the wavelength of the sound at the frequency of vibration. Similarly, in the fruitfly courtship includes a song transmitted by wing vibrations at 166 Hz that is detected by particle motion in the near field (Bennet-Clark 1971). On caterpillars certain hairs are sufficiently sensitive to air movements to detect a flying wasp half a meter away (Markl 1977), and spiders have similar hairs (Barth 1982). The well-known ability of cockroaches to avoid a rapidly approaching newspaper is based on the detection of air movements ahead of the newspaper, an ability that also allows them to detect and avoid the strike of a toad 10 cm away (Camhi 1978, 1980).

SURFACE WAVES

As is often observed, small animals moving on or near the air–water interface generate surface waves that propagate away from the source in concentric circles or ripples. A few species of a wide variety of animals, such as spiders, insects, fish, frogs, and bats make use of these waves to obtain information about prey or predators or to communicate with conspecifics (Wilcox 1988). Small animals produce waves with amplitudes of a few micrometers to a millimeter and frequencies in the range of 10 to 100 Hz (Bleckmann 1988; Wilcox 1988) (Figure 10-1). (Wind driven waves have lower frequencies.) Surface waves differ from sound waves in that their much lower velocity of propagation varies significantly with frequency. They have a minimum speed of 18 cm/s at 6 Hz (λ = 5 cm), which increases to 53 cm/s at 100 Hz (λ = 0.36 cm). Although many animals are quite sensitive to surface waves (see Table 10-1), attenuation is severe, with the attenuation length ($1/\alpha_e$) decreasing from 2.6 cm at 10 Hz to 0.9 cm at 100 Hz (Lang 1980). The useful ranges of transmission are therefore limited to a fraction of a meter. A backswimmer, for instance, can probably detect very small prey like aphids at distances only of 1.5 cm, ants at 6 cm, but bees at more than 14 cm (Lang 1980).

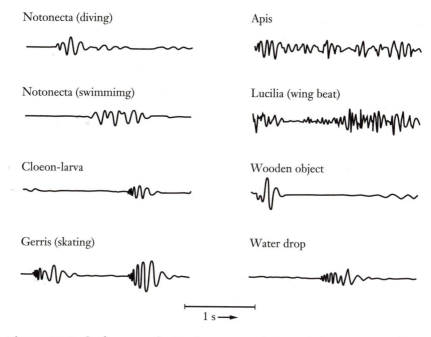

Notonecta (diving)

Notonecta (swimmimg)

Cloeon-larva

Gerris (skating)

Apis

Lucilia (wing beat)

Wooden object

Water drop

1 s →

Figure 10-1 Surface waves from various sources. Some typical wave patterns from water striders, backswimmers, their prey, and other objects. (After Wilcox 1988.)

Directional discrimination with surface waves is quite good (Wilcox 1988). A backswimmer orients itself to within an accuracy of about 2° for a source in front, the error increasing to 20° in back. Fishing spiders orient to an average accuracy of about 10° in all directions, and a frog orients itself with an accuracy of 5°.

A surface-feeding fish has a maximum sensitivity of about 0.01 mm amplitude at 100 Hz and can estimate the distance as well as the direction to a source (Bleckmann 1988). The distance is assessed from the degree to which the arrival of low frequencies lags behind those of high frequencies, because of the unequal propagation speeds and the curvature of the wave-front. A fishing spider, which has a maximum sensitivity of 1 μm at 70 Hz, identifies its likely prey by selecting sustained disturbances containing high-frequency components in the surface wave spectrum, correcting for excess attenuation of the high frequencies as dependent on the distance of travel (Bleckmann 1988). Similarly, a backswimmer attacks stimulus sources having frequency components above 50 Hz (Wilcox 1988). The amplitude of surface waves does not seem to be a factor in distinguishing prey (Wilcox 1988).

Table 10-1   Frequencies and Thresholds for Detecting
Surface Waves

Animal	Best Frequency (Hz)	Threshold (μm)
Bryozoan	40–80	30
Backswimmer	≥ 1,000	0.5
Water strider	20–40	2.5
Whirligig beetle	150	0.5
Mosquito larva	10–20	5
Fishing spider	≥ 1,000	0.1
Surface-feeding fish	70–140	0.04
Frog	14	0.2

Source: Wilcox 1988.

Surface waves are used for communication by some surface-dwelling species like water striders (Wilcox 1988). In them the females can be attracted and induced to oviposit simply by reproducing the surface-wave signal of the male. But a female will be attacked by a male if made to generate the male signal. Male Siamese fighting fish protect their young and generate surface waves to attract them from up to 40 cm away.

Surface waves also occur at air-to-solid-surface interfaces. Desert sand, for instance, conducts surface waves at 40 to 50 m/s. Scorpions can detect insect prey from 50 cm away and estimate their direction and distance from the surface waves they generate (Brownell 1977; Brownell and Farley 1979).

THE VIBRATION OF SOLIDS

Solids are distinguished from fluids in having a tendency to return to their previous shape after undergoing a small bend or twist. They can thus support a variety of transverse waves as well as the compression waves of sound (Markl 1983; Gogala 1985). Rod- or plate-shaped solids tend to vibrate in particular modes, depending on their shape. Any of these vibrations can provide organisms with information. In fact, many insects do use them for communication (Bailey 1991), and they may be the primary mode of communication within the nests of social insects (Markl 1983).

The sources of vibration vary enormously in their amplitude and energy, as shown in Table 10-2.

Table 10-2 Vibration Amplitudes and Power

Animal	Amplitude (m s^{-2})	Power (W)
Fruit fly	?	10^{-16}
Stridulating ant	0.1	10^{-8}
Drumming ant	10	$10^{-5} \leftrightarrow 10^{-4}$
Fiddler crab	100	$10^{-1} \leftrightarrow 1$
Drumming woodpecker	High	High
Foot-stamping ungulate	High	High

Source: Markl 1983.

Animals' sensitivity to vibration is typically on the order of 10^{-2} m s^{-2}, with an established low of 10^{-4} m s^{-2} at the best frequencies of 100 Hz to a few kHz (Markl 1983).

The attenuation of vibrations depends on the shape and composition of the substrate and the mode of vibration in a frequency-dependent fashion. As a result it is difficult to make general predictions. Some vibrations have attenuation lengths of less than a centimeter (Table 10-3), although communication by vibration has been observed over distances of several meters (Barth 1982; Markl 1983). Markl (1983) has even observed carpenter ants retreating into their nest when a woodpecker hammers on their tree more than 10 m away.

Plant stems and leaves transmit vibrational information most efficiently as bending waves, which have a large dispersion—the variation in propagation velocity with frequency. In *Vicia faba*, for instance, $v = 36$ m/s

Table 10-3 The Attenuation of Vibrations in Solids

Material	Mode of Vibration	Frequency of Vibration (Hz)	Attenuation Distance, $1/\alpha_e$ (cm)
Spider silk	Longitudinal	50–2,000	2.9
Spider silk	Transverse	20	2.2
Spider silk	Transverse	160	1.6
Wood	Surface waves	4–5 kHz	2.3
Soil	——	1–3 kHz	0.7

Source: Barth (1982), converted to α_e by Equation 9-12.

at 200 Hz and 120 m/s at 2 kHz. Accordingly, differences in the time of arrival for different frequencies are a sensitive measure of distance to a source. Because objects having specific shapes and mechanical properties tend to vibrate in specific modes with nodes of zero amplitude at certain locations, communication becomes more reliable if a broad band of frequencies is transmitted, as is found in many land bugs (Gogala 1985).

Solids are also distinct from fluids in that they support tension, which provides the restoring force for spider webs. The web-building spiders probably use substrate vibrations for a wider variety of information-gathering purposes than does any other animal. These spiders detect, identify, and localize prey within their webs by its vibrations and also communicate with their conspecifics through web vibrations (Masters et al. 1986). Wandering spiders also use substrate vibrations to locate prey (Hergenröder and Barth 1983).

STEADY FLOW

Many animals respond to the flow of the surrounding medium, whether it is air or water. Fish and other aquatic animals in a stream tend to orient themselves with their heads upstream (Fraenkel and Gunn 1961, 250; Dijkgraaf 1962), an attitude that reduces drag and allows them to swim against the current more easily to maintain position. Similarly, swarming insects fly upwind to maintain a constant position. Male moths, having detected the female pheromone, fly upwind to find its source. In other situations, however, animals may fly downwind, which lets them move farther with less effort (Gillies and Wilkes 1974; Dusenbery 1989c). Such responses, oriented to a flow, are frequently described as **anaemotaxis** in air and **rheotaxis** in water (Fraenkel and Gunn 1961, 249–260).

A fundamental feature of uniform flow is that the flow is not detectable by an organism immersed in it, unless it makes reference to external stimuli (Lyon 1904). Thus, high-flying birds cannot determine the wind's direction, unless they can see the ground well enough to detect their motion or can perhaps detect their motion relative to the earth's magnetic field. Fish in open water and night-flying insects face similar problems.

In contrast to the situation under uniform flow, determining flow direction is straightforward when an animal is close enough to a surface to have receptors in different parts of the velocity gradient formed by drag on the surface. Thus, fish having extensive lateral-line systems are well equipped for this task. It has been suggested that night-feeding mosquitoes can orient themselves to the wind in complete darkness and might obtain the information they need by sudden dips to levels with lower wind speeds and by detecting the direction of acceleration (Gillett 1979). However, it is not clear if this mechanism can in fact work. How does a mosquito determine the vertical direction, for instance, and can it really distinguish accelerations caused by its flight from those caused by lowered wind speed?

Sporangiophores of the mold *Phycomyces* grow into a wind, which seems adaptive, but the mechanism for this is not yet known, although it has been much studied (see Section 12-6).

TOUCH

Information is often provided by forces that arise from direct mechanical contact with an object. Humans judge the ripeness of fruit by its softness, the sharpness of a blade by its feel, and the type of material from its weight and feel.

Even some plants make use of tactile information. The Venus flytrap recognizes the presence of an insect in one of its traps by its sense of touch and responds by closing the trap. Also, many orchids have pollinating mechanisms that are triggered by the forces exerted by an appropriate insect once it has been attracted to the flower. This should probably be considered an information-mediated response, even if all the linkages between the stimulus and the response are mechanical, because the forces exerted by the insect act as signs that the insect is present. This response may also be a case involving physical interaction, if the insect provides the energy to move parts of the pollinating mechanism. These considerations demonstrate that there can be a significant gray area in which it is not obvious whether an interaction is physical or informational.

In animals with internal fertilization, mating requires close contact and touch becomes a universal mechanism of communication, as in humans, arthropods (Markl 1983), and nematodes (Chalfie and White 1988). Similarly, walking animals must use information from contact with the substrate to control their locomotion. Some dramatic examples are the righting behaviors of a variety of invertebrates that come into effect when their legs lose contact with the substrate, as well as the observation that insects usually stop flying if their legs contact a substrate (Fraenkel and Gunn 1961, 244–245).

In spite of their importance, touch stimuli are very difficult to measure or manipulate, which makes them difficult to study (Markl 1983). In addition, they are probably quite specific to a given situation, so that there are probably few generalizations that can be made about them as a whole.

10-2   GRAVITATION

Gravity is a unique stimulus in that it is constant. No organism is thought to use the small change in the force of gravity that occurs with a change in elevation; only its direction is useful. Another unique characteristic of gravity is that a gravity receptor cannot distinguish between a gravitational force and a force caused by inertia during acceleration. Animals capable of moving rapidly must deal with this ambiguity.

Gravity receptors in animals usually consist of an object called a **statolith** with a density greatly different than that of the surrounding fluid that is also in contact with certain mechanoreceptors. The direction in which gravity pulls on the statolith determines which of the mechano-receptors are stimulated. A statolith is usually of high density and is formed as a mineralized secretion or is acquired from the environment, as perhaps a sand grain, although in some aquatic insects an air bubble is employed for this purpose (Markl 1977; Budelmann 1988).

Gravity exerts force on all organisms that are not buoyant, in proportion to their size. Large, mobile, nonbuoyant organisms (such as familiar animals) use information about the direction of gravity to help them control their locomotion. Even buoyant animals (such as fish) find it useful to obtain information about the direction of gravity to use as a reference. Indeed, all arthropods and vertebrates have gravity receptors. The large, sessile, nonbuoyant organisms (such as familiar plants) use gravitational information to orient their growth toward sunlight and perhaps to resist gravitational forces. In plant roots, the statolith is thought to be starch grains acting inside the cells (Iversen and Rommelhoff 1978).

Locomotion oriented to gravity is often called *geotaxis*, but this term can be misleading. Although many small aquatic animals tend to orient themselves vertically (Budelmann 1988), most of such cases are likely to be simply the result of gravitational force acting on an animal that is unequal-ly buoyant in the various parts of its body. The clearest example comes from nematodes, none of which have been shown to have gravity receptors or definitely to respond to gravity (Croll 1970, 77) except for the vinegar eel, which swims upward, because its posterior is denser than its anterior

(Peters 1952). Thus, this nematode's vertical orientation is not based on a stimulus–response mechanism, and the force of gravity supplies the energy for the rotation to the vertical.

10-3   ELECTRIC FIELD

More than two hundred years ago the ampullae of Lorenzini of certain fish were described anatomically, but only within the last thirty years has it been recognized that their principal function is to sense low-frequency electric fields (Kalmijn 1988). Weakly electric fish possess other types of electroreceptors, which have a high frequency response (Hopkins 1988; Zakon 1988). In general, the lateral-line system in vertebrates contains electroreceptors as well as mechanoreceptors. This situation is probably the primitive condition, although various types of receptors have been both gained and lost along several evolutionary lines (McCormick and Braford 1988). No invertebrate is known to possess electroreceptors (Zakon 1988). The collection of papers in the volume edited by Bullock and Heiligenberg (1986) is a useful source of information on this topic.

Freshwater fish have skins with high electrical resistance (for osmotic control) and most of any external electric field appears across the skin, leaving the internal tissues all at the same electrical potential (Kalmijn 1988; Zakon 1988). In these fish the electroreceptors are localized in, and detect the potential across, the skin. In contrast, the marine elasmobranchs (sharks and rays) have skin with a lower resistance, and external fields can cause significant potential differences within the animals. Their electroreceptors have long canals connecting many pores through the skin to a few central clusters of ampullae where thousands of receptor cells are located. These long canals have the electrical properties of a highly conductive cable surrounded by a good insulator and conduct the electric field from the site of the pore to the central cluster. In both the freshwater and the marine cases, individual ampullae respond to the electric field potential difference between some point on the surface and a common potential (either of all the internal tissues, in freshwater fish, or of a cluster of ampullae, in the elasmobranchs). The pores in sharks are distributed over the head, whereas in rays they are more widely distributed over the body. The electrical properties of the canals limit the response to frequencies of less than a few hundred Hz for the shortest canals to less than 3 Hz for the longest ones. Because the receptor cells seem to be insensitive to steady fields, stationary animals in stationary fields would sense nothing. The tuberous receptors of weakly electric fish are usually tuned to the major frequency components of the electric discharges of their own species, which are usually in the range of a few hundred Hz but run as high as 18 kHz in some species (Zakon 1988).

In principle, electric fields can provide a wide range of information to aquatic organisms (Kalmijn 1988). In fresh water, significant electric field sources at low frequencies are geochemical reactions (Kalmijn 1988) and distant lightning at high frequencies (Hopkins 1988a). In saltwater, ocean currents flowing through the earth's magnetic field are a significant source of electric fields. In all aquatic environments, biological sources are important within a range of one meter.

Most biological tissues, unless insulated by skin or cuticles, produce significant electric fields. Typically, the electric field around an animal is bilobed and similar to that of a dipole source (meaning a source and sink of equal strength), as the result of current flowing out of one opening and into another. An electric field's strength around a dipole decreases with the cube of the distance from the source. A field of 200 nV/cm was recorded 10 cm in front of a saltwater flounder. In fresh water the electric fields that are of biological origin tend to be of a higher potential (voltage) and lower current than in saltwater because of the greater electrical resistance of fresh water as compared to seawater. In tropical fresh waters a significant source of electric impulses is often fish with specialized organs for generating electric fields that the fish use for sounding (Chapter 12), communication (Chapter 13), or, in a few cases, for stunning their prey (Hopkins 1988b).

Ocean currents passing through the earth's magnetic field generate electric fields that are oriented perpendicular to the plane containing the flow and the magnetic field. The strengths of these fields are from 5 to 500 nV/cm. These fields could provide useful information about currents in which an animal is being carried.

Lightning produces electric-field pulses that can be significant signals to weakly electric fish that detect high frequencies (Hopkins 1988a). The electric storms can be distant, and the pulses as frequent as several per second. These stimuli presumably represent noise interfering with communication rather than providing useful information. Noise from this source is about 100 times higher in fresh water than in saltwater (Hopkins 1988b).

Stingrays can detect electric fields down to 5 nV/cm. It therefore appears that at least some fish are sufficiently sensitive to detect ambient electric fields, which can often provide them with useful information. The animals in which this is best understood are the elasmobranchs in saltwater and fresh water catfish and weakly electric fish (Kalmijn 1988).

Behavioral experiments clearly indicate that these animals (Kalmijn 1988) and salamanders (Himstedt et al. 1982; Zakon 1988) use electric fields to locate their prey. Weak, steady dipole fields alone are sufficient to induce approach and a biting response that is accurately aimed at the dipole. Responsiveness declines with alternating fields above a few Hz. The thresholds observed were 5 nV/cm in saltwater species and 5 μV/cm in fresh water species. Communication among the weakly electric fish by brief electrical pulses has also been clearly demonstrated (see Section 12-3). The maximum range of detection is within a factor of two of one meter (Hopkins 1988a).

Electric fields have some special advantages for carrying information. For one, electrical signals in the external environment spread much faster than signals in the nervous system. In addition, the high speed of propagation of electromagnetic waves (10^8 m/s) means that the highest frequencies that organisms can generate (10^3 Hz) would involve wavelengths so large (10^2 km) that they would be generated quite inefficiently by the much smaller organisms and would not form shadows or reflections of the much smaller objects in the environment. Thus, any detected variations in their time course would correspond accurately to actual variations in the source, and electrical signals are much more efficient at carrying temporally coded information than is sound, in which temporal patterns are smeared by reverberations from the environment (Hopkins 1988b). Experiments have demonstrated that the time-domain patterns of one-millisecond duration electric discharges can in fact be discriminated even when the power spectrum is identical.

10-4 MAGNETIC FIELD

The earth's magnetic field is potentially quite a useful source of information about one's location (Gould 1980). The earth acts as a big magnet, with magnetic field lines traveling from one pole to the other. These lines are horizontal near the equator but become more and more steeply inclined, and concentrated, as the poles are approached. The earth's magnetic field can as a result not only provide information on direction to the poles but also on latitude, from the inclination or field strength. In addition, local geological formations may perturb the field and provide characteristic magnetic landmarks.

In spite of these advantages, exploiting the earth's magnetism leads to complications. One is that its magnetic poles do not coincide with the earth's axis of rotation—in fact, the magnetic poles actually move, over geological time spans. For instance, in 1984 the north magnetic pole was at latitude 79° N instead of 90°, and the south magnetic pole was at 65°S (Bowditch 1984, 24). This wandering of the poles produces changes in the angle between the geographic and magnetic poles that are on the order of 1° per decade over much of the earth's surface. Over millions of years, the magnetic field even reverses the positions of the north and south poles (Skiles 1985). Moreover, small changes in magnetic field strength occur on a daily basis and also with changes in solar activity.

It is only within the last two decades that convincing evidence has been obtained that some organisms have access to this information, and plausible mechanisms for its reception have been proposed.

Just as an ocean current flowing through the earth's magnetic field induces an electric field, a swimming animal induces a field across its body. In seawater a speed of only 1 cm/s is sufficient to produce a threshold stimulus of 5 nV/cm (Kalmijn 1988). This electric field, which is oriented perpendicular to the plane containing the direction of motion and the magnetic field, has its maximum strength if the motion runs perpendicular to the magnetic field. Thus a swimming animal can, in principle, find any compass heading by changing its swimming direction until its self-generated electric field has an appropriate direction and strength. In addition, if the animal can determine what direction is vertical, by using gravity or the visual horizon, it should be able to estimate its latitude by how much the induced electric field is inclined. Although the experiments in this area have been few, it has been demonstrated that stingrays can determine magnetic field direction, presumably by reference to induced currents (Kalmijn 1988).

It has often been suggested that animals may be able to detect the earth's magnetic field directly (Presti 1985), thus providing terrestrial animals with access to this same information. For a collection of recent discussions of the problem see the volume edited by Kirschvink et al. (1985). Evidence has been obtained suggesting that a variety of animals can detect magnetic fields (see below), although the effects on their behavior are subtle. These claims gained support from the finding that certain bacteria synthesized the naturally occurring magnetic mineral, magnetite, which could provide the basis for a sensory mechanism. Magnetite has subsequently been found in a variety of animals, but despite numerous attempts no one has yet succeeded in finding it associated with neurons or identified it in relation to any other sensory mechanism. Thus, the hypothesis that terrestrial animals may be able to obtain information from the earth's magnetic field remains unproven in the eyes of some researchers (Griffin 1982).

Some of the clearest apparent examples of the use of magnetic fields by terrestrial animals come from honeybees (Towne and Gould 1985). The direction of their dances (see pages 336–347), which are easily measured, are often influenced by magnetic fields. In addition, when a swarm of bees establishes a new colony the bees build their vertical sheets of honeycomb in a parallel array that is not oriented to the side of the nest cavity (Figure 10-2). Evidence indicates that its orientation is chosen to correspond with the orientation of the parent nest, and that the earth's magnetic field is used as the reference to determine this orientation.

Early experiments by several different research groups suggested that the orientations of homing pigeons could be influenced by placing magnets on their heads (Gould 1980). That evidence suggested that they could not determine the north–south polarity but could distinguish the direction of north by the inclination of the magnetic field. However, a recent analysis of a large number of these experiments casts doubt on the conclusion that such magnets in fact disrupt the homing process (Moore 1988). Other evidence suggests that the flight orientation of migrating birds as well as homing pigeons is altered by magnetic storms and local magnetic anomalies, which alter the magnetic field's strength by only a few percent. Correlating behavior with the field strength of the perturbations of the normal field strength suggests a sensitivity to changes of less than 0.1 percent. Such a level of sensitivity is sufficient to use changes in field strength to locate one's latitude within a few kilometers, and homing pigeons with obscured vision often do get to within a few kilometers of their loft. This, and numerous other observations of specific effects at different release sites have led to the suggestion that birds may have some kind of map sense that may be based on magnetic fields (see pages 469–72).

There is also evidence that newts use magnetic fields to locate their home shoreline (Phillips 1986). They seem to obtain directional information from the inclination of the magnetic field and seem to gain map information as well. The location of whale strandings appears also to correlate with the locations of local minima in the earth's magnetic field (Kirschvink et al. 1986). And there is a variety of evidence that honeybees are extremely sensitive to magnetic field strength in a variety of ways (Gould and Gould 1988). They can, for instance, probably use diurnal

Figure 10-2 The natural orientation of honeycombs. A view upward into a screech-owl nest box 20 cm square internally, which had been taken over by a swarm of honeybees. The slabs of honeycomb are parallel to each other, but not to the sides of the box. (Photo by the author.)

fluctuations in magnetic field strength to maintain a sense of the time of day.

Finally, as mentioned in previous chapters, certain bacteria are clearly affected by magnetic fields (Blakemore 1982), but it is now clear that this is the action of a physical force acting on the magnetic material they contain, not the action of a stimulus–response mechanism, and the process does not involve an exchange of information. The magnetic field provides

Table 10-4 The Relative Advantages of Different Stimuli

Characteristic	Stimulus					
	Chemical	Light	Sound	Gravity	Electrical	Magnetic
Speed	Slow	Instant	Fast	Constant	Instant	Slow
Localization	Hard	Easy	Possible	Easy	Indirect	Possible
Obstruction	Little	Common	Some	None	Little	Some
Variation	Much	Much	Much	None	Little	None
Patterns	Mixtures	Color	Pitch	None	None	None
Generating cost	Low	Low	High	None	High	None

the energy to rotate the cell. The importance of these findings in the present context is their demonstration that organisms are capable of acquiring magnetic materials, which makes it more plausible that other organisms do in fact have magnetic receptors. It has indeed been demonstrated recently that honeybees and pigeons contain the mineral magnetite, the lodestone of which the first compasses were made, and calculations have demonstrated a potential sensitivity even greater than that observed in the behavioral experiments (Yorke 1985). The possibility of animals having a magnetic sense has been further buttressed by reports of nerve responses to magnetic stimuli (Semm and Beason 1987). Nonetheless, many researchers remain skeptical.

Table 10-4 summarizes the relative advantages of the various types of stimuli.

CHAPTER 11

Stimulus
Patterns

■ ■

Single stimuli by themselves often do not convey sufficient information to make reliable decisions, but patterns of stimuli do. A *stimulus pattern* is some kind of predictable variation of a single stimulus in either space or time, or the conjunction of a particular combination of stimuli. In this very general sense, being able to recognize a pattern is almost always the key to obtaining useful information from the environment. This chapter includes a few examples of patterns selected for their general importance or because they demonstrate the use of a complex pattern by a relatively simple organism.

11-1 SPATIAL PATTERNS

VERTEBRATE EYES

Complex spatial patterns of light are detected only by animals such as vertebrates, arthropods, a few mollusks with image-forming eyes. The patterns that are most easily detected and recognized are those that have boundaries with strong contrasts and a high degree of symmetry.

One of the most easily recognized visual patterns in nature is that of the vertebrate eye, which consists usually of a rounded shape with a sharply defined black disk, the pupil, in its center (Cott 1941, 82). Its image includes both strong contrast and a high degree of symmetry. The eye's inherent recognizability is attested to by the frequency with which it is disguised by disruptive coloration and also by the use of patterns resembling eyes for defense.

Eye camouflage is especially prevalent in the vertebrates lacking eyelids, the fish and snakes (Cott 1941, 88). In them the eye is usually disguised by incorporating the pupil into a black stripe or **eye mask** (Figure 2-2). The stripe frequently is positioned so that the edge of the pupil coincides with one or both edges of the stripe (Cott 1941, 84). This patterning probably helps conceal the pupil by not presenting it against a uniform background. The high contrast at the edge of the stripe interferes with perception of the round disk.

On the other hand, many moths and butterflies possess eyelike patterns on their wings, which they suddenly expose when they are disturbed (Figure 2-3). Experiments (Blest 1957) have confirmed that displaying these eyespots discourages birds from preying on the insects. A comparison of artificial models has indicated that circular patterns were more effective than designs using straight lines. Concentric circles were superior to a simple ring, but the most effective was a pattern with highlights suggesting a three-dimensional eye.

Although a pattern like that in Figure 2-3, which is revealed suddenly when the moth is disturbed, certainly seems to function by suggesting a pair of eyes belonging to a large vertebrate, it is difficult to demonstrate that such patterns are indeed perceived as eyes, and it is important to guard against assuming that other animals perceive things as humans do. One caveat is that different species of butterflies and moths have patterns that cover a wide range of resemblances to eyes (Blest 1957). Furthermore, the smaller and simpler patterns may resemble food items like berries and seeds and cause a bird to peck at the periphery of the wing, allowing the insect to escape (Owen 1980, 78–79).

CHEMICAL TRAILS

A fundamental question about chemical trails is whether it is possible to determine the direction in which the trail was made and thus identify the direction in which to follow the trail in order to find its source. Some snails can apparently determine the direction of their own previously made trail, but their mechanism for doing so is not known (Wells and Buckley 1972).

In general, it seems that such information is not available to most animals, including ants (Wilson 1962). One might imagine that when animals lay down a pheromone trail for conspecifics they would deliberately provide some clues to direction, perhaps as some sort of spatial patterning. However, there does not seem to be any evidence that this occurs.

One example in which the mechanism for determining direction has been established, however, is in snakes (Ford 1986). Male snakes frequently follow the trails of conspecific females in the breeding season. They apparently use their forked tongues, which can spread to twice the width of their heads, to provide spatial comparisons, and thus use tropotaxis (see Section 17-3) to follow the trails with great accuracy. More impressively, a snake encountering a trail can apparently determine the direction in which it was made. This analysis probably occurs by using the pattern of pheromone distribution on objects projecting above the surface to determine which side the female pushed against during locomotion (Ford and Low 1984). As can be seen in Figure 11-1, this provides a clear clue as to the direction of the female's locomotion.

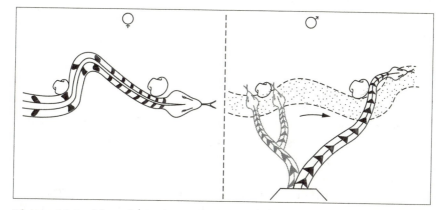

Figure 11-1 A hypothesized mechanism for snakes to determine the direction of a trail. *Left*: As a female leaves a trail of pheromones, her body makes more contact with objects she pushes against on the side toward the direction of travel. *Right*: A male snake, after encountering the pheromones, explores for chemical stimuli on both sides of nearby objects and recognizes which side the female has pushed against. The stippling indicates the locations in which the pheromone is probably deposited. (After Ford and Low 1984.)

11-2   TEMPORAL PATTERNS

SOUND

To humans the most meaningful temporal patterns are those of sound. Speech and music are temporal patterns par excellence. In other animals specific temporal patterns of sound known as songs are frequently used for communication (see pages 347–350). This is common in insects, frogs, birds, and mammals, including the recently discovered songs of whales.

The first animals to employ sound for communication could have used a relatively simple signal. However, as more and more species employed the strategy, it would become necessary to use more complex patterns for each species to distinguish its own song. Sound has two parameters that can be modulated to produce a pattern: amplitude (intensity) and frequency (pitch). Birds in fact employ both **amplitude modulation** (AM) and **frequency modulation** (FM) in generating and recognizing their many distinctive songs (Greenewalt 1968; Lewis and Gower 1980, 58). Likewise, human speech requires the production of many symbols, and it also uses both parameters to form the distinctive units called *phonemes*. Insects, however, employ only amplitude modulation. The simplest pattern they have is a continuous *trill*, the more complex being a series of *chirps* (Lewis and Gower 1980, 137).

LIGHT

The use of light to obtain information usually requires good temporal resolution and the ability to recognize temporal patterns. Because vision provides good spatial resolution as well, most visual patterns that are of relevance to animals with sophisticated eyes are the spatiotemporal patterns discussed below.

A common pattern that is purely temporal is the "shadow response" that initiates defensive reactions when there is a sudden reduction in light intensity. Another pattern that is mostly temporal is the flashing of light reflected from the wings of flying butterflies, which is used for mate recognition (Lewis and Gower 1980). The flashing bioluminescence of fireflies is a similar situation. Different species have their own distinct patterns of flashing (Lee 1989).

CHEMICALS

There has been some speculation that animals may transmit information through the temporal patterning of a chemical release (Bossert 1968). This possibility gained interest when Conner et al. (1980) discovered that a moth released a pheromone sex attractant in a pulsed manner. However, given the turbulence of air in natural environments (see pages 75–86), it is doubtful that a pulsed pattern would be detectable very far from its source, though it might provide some information at close range. On the other hand, it has been demonstrated that a pulsed release of chemicals could function to extend the range of detection of a given amount of pheromone (see Section 7-4) (Dusenbery 1989a), which seems a more parsimonious hypothesis. Besides this possible case, there seem to be no established examples where a purely temporal pattern of a chemical stimulus provides information. The interesting case of slime-mold aggregation is discussed in the next section.

11-3 SPATIOTEMPORAL PATTERNS

MOVEMENT

An important type of pattern is the spatiotemporal pattern caused by the movement through an environment of a physical object such as another organism. The visual systems of many animals seem to be adapted to detecting this type of pattern, which is important in a variety of ways. In humans this tendency is so strong that a pair of alternately flashing lights with appropriate spacing and frequency can generate a perception that there is only a single light moving back and forth. When the individual stimuli are not as clear-cut, however, a more continuous pattern of motion is required for detection. Think of the sounds of creaking floorboards, rustling leaves, or snapping sticks at night. Although an individual sound may attract some interest, a pattern of such sounds that follows a coherent path will rivet our attention if we are not expecting someone. The individual sound could have many causes—thermal expansion, wind, or falling limbs—but a coherent pattern is likely to be produced by a walking animal. Marr (1982, 159–215) provides a critical discussion of this subject with respect to vision. Some animals disrupt this pattern by moving in an irregular manner, such as a quick movement followed by a long pause (O'Brien 1979).

An animal that is itself in motion and has a good visual system will experience a spatiotemporal pattern that is often described as an **optical flow field** (see pages 192–195) (Wagner 1982). The orientation and speed of this flow provide the animal with information on the orientation and speed of its movement through the environment, which is of particular importance to flying animals.

Communication between animals often depends on the spatiotemporal pattern in the image of a moving appendage. For example, male fiddler crabs wave their large claw as a signal to conspecifics (Figure 11-2), and male spiders that are of hunting species with good vision communicate with females by a patterned movement of their pedipalps or forelegs (Lewis and Gower 1980). And the mating displays of many birds primarily involve spatiotemporal visual patterns.

SLIME-MOLD AGGREGATION

The best example of a spatio-temporal pattern of a chemical stimulus is in the aggregation of ameboid cells of the slime mold *Dictyostelium discoideum*, to form a fruiting body (Gerisch 1982). In this unusual situation, each cell has on its surface an enzyme that degrades cyclic AMP but responds to an increase in the concentration of cyclic AMP in its environment by releasing more of this compound for several minutes (Devreotes and Steck 1979). A population of cells sufficiently close to one another will consequently propagate waves of changing cyclic AMP concentration (Tomchik and Devreotes 1981). In these waves the concentration of cyclic AMP varies from about 10^{-8} to 10^{-6} M. The period varies from 5 to 10 min, and the wave propagates at a speed of roughly 100 μm/min. Individual cells make use of the information from this dynamic pattern to aggregate into clumps containing 10^4 to 10^5 cells (Raper 1940). Exactly which aspects of the wave are used to provide directional information is not clear, but the sensing of both temporal and spatial gradients is likely (Soll 1990).

11-4 CHEMICAL MIXTURES

The leech sucks blood from its vertebrate hosts to obtain nourishment. Investigations of the stimuli that initiate feeding indicate that the leech recognizes blood as a mixture of 150 mM NaCl and 0.1 mM L-arginine.

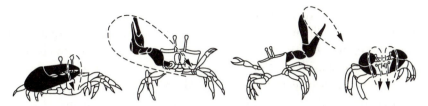

Figure 11-2 Fiddler crab displays. Male fiddler crabs display to females by waving their large, conspicuously colored cheliped (shown in black) in species-specific patterns. Each crab represents a different species, with the complete pattern of motion indicated by the dotted line: From left to right, *Uca rhizophorae; Uca lactea; Uca maracoani,* faces away from the female; *Goniopsis cruentata,* which waves both chelipeds simultaneously. (After Wehner 1981, 510.)

This simple mixture is as effective as whole blood in eliciting all the aspects of feeding, including ingestion (Elliott 1986).

During extensive research on the sex-attractant pheromones of flying insects it became increasingly clear that many if not most pheromones are blends (Tamaki 1985). Initial suggestions that the individual components might act as discrete signals have now been supplanted by the supposition that the complete blend acts as a unit (Linn et al. 1984, 1986, 1987). Consequently, it is the pattern of the chemicals that is recognized. Using mixtures to provide specificity is probably easier to evolve than mechanisms for synthesizing and recognizing a new chemical (see page 138).

11-5 COLOR

Having a sense of color requires the ability to distinguish between light of different wavelengths. This sense is widespread among organisms, though far from universal. For example, the cells of certain bacteria called *Halobacterium* are attracted by yellow-green (565 nm peak) light, which is used for energy, but are repelled by ultraviolet light (370 nm peak), which has the potential to damage the cells (Hildebrand and Dencher 1975). Plants can also discriminate colors by using the phytochrome system described in Section 8-1 (Bradburne et al. 1989; Ballaré et al. 1990).

On the other hand, many if not most animals have little sense of color. Nocturnal animals sacrifice color vision in favor of increased sensitivity. In particular the mammals, except for primates, generally lack color vision, which accounts for the lack of color on their body surfaces. In contrast, most birds have a keen sense of color, and colorful coats. In addition, flowers that are pollinated by birds are brightly colored, often looking red to humans (Cott 1941, 191; Waser 1983), but lacking scent, to which birds are relatively insensitive (Hinton 1973). Many reptiles, which are closely related to birds, also seem to have a good sense of color (Hinton 1973), which is consistent with most of them being diurnal and many of them having colors on their body surfaces. Color sensitivity and display also occur among the surface-dwelling diurnal fish. Fish may have fewer types of cones, or more, than the three in primates. Their color sensitivity often seems to be related to the dominant or important colors in their environment. In some fish eyes there is a different ratio of cones for looking upward as compared to downward. At least some diurnal insects also have a good sense of color. For example, aphids use the yellow–blue ratio to discriminate the sky from foliage (Kennedy et al. 1961), and bees and butterflies use color to discriminate between one flower and another (Waser 1983).

In primates, color information is provided by three kinds of receptor cells or cones that are sensitive to light of different colors: yellow, green, and blue. The cone type that is sensitive to the longest wavelengths has its maximum sensitivity to yellow-green light but is often called red sensitive, because stimulating it without stimulating the other cone types produces a sensation of red. Because color receptors are less sensitive than rod receptors, we lose our sense of color in dim light.

Genetic variation in humans is such that about 8 percent of men lack red- or green-sensitive cones (Nathans et al. 1986), which restricts the colors they can perceive. However, this lack may be beneficial in some circumstances. For example, it has been said that such persons can see through camouflage that the typical person cannot (*Scientific American* 1991). This population variation might be beneficial because a group of hunters with a variety of perceptual tendencies might collectively obtain more information about their environment than if all individuals had the same perceptual abilities.

Because color is a powerful stimulus for humans, describing colors is of great practical importance and much effort has been devoted to devising methods to describe them accurately. This undertaking has proven to be surprisingly complicated. Some of the most famous names in science have worked on the problem—Newton (1730), Maxwell (1860), Helmholtz (1852–1911), Schrödinger (1920–1925)—but are better known for their contributions to other areas of science (see Judd and Wyszecki 1963).

Light with a given spectral distribution will produce a specific sensation of color, but the converse is not true—a given color sensation in an individual can be produced by any number of spectral distributions. This situation occurs because color is determined only by receptors of three different spectral sensitivities. As a result, color has three independent dimensions, often described as **hue** (green, red, etc.), **saturation** (the degree of gray), and **brightness** (intensity) (Levi 1968, 11). Other triplets are also used. For example, printers use the three colors cyan, magenta, and yellow, and video technology uses red, green, and blue.

The rules regarding the perception of color may be summarized by Grassmann's Laws (Levi 1968, 10):

1. The eye can discern only three types of color variation, expressible as hue, brightness, and saturation.
2. If, in a mixture of two unequal colors, the proportion is steadily changed, the color of the mixture changes.
3. When lights of two given colors are mixed, the result is always the same, regardless of the particular spectral compositions that produce the two colors in the mixture.
4. When two lights are mixed, the luminous flux of the mixture equals the sum of the luminous fluxes of the components.

What we commonly mean by color is actually the relative spectral distribution, independent of intensity, which depends on hue and saturation. This accords with the view that the function of vision is to provide information about objects, whose surface reflectance spectra are primarily related to hue and saturation. We more commonly ascribe color to objects than to light. With our interest focused on only two dimensions, it is possible to plot on paper the relationships between all the perceivable colors.

The most commonly used system is that defined by the Commission Internationale de l'Eclairage, known as the CIE. Since individuals vary significantly in how they see colors, the CIE defined a "standard observer" in addition to the coordinate system (Levi 1968, 20ff). The plot they designed is shown in Figure 11-3. Hue changes with position around the periphery, saturation with position between the central white or gray point and the periphery. Brightness is represented in a third dimension that is perpendicular to the paper. On this type of diagram, mixing two colors in different proportions produces colors that fall on a line between the two parent colors. A line passing from the white point to the periphery includes all the colors with an identical dominant wavelength. The hues of

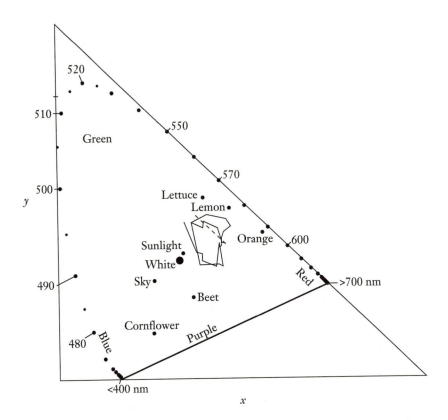

Figure 11-3 The CIE color plot. The *x* and *y* axes represent parameters varying from 0 to 1. The corners of the triangle represent theoretical designations for blue, green, and red. The outermost ring of points represents the perceived color of pure monochromatic light for the wavelengths indicated, in nanometers. Only the colors within the area defined by these monochromatic colors are physically achievable. The points plotted in the interior represent the perceived colors of sunlight, blue sky, and specific plant parts. The solid lines represent the color change of leaves in the fall. The dotted line is for the ripening of tomatoes. The large point labeled "white" is for light with equal energy in all its wavelength bands. This is not a special distribution from the physical point of view; equal numbers of photons per frequency band would be the best physical definition of a uniform distribution. Specifying a distribution of equal energy per wavelength band for white was probably done because it matches the distribution from a uniformly reflecting surface illuminated by a mixture of sunlight and blue sky, as indicated by the white point's being nearly on the line between sun and sky light. (Data from Le Grand 1968, 166, 492; Rossotti 1983, 164; and Gates et al. 1965.)

the colors on opposite sides of the white point, which are called complementary colors, may be mixed to generate white. The farther away a color is from the white point, the greater its saturation. The regions of the plot beyond the monochromatic colors are not accessible.

It may be noticed on the diagram that equally distinct colors are not equally far apart. For instance, the various green colors occupy a much larger area than shown for the combined area of reds and yellows. In fact, there is a twentyfold increase in the green region compared to the violet one, and attempts to define a coordinate system in which a given distance represents an equal change in color for all regions have not been very successful (Levi 1968, 27).

On such a trichromatic plot the majority of a sample of flowers would have colors falling on a line from yellow to white (Kevan 1972, 1978). If, however, the same reflectance spectra are plotted on a trichromaticity plot appropriate for the honeybee, the flower colors become much more widely dispersed (Figure 11-4). This situation is consistent with the hypothesis that flower colors have been selected to be distinctive to pollinating insects.

A remarkable feature of color is that how it is perceived is highly relative to its surroundings. Although most people do not often recognize this fact, artists have long realized it, and Edwin Land (1974) has done much to promote the idea. The basic point is that how the color of a particular object is perceived depends on how its color contrasts with the colors of surrounding objects and is relatively independent of the spectral distribution of the illuminating light, which varies with the time of day, the cloud cover, and in relation to reflecting objects outside the field of view. This characteristic of information processing in our visual system can be understood as being adaptive if we adopt the view emphasized by David Marr (1982) that the function of vision is to provide information about the physical objects in the surrounding environment rather than about the nature of light in the environment. Consequently, the important information about color is not the ratio of red, green, and blue photons currently being received but what this ratio would be under some standard lighting condition and thus whether the object being viewed has the same reflective qualities as certain other objects. The human visual system accomplishes this to a remarkable degree, but the strategies by which it does so are not understood (Marr 1982, 250–258).

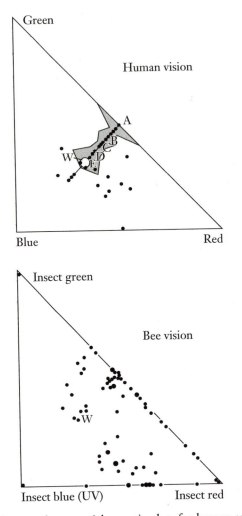

Figure 11-4 Flower colors on trichromatic plots for human and bee visual systems. The points are the flower colors measured for one or more species or flower parts, from weeds growing in Canada. *Top*: This plot for human color vision shows most flower parts distributed along the line from points A (yellow) to W (equal-reflectance white). The width of the shaded area is a histogram for the number of observations falling at positions along the line; A:37 flower parts, B:12, C:5, and E:22. Specific points are labeled A–E. Other points are for single observations. *Bottom*: On this plot, the same data are presented on a trichromatic plot appropriate for honeybees. The diameter of each spot is proportional to the number of observations falling within it. This sample of flowers is spread much more widely over the range of colors perceivable by bees than over the range seen by humans. (From Kevan 1978.)

11-6 MULTIMODAL PATTERNS

In many situations, animals employ a variety of different sensory modalities to identify an appropriate object or condition. In few cases has there been sufficient investigation to suggest that we have a relatively complete understanding of the information used. Two cases involving insects' recognition patterns may serve to illustrate.

The onion fly normally deposits its eggs at the base of onion plants. A series of experiments has indicated that this fly's recognition of an appropriate egg-laying site involves at least the following stimuli (Miller and Harris 1985):

The odor of onions (10-100 ng/s n-dipropyl disulfide)

The sight of an onion stalk (a yellow or green vertical cylinder 4–8 mm in diameter and at least 10 cm high)

Even this set of stimuli is not sufficient to substitute completely for real onion plants. The fly tests these stimuli over a ten-minute period of walking back and forth on the "stem" and the substrate. This test period may provide the fly with additional information through contact chemoreception or mechanical mechanisms.

Breathing on a fungus beetle causes it to secrete irritating chemicals that will likely cause a predatory mouse to drop it. And experiments on the cues used to recognize mammalian breath (Conner et al. 1985) indicate that an unsteady flow of warm air and high humidity produced the response reliably. Interestingly, temperatures above mammalian body temperature were superior to that of body temperature and employing carbon dioxide did not significantly increase the effectiveness of the stimulus. Thus, the beetle does not use all the cues it might, but it appears to use enough for satisfactory recognition in its natural environment.

PART 3

Stimulus Generation

■ ■ ■

There are certain behaviors in which the deliberate generation of a stimulus plays a fundamental role. Some organisms probe their environment by producing a stimulus, then detecting any modifications made in it by the environment. This general strategy, called *sounding*, is the subject of the next chapter. The other main class of behaviors in which stimuli are generated is that in which one individual sends a message to another establishing *communication*, the subject of Chapter 13.

CHAPTER 12

Sounding

Probing the Environment

■ ■

Heave oft the lead, and mark the soundings well.
The Shipwreck, quoted by Sir Walter Scott

Sounding is an old nautical term for the act of determining the depth of water by lowering a weight attached to a line until the weight reaches bottom. This technique is representative of a class of strategies organisms have for gaining information about their environment by sending out some kind of probing stimulus. As we shall see, this strategy is used by a wide variety of organisms employing all types of stimuli. The strategy is most commonly used where vision is limited by darkness or turbidity. The term sounding is used here to cover all the cases where this strategy is used, irrespective of the nature of the stimulus.

12-1 GENERAL ASPECTS

TECHNOLOGICAL APPLICATIONS

The strategy of sounding has been found to be quite useful in human technology. One early development was probably the use of a torch to provide light when the sun was not present. A common contemporary example is radar, in which pulses of microwaves are sent out, usually in a beam, from an antenna and their reflected waves or echoes are collected by the same antenna and detected by a receiver (Eaves 1987). Information concerning the direction of the reflecting target can be determined from the orientation of the antenna. The distance to the target is determinable by the time delay between the outgoing pulse and the received echo. Radar is not used by other organisms because they do not have mechanisms to produce or detect microwaves.

Another technological example of this strategy is found in active sonar (Cox 1974, 3, 5). In this technique, pulses of sound in water are used to detect submarines or fish and to map the floor of the ocean. More recently, some snapshot cameras have been fitted with active sonar devices to provide automatic focusing. Many animals can generate and detect sound, so it is not surprising that some have developed similar capabilities, which are usually termed *echolocation*.

DISTANCE LIMITATIONS

In order to gain insight into the physical constraints on the strategy of sounding, let us consider some simple idealized cases for the basic types of stimuli: propagating stimuli (of sound or light), diffusing stimuli (of temperature or chemicals), and electric fields.

PROPAGATING STIMULI   For propagating stimuli the simplest model is to assume that a stimulus is transmitted uniformly in all directions from the sounder, then intercepted by a spherical target of effective radius r_T at range r_{ST}, reradiated equally in all directions, and finally intercepted by a receiver of effective radius r_R back at the location of the sounder. This model would be appropriate for using sound to detect a small gas bubble in water, because the sound will cause the whole bubble to pulsate and reradiate the sound in all directions. In this situation, as shown in Figure 12-1, the fraction of transmitted energy that is received in the echo is simply

$$\frac{E_R}{\mathcal{J}} = \frac{r_T{}^2 \, r_R{}^2}{16 \, r_{ST}{}^4} \qquad (12\text{-}1)$$

The signal is proportional to the effective area or crosssection of the target and the effective area of the receiver. More importantly, the strength of

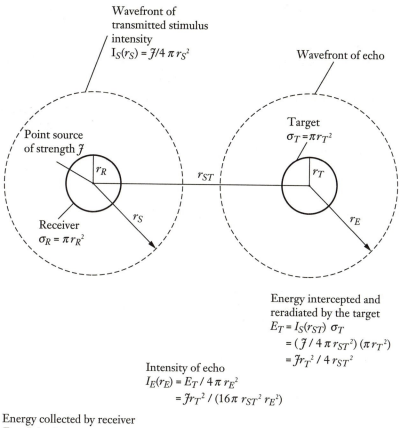

Figure 12-1   The intensity of an echo from a uniformly radiating source and target. For any uniformly radiating source, including this target, the energy is spread uniformly over spheres centered on the source. The intensity at distance r is simply the total energy divided by the area of the sphere with a radius equal to the distance from the source ($A = 4\pi r^2$). The energy received is calculated as the intensity at the position of the receiver times its cross section (σ). The cross sections of the source and the target are indicated by the heavy circles, particular wavefronts by dashed circles. The cross sections are assumed to be spherical. Equations describing the intensity of the stimulus as it progresses are presented from the top downward. Attenuation is assumed to be insignificant.

the echo declines with the fourth power of the distance to the target, because the target acts as a new source, and the inverse square law applies in both directions. This relationship places severe limitations on the range at which this strategy is effective. When attenuation is important, the ratio in Equation 12-1 is multiplied by a factor of exp $(-2\,\alpha\,r_{ST})$, where α is the attenuation coefficient (see pages 58–59).

The relations in Figure 12-1 can be solved for the maximum range (r_{max}) of detection of a target when the sounder has a threshold energy (E_{Tb}) or intensity (I_{Tb}), giving

$$r_{max} = \frac{1}{2}\left(\frac{\mathcal{J}}{E_{Tb}}\right)^{1/4}(r_T\,r_R)^{1/2}$$

$$= \frac{1}{2\,\pi^{1/4}}\left(\frac{\mathcal{J}}{I_{Tb}}\right)^{1/4} r_T^{\,1/2} \tag{12-2}$$

For comparison, the range of detection by a target with an effective detector radius of r_T and threshold energy E_{Tb} or intensity I_{Tb} is

$$r_{max} = \frac{1}{2}\left(\frac{\mathcal{J}}{E_{Tb}}\right)^{1/2} r_T$$

$$= \frac{1}{2\,\pi^{1/2}}\left(\frac{\mathcal{J}}{I_{Tb}}\right)^{1/2} \tag{12-3}$$

Thus, a sixteenfold increase in energy output would increase the range of detection by another receiver by fourfold, but the increase in the range at which an echo was detected would be only twofold (equation 12-2).

If the attenuation factor is included in either of these two equations they cannot be solved for the maximum range, but it can be determined by trial and error substitution of approximate values for r_{ST} or by plotting the equation as a function of r_{ST} as shown later in Figure 12-6.

In most real situations the echo will not have equal intensity in all directions, and the relations above will be modified by numerical factors. These factors depend on the shape and materials of which the target is composed, as well as on its size in comparison to the wavelength (λ) of the stimulus. For the simplest shape, which is a spherical target, echo intensity depends on wavelength, as follows (Urick 1983, 299):

$\lambda \gg r_T$	$I \propto \lambda^{-4}$	Scattering of a wave
$\lambda = 2\pi r_T$	I maximum	Constructive interference
$\lambda \approx r_T$	I variable	Interference
$\lambda \ll r_T$	I independent of λ	Reflection of a ray

The situation where there is reflection from a large, smooth, flat target is shown in Figure 12-2. Specific formulas for determining the fraction of energy emitted that is returned in the echo are presented in Table 12-1 for the simple cases of a large flat surface and of a sphere that

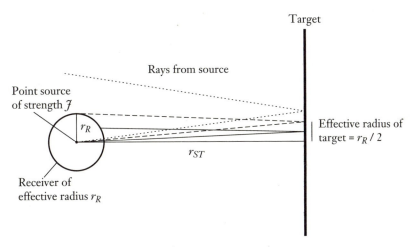

Energy reflected by the target to the receiver
$E_T = \mathcal{J} \times$ effective area of target $\div$ area of sphere that energy is spread over at the range of target
$\quad = \mathcal{J}\,(\pi\,(r_R/2)^2)\,/\,4\pi\,r_{ST}^2$
$\quad = \mathcal{J}r_R^2\,/\,16\,r_{ST}^2$

Figure 12-2 The intensity of an echo from a large, smooth, flat, reflecting target like a mirror. The stimuli are assumed to be transmitted equally in all directions. For a mirror target only the rays intercepting the target in the small area around the point on it closest to the source will be returned to it. From the law of equal angles of incidence and reflection (Section 4-1), the effective area of the mirror will have half the effective radius of the receiver, which is assumed to be spherical. This leads to a relationship in which the energy received in the "echo" is proportional to the square of the effective radius of the receiver and inversely proportional to the square of the distance to the mirror target.

Table 12-1 Summary of Echo Energy Returned from Different Types of Targets

Target Type	Type of Interaction Between Stimulus and Target	
	Specular Reflection	Diffuse Scattering
Flat	$\dfrac{r_R{}^2}{16\, r_{ST}{}^2}$	$\dfrac{r_R{}^2}{12\, r_{ST}{}^2}$
	Smooth at λ	Rough at λ
Spherical	$\boxed{\uparrow r_T \gg r_{ST}}$ $\dfrac{r_T{}^2\, r_R{}^2}{16\, r_{ST}{}^2\, (r_{ST}+r_T)^2}$ $\boxed{\longrightarrow \\ r_T \ll r_{ST}}$ $r_T \gg \lambda$	$\dfrac{r_T{}^2\, r_R{}^2}{16\, r_{ST}{}^4}$ $r_T \ll \lambda$

This table gives the fraction of transmitted energy returned to the sounder in an echo. The calculations assume that the target is distant compared to the size of the receiver ($r_R \ll r_{ST}$). Diffuse scattering assumes equal scattering in all directions. For diffuse scattering from a sphere, r_{ST} is the distance to the center of the sphere. In other cases it is the distance to the surface of the target. For the intensity of the echo as a proportion of the power emitted, replace r_R2 with $1/\pi$. Where attenuation is important, multiply each term by exp $(-2\, \alpha\, r_{ST})$.

are both completely smooth (specular reflection) or are diffusely scattering (equal reflection in all directions).

Note from Table 12-1 that for large, flat targets the echo intensity declines with the second power of the distance, as compared to the fourth power with a spherical target. Most nonabsorbing flat surfaces should fall in between the similar relations for the mirror and the diffuse models. However, some surfaces are exceptional. Certain shapes called *retroreflectors*, reflect back in the direction of the source—even if that direction changes. The surfaces of highway signs are frequently coated with retroreflectors, and the eyes of nocturnal animals often act as retroreflectors. There are, however, few if any organisms known to make use of retroreflection.

DIFFUSING STIMULI   Although propagating stimuli have severe enough limitations for sounding, diffusing stimuli are even more difficult to use. One reason for this is that diffusing stimuli return to their source to some extent even in the absence of a reflecting target. In addition, the calculations for diffusion are more complicated than those for propagation.

In situations where a stimulus, whether of chemical or heat, diffuses through a homogeneous, isotropic, stationary medium without sources or sinks, the steady-state distribution of the stimulus concentration or temperature is the solution of Laplace's equation (Carslaw and Jaeger 1959, 9, 28) that fits the boundary conditions of the particular situation. There is no general method of finding solutions to this differential equation, but in certain situations simple methods suffice. One of the most important such conditions is that of a point source in the vicinity of a plane reflecting boundary of infinite extent. In this case the distribution of the stimulus on the side of the boundary toward the source is identical with the distribution that would be produced were the reflecting boundary to be replaced by an *image source*, meaning an identical point source located where the image of the real source would be if the boundary were to reflect light. Specifically, an image source is on a line from the real source perpendicular to the boundary and at the same distance from the boundary as the real source—but on the opposite side. This mathematical trick of using image sources allows many problems involving point sources and plane boundaries to be solved easily.

This technique can be applied to the problem of how an organism might obtain information about the distance of a target consisting of a large planar reflecting surface, by analyzing the increased intensity of a stimulus diffusing from a small source. In the space around the source, the intensity distribution would be identical to a situation in which the reflecting plane were to be replaced by an image source. The symmetry across the position of the reflecting plane ensures that there will be no net flux of stimulus across it, which has the same effect as the reflecting plane. Consequently, the additional intensity observed at the source is just the intensity coming from the image source. Thus, a hypothesis that a given response is due to the reflection of a diffusing stimulus from a plane could be tested by using an identical source at twice the distance of the plane.

Quantitatively, it can be seen from Figure 12-3 that a large reflecting target at a distance r_{ST} from a steady source causes an increase in intensity of

$$\Delta I = \frac{\mathcal{J}}{8 \pi D r_{ST}} \tag{12-4}$$

where $\mathcal{J}$ is the strength of the source and D is the diffusivity (Table 4-3). An absorbing target would cause a decrease of the same magnitude. In either case, the signal would decline only in proportion to the distance.

If an organism makes a spatial comparison of a given stimulus's inten-

Actual situation modeled:

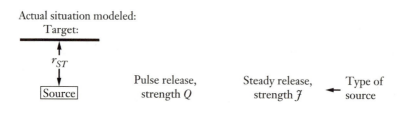

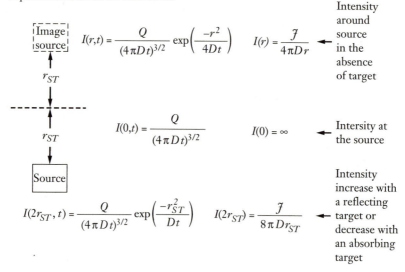

Figure 12-3 The intensity of an echo from a totally reflecting plane surface for a diffusing stimulus. The sources are treated as points, which is appropriate as long as their size is small compared to the distance to the target. In the mathematical model (*bottom*), the intensity is infinite at the point source during emission, whereas real sources must have a finite size and intensity. The equations at the bottom show the intensity change caused by a target at a distance r_{ST}, for both a pulsed and a steady release. The intensity increases this much with an ideal reflecting target and decreases an equal amount with an ideal absorbing target.

sities, the gradient of the intensity becomes the important parameter. Then, taking the derivative of Equation 12-4,

$$\frac{dI}{dr} = \frac{J}{4 \pi D r_{ST}{}^2} \qquad (12\text{-}5)$$

This equation is applicable only to situations where the gradient is measured over a distance that is small compared to that to the target.

For a pulsed release a similar argument will show that the change in intensity in the presence of the target is figured as

$$\Delta I = \frac{Q}{(4 \pi D t)^{3/2}} \exp\left(- \frac{r_{ST}^2}{Dt}\right)$$

(12-6)

Here the signal varies exponentially with the distance from the target.

In this case we can also consider the ratio ($\Delta I/I$) of the change in intensity in the presence of the target to the intensity in its absence at the location of the pulsed source:

$$\frac{\Delta I}{I} = \exp\left(- \frac{r_{ST}^2}{Dt}\right)$$

(12-7)

This ratio simply increases steadily with time, from very small toward a limiting value of 1. Thus, the intensity of the signal in the presence of the target is never more than twice the intensity in its absence. For small targets the signal would be even smaller. Calculations have also been performed for sounding by diffusing chemicals in various other geometries (Meyer et al. 1987).

Currents also influence the concentration of a diffusing stimulus, thereby allowing chemical or thermal sounding to be used to detect currents. Basically, a current acts to carry a stimulus away from a source and reduce its concentration near the source. A familiar example is the way wind speeds up the evaporation of water by increasing its concentration gradient in the air near the water surface. Man-made instruments for measuring currents are often designed to determine the cooling of a heat source. Unfortunately, predicting the magnitude of such effects is quite complicated. Extensive discussions are available, however, in the context of gas exchange with plants (Campbell 1977, 61–72).

ELECTRIC FIELDS   Electric fields can also provide a stimulus useful for sounding, and some species of fish have developed the ability to exploit this strategy. Like diffusing stimuli, the electrical potential in a homogeneous, isotropic medium in a steady state obeys Laplace's equation, and using image sources the distribution of potential is easy to determine for point sources of electrical current and planar boundaries between media of differing electrical conductivity. If the boundary is with an ideal conductor—metals come close—the image source has a polarity opposite that of the real source. If the boundary is with an ideal insulator—air comes close—the image source has the same polarity as the real source.

The lines of electrical potential and those of current flow are mutually perpendicular. Thus, a point source of current is surrounded by concentric circles of electrical potential. This pattern is modified by the presence of objects of higher or lower conductivity than that of the surrounding medium. A conductor tends to push potential lines away from it, which tend to run parallel to its surface. A current thus tends to flow perpendicular to the surface of a conductor. Conversely, an insulator tends to pull potential lines toward it, and they tend to run perpendicular to its surface. As a result, current tends to flow parallel to the surface of an insulator.

To a first level of approximation, the fields around an electric fish are similar to those generated by a point source and a point sink (a negative source). Electric current flows out of one point and into the other. The source and the sink are generally located near the gills and at the tail.

The effect that having a nearby target with a large planar boundary differing in its conductivity from the surrounding water has on the distribution of electrical potential has been calculated by using the method of image sources shown in Figure 12-4. When the target is either ahead or behind, it can be detected as a shift of the potential along the axis of the body or as a change in the potential level at the end closest to the target. If the target is alongside, there is no potential change at the point midway between the source and the sink of current, but a relatively large change in the potential level occurs one-quarter of the way between them. A cross section at this level shows relatively large changes in potential on the side nearer the target. In summary, a variety of changes in the distribution of the electrical potential exist that might provide useful information.

However, the range of these changes is limited. In Figure 12-4, the distance between the target and the source or axis is only one-quarter of the distance between the source and the sink. Even for such a large, close target the changes in potential are relatively small. Sounding with electrical stimuli can thus be seen as limited to very short ranges, less than the length of the animal. The range is comparable to that of sounding with appendages, as discussed on pages 311–12.

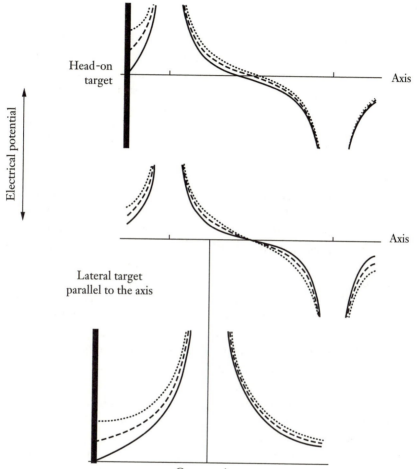

Figure 12-4 The electrical potential around a point source and sink. The source and the sink lie on the axis at the positions of the tics. The target (*heavy lines*) is a large plane, at a distance of one-quarter the distance between the source and the sink. The cross-section is taken one-quarter of the way between the source and the sink. The curves demonstrate the variation in electrical potential for (*a*) a target acting as an ideal insulator (*dotted line*); (*b*) a "transparent" match to the surrounding medium, which is equivalent to having no target (*dashed line*); or (*c*) an ideal conductor (*solid line*). For a target having this geometry no material can produce larger changes in its electrical potential than those shown here.

Top: a situation similar to a fish facing a flat target at a distance of less than one-quarter of the fish's length. *Center and bottom*: a situation similar to a fish oriented parallel to a large flat target at a distance of less than one-quarter of the fish's length.

DETECTION

Sounding provides an opportunity for using special strategies to detect a signal against noise, which helps to detect weak echoes. Because the receiver generates the outgoing signal, it can optimize the characteristics of that signal. For propagating stimuli a common strategy is to compress the outgoing stimulus into a series of short, intense pulses. If the period between the pulses allows sufficient time for an echo to be received, then when a target is present at a certain distance there should be an echo after a certain delay following each pulse.

This relationship can be exploited by using multiple-pulse integration (Echard 1987), which correlates the signal intensity with the time after a pulse is transmitted. This strategy requires having a series of memory elements each able to integrate the signal present during a certain time window after a pulse with the signal in the same window after previous pulses. The series of elements is arranged so that its windows correspond to a succession of time intervals after a pulse. With this system it is possible to integrate the stimulus in each of several windows over many pulses, making possible a great improvement in the signal-to-noise ratio (Figure 12-5). This technique has been employed in a general-purpose device called a lock-in amplifier, and many variations on it have been developed for use with radar (Barrett 1987). However, it is not clear to what extent organisms have utilized this strategy.

Another strategy often employed is to focus a stimulus in space as well as time. Thus, special structures like antennas in radar and facial features in bats are used to focus the transmitted pulse into a narrow beam, and the same or other structures like the ears of bats are used to make the detector more sensitive to signals from the same direction. This method can greatly improve the signal-to-noise ratio for a given pulse of energy and for a target in the appropriate direction. In addition, it allows more targets to be distinguished one from the other.

PREY FINDING

In general, it is difficult to use sounding to find prey. For one thing, if predator and prey have similar sensitivities to the probe stimulus, the prey can detect a probing stimulus at much greater distances than the predator can detect its echo. This occurs because a target usually reflects only part of the impinging stimulus and also disperses the propagating stimuli into

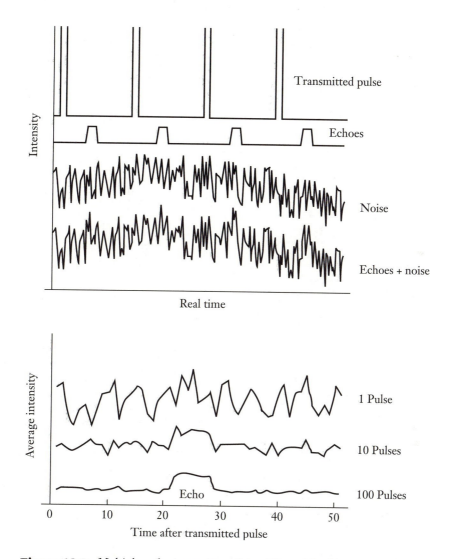

Figure 12-5 Multiple-pulse integration. *Below*: The results of averaging the echoes plus noise (*above*) over 1, 10, or 100 pulses. The curves have been shifted vertically for clarity.

new directions, and the stimulus has to make a return trip. As we have seen, intensity falls as the distance squared from the source, which, combined with having the target disperse the propagating stimuli, will let the strength of the return signal fall as the fourth power of the distance to the target, as seen in Equation 12-1 and shown in Figure 12-6. For example, certain moths can probably detect the sound of a bat at 35 m and take evasive action, but the bat can detect the moth only when it is within 3 to 6 m (Roeder 1967, 57, 77; Miller 1983).

Clearly, it can be beneficial for prey to develop mechanisms to detect the sounding stimuli used by their predators. For example, moths have

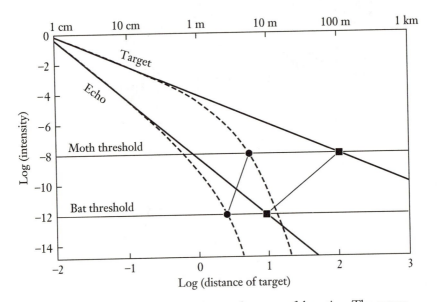

Figure 12-6 The effect of attenuation on the range of detection. The curves, with the distances measured in meters and the power in watts, are appropriate for bats and moths. The horizontal lines show the threshold intensities for moths and bats. The heavy lines show the decline in intensity with distance between the source (the bat) and the target (the moth). The lines labelled "Target" are the intensity at the target; the lines labeled "Echo" show the intensity of the echo to the bat. The curved dashed lines show how intensity is modified by an attenuation of 1/meter, as is appropriate for 100 kHz sound in typical conditions. The maximum ranges for the bat and the moth are indicated with solid circles for the case of attenuation and solid squares for no attenuation. The relative advantage that the prey has decreases with attenuation. Without attenuation, the maximum ranges are 8.4 m for the bat to detect the moth and 89 m for the moth to detect the bat, a relative advantage for the moth of 10.6 : 1. With an attenuation length of one meter the maximum ranges become 2.5 and 5.5 m, for a relative advantage of 2.2 : 1. The assumed parameters in the equations in Figure 12-1 were $J = 1$ mW, $r_T = 1.6$ cm, and $\alpha = 1$ /m.

developed a great variety of mechanisms to detect the ultrasonic cries of bats (Pye 1983). In contrast, it is thought that fish have not developed mechanisms to detect the high-frequency sounds used by dolphins to locate fish. In general, sounding is likely to be of continuing success over evolutionary time spans only when the predator can move much faster than the prey.

In fact, this seems to be the case, but even bats hunting slow-flying moths have adapted strategies to avoid early detection (Pye 1983). One obvious technique is to shift to a stimulus the prey cannot detect as well as some bats do in employing very high frequencies that may be difficult for moths to detect (Surlykke 1988). In addition, an increase in attenuation gives the predator a relative advantage, as in Figure 12-6. Since attenuation in air is significant over a few meters at high frequencies (see Table 9-3) and increases with frequency, attenuation may provide an additional advantage of using high frequencies for bats.

Emitting low intensities may also improve the situation for the predator. Because echo intensity for predators declines as the fourth power of the range to the prey but declines only as the second power for detection by the prey, stronger transmissions increase the range at which the prey can detect the predator more than they do the range at which the predator can detect the prey. To see this take the ratio of equations 12-2 and 12-3:

$$\frac{r_{\text{prey}}}{r_{\text{predator}}} = \left(\frac{r_T{}^2 \, \mathcal{J}}{\pi \, I_{Tb}}\right)^{1/4} \tag{12-8}$$

The relative advantage of the prey increases with the fourth root of the power $\mathcal{J}$ emitted. The situation with attenuation is depicted in Figure 12-7.

The loss of the range of detection with the lower energy of transmission may be offset by not alerting the prey so far away. This strategy is most appropriate when the prey occur in high density, and the range of detection becomes less important in searching.

Another consideration in using sounding to hunt prey is the possibility that the prey can disrupt the hunt by producing a stimulus similar to an echo, to interfere with the predator's analysis of the true echo. Such signal **jamming** is a common strategy in military radar. In fact, some species of moths are known to emit ultrasonic clicks when disturbed or when they hear ultrasonic pulses similar to those of the bats that prey on them (Miller 1983). These observations suggest the use of a jamming strategy. Since the moth clicks are often produced only when a bat is very close, they may

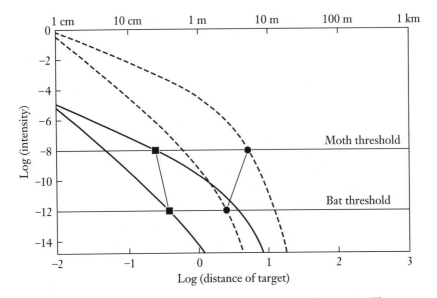

Figure 12-7 The effect of power output on the range of detection. These curves, with distances measured in meters and power in watts, and with an attenuation length of one meter, are appropriate for bats and moths. The horizontal lines show the threshold intensities for moths and bats. With a power emission of 1 mW (dashed curves), the intensity at the target is shown in the upper such curve, the intensity in the echo in the lower one. The solid curves show the same thing, except that the power output is reduced to 10 nW. The maximum ranges for the bat and the moth are indicated at high power levels (solid circles) and for low (solid squares) levels. The relative advantage of the prey decreases with decreasing power. Because of the bat's greater sensitivity, at sufficiently low power it can even detect the moth before the moth can detect it.

function by a startle effect (see Section 2-5) (Stoneman and Fenton 1988). Specifically, the bat suddenly receives what apears to be an intense echo indicating that there is a large target ahead. However, bats hunting by vision are just as disrupted by moth sounds as are those hunting by echolocation (Stoneman and Fenton 1988). Most of these moth species have chemical defenses, and many are brightly colored, suggesting their use of warning coloration, as discussed in Section 2-5. Thus, the moths' clicks may function as a sonic warning to the bats rather than as a mechanism of jamming. No other possible examples for jamming are known.

12-2 ECHOLOCATION

The prime example of using sounding is that of echolocation, in which sound serves as the stimulus probe. The peculiar mammals, bats and whales, both rely heavily on echolocation, which allows bats to fly in darkness and even catch flying insects. Echolocation permits whales, which swim at high speeds, to obtain information about their environment at much longer distances than would be possible from vision in even the clearest water. (The attenuation distance is > 100 m for sound, as compared to 15 m for light; see Tables 9-3 and 8-6.) Developing a sounding technique has probably been essential to these animals in occupying the niches they do as high-speed predators in environments where sight is limited.

A few species of birds have developed more limited echolocation capabilities, which allow them to nest deep in caves (Griffin 1953; Medway 1967; Pye 1983). There is also recent evidence that seals may use echolocation (Evans 1988). It is interesting to note that all the species thought to employ echolocation are warm blooded. Pye (1983) has suggested that precise temperature regulation may be necessary for the accurate timing required to extract useful information. However, mammals also have more highly developed nervous systems than other animals, which may likewise be required. Furthermore, because homeothermy permits high-speed locomotion in the absence of warming sunlight, homeotherms may have more use for echolocation.

Other types of animals may well employ sound as an environmental probe, in less sophisticated ways. For example, the sea catfish is reported (Tavolga 1976) to use low-frequency (100 Hz) sounds to detect invisible barriers. Since the 15 m wavelength is much greater than the size of the animal (0.3 m) it cannot efficiently project sound to any distance. In fact, this fish can only detect barriers within 5 cm. Clearly, it is relying on near-field acoustics (see pages 250–52), so it is doubtful that this behavior should be considered echolocation.

The rest of this section presents certain general characteristics of echolocation, followed by specific observations of bats, whales, and humans.

ECHO STRENGTH

In addition to the general relations already established for propagating stimuli on pages 284, 286–88, more specific predictions for acoustic stimuli interacting with a small ($r_T \ll \lambda$), homogeneous spherical target can be based on Equation 9-10. For the ideal rigid and fixed or high-density target, using the relations in Figure 12-2, the fraction of energy returned to a uniformly radiating sounder is

$$\left(\frac{5}{3}\right)^2 \pi^4 \frac{r_T{}^6 r_R{}^2}{\lambda^4 r_{ST}{}^4} \qquad \begin{array}{l} \lambda \gg r_T \\[4pt] r_{ST} \gg r_T, r_R \end{array} \qquad (12\text{-}9)$$

where r_T is the actual physical radius of the target. In this situation the intensity of the echo decreases with the fourth power of the wavelength and increases with the sixth power of the target's radius. Spheres that are not rigid and fixed will differ in intensity only by fixed numerical factors, and these relations will still hold. From Table 9-6, it can be seen that for mineral targets in air or water or for water drops in air the factors do not vary much from those for the fixed, rigid model. However, for gas bubbles in water the echo's intensity is increased 100 million times over those of similarly sized targets of the other compositions. A bubble 4 percent the size of a dense, rigid sphere produces the same echo intensity, suggesting that echolocation fish with swim bladders is a particularly attractive method. On the other hand, echolocation of aquatic organisms consisting only of soft tissues that resemble water in their density and rigidity would be difficult.

Small targets are also difficult ones for echolocation. The maximum frequencies available are about 100 kHz, corresponding to a wavelength close to a centimeter (see Table 9-3). Targets much smaller in size than this wavelength are difficult to detect, because of the decreasing efficiency of scattering that occurs from smaller targets (see pages 284, 286–88).

In many cases of interest the targets are about the same dimensions as the wavelength. In these cases the calculations are complex, but some measurements are nevertheless available, particularly for fish, which are of interest for fishermen using sonar. The data are usually reported as target strength (*TS*) (Urick 1983), related to the effective radius of the target (r_T) as shown in Table 12-2.

Some examples of biological targets are shown in Table 12-3. For a wide range of sizes the effective radius of organisms in water is about one-tenth their length. For an organism in air the effective radius is close to its physical radius.

Table 12-2 The Relationship Between Target Strength and Effective Radius

$$TS \equiv 10 \log \frac{I_E \,(1 \text{ m})}{I_S}$$

$$= 10 \log \frac{E_T}{4\pi \, I_S} \quad \text{From Figure 12-1}$$

$$= 10 \log \frac{I_S \, \pi \, r_T{}^2}{\pi \, I_S} \quad \text{From Figure 12-1}$$

$$= 10 \log \frac{r_T{}^2}{4}$$

$$= 20 \log \frac{r_T{}^2}{2}$$

$$r_T = 2 \times 10^{(TS/20)}$$

The symbols have the same meaning as those in Figure 12-1.

TARGET DISTANCE

Sound waves travel at speeds of 340 m/s in air and 1,500 m/s in water, which are high enough that they cover biologically relevant distances (meters) in the time intervals (milliseconds) that the nervous system acts on. A similar relationship exists in radar between the speed of microwaves (3×10^8 m/s), the length of important distances between vehicles (kilometers), and the time resolution of electronic equipment (microseconds). In these situations the distance to a target can be estimated simply from the time delay between the transmission of a pulse and the reception of its echo.

The uncertainty in determining the distance to a target is limited by how short a pulse is transmitted. A simple model suggests that the uncertainty in the time of arrival is roughly equal to pulse width (t_w), which translates into an uncertainty about the distance (Δr) of (Pye 1983)

$$\Delta r = \frac{v \, t_w}{2} \qquad (12\text{-}10)$$

Table 12-3 Some Examples of Target Strength

Target Strength	Effective Radius (r_T)	Examples*
−80 dB	0.02 cm	0.3-cm Copepod in water
−60 dB	0.2 cm	3-cm Crab in water
−50 dB	0.6 cm	5-cm Fish in water
−40 dB	2 cm	Moth in air[†]
−30 dB	6 cm	50-cm Fish in water
+7 dB	4.5 m	45-m Whale in water

* Data for frequencies in the range 10 ↔ 100 kHz.
[†] Data from Miller 1983; all other data from Urick 1983, 316–17.

The shortest pulses (≈ 0.1 ms) are generated by dolphins and porpoises, suggesting a distance resolution on the order of 10 cm.

One problem with the strategy of using short pulses is that the energy declines in proportion to the pulse's duration, unless the intensity increases, and there are limitations to maximum intensity. A possible solution is to use a long pulse in which the frequency is swept through a whole range, a technique that has been used in radar (Holm 1987) as well as by bats (Pye 1983).

APPROACH VELOCITY

When either the source or the receiver of a propagating wave moves, there is usually a change in frequency called the *Doppler shift*. This effect causes the familiar descending tone from the whistle of a train or the horn of a car as it passes. In sounding, such a shift in frequency can be used to determine the velocity of the target along the line to it, which is to say its **approach velocity.**

If the transmitter generates a frequency of f_0, the velocity of sound is v_s, and the approach velocity is a much smaller value v_a, the received frequency is approximated by

$$f \cong f_0 \left(1 + \frac{2\,v_a}{v_s} \right) \quad v_a \ll v_s \tag{12-11}$$

and the frequency shift is simply proportional to the approach velocity

$$f - f_0 \cong \frac{2 f_0}{v_s} v_a \qquad v_a \ll v_s$$

(12-12)

The precision with which frequency can be measured is limited by the bandwidth and is approximately the reciprocal of the pulse width (t_w) (Pye 1983). Therefore, the precision of estimating the approach velocity is

$$\Delta v_a \cong \frac{v_s}{2 f_0 t_w} \qquad v_a \ll v_s$$

(12-13)

High frequencies and long duration pulses are desirable. However, long duration pulses of small bandwidths are unfavorable for estimating the distance to a target, which probably explains why many echolocating animals emit a sound containing both a long-duration, constant-frequency component and a frequency sweep. Other animals employ different types of pulses, according to their needs. Another possible solution is to transmit pairs of short pulses, to reveal the approach velocity by changes in the time between the corresponding pairs of echoes. Some bats and echolocating birds have been observed to use pairs of pulses (Pye 1983).

Attempts to minimize uncertainties about distance and approach velocity are frustrated by the mathematical restriction that bandwidth and pulse duration are inversely related (see Figure 5-10), meaning that a reduction in one causes an increase in the other. Detailed studies by Wiersma (1988) of the sound pulses emitted by several species of dolphins indicate that they all use waveforms that minimize the product of the bandwidth and the duration. This method should maximize the detectability of a signal in the presence of unpredictable noise.

THE NATURE OF THE TARGET

An echo may contain information about the size and composition of the target. For targets that are approximately the same size as the signal's wavelength or larger, interference will occur between waves scattered by different parts of the target, so that the echo may carry information about the shape of the target.

Information may be obtained by one of two different types of analysis of the echo. A single frequency may be analyzed in terms of its intensity variations over time as reflections from different distances are received.

Alternatively, a broadband emission may be analyzed in terms of changes observed in its spectral frequency distribution.

The biological literature often treats the propagation of sound by ray tracing, even when the structures involved are close in size to the wavelength of sound, but this practice is clearly unwarranted. Even when high-intensity pulses or glints can be observed in the echo, these are just as likely to result from constructive interference between waves scattered from different parts of the target as from specular reflection off a surface perpendicular to the path of the sound. A critical test to distinguish between these two possibilities is to determine if the intensity pattern changes with the frequency. If ray tracing is applicable, frequency changes will have no dramatic effects, but interference would be highly dependent on frequency.

SOME EXAMPLES OF ECHOLOCATION

BATS   Most bats prey on nocturnal insects, especially flying ones (Surlykke 1988). Different species of bats specialize in hunting in specific environments, which vary in their proximity to obstacles (Figure 12-8).

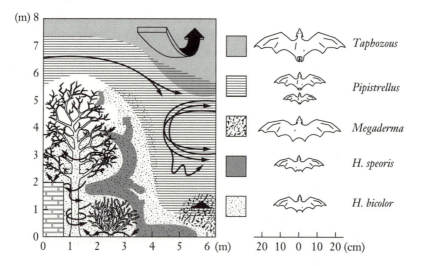

Figure 12-8 Bats' hunting environments. A depiction of echolocating bats and their preferred distinctive hunting environments in Madurai, India. (From Neuweiler 1983.)

The sound pulses bats employ vary widely in their frequency content and whether the frequency is constant or swept (see Figure 12-9). The differences in their sound patterns appear to be related to the environments in which bats hunt.

Most of these bats exhibit three phases of echolocation behavior in hunting (Simmons and Kick 1983; Surlykke 1988). While they are search-

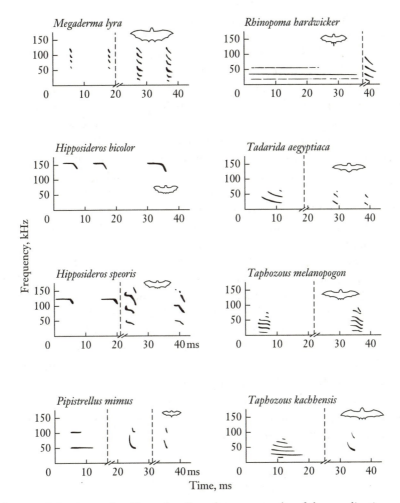

Figure 12-9 A sample of bat cries. Sound spectrographs of the vocalizations of different bat species in Madurai, India. *Left of vertical broken lines*: For each species the sounds emitted during cruising in the open are shown. *Right*: The sounds emitted close to targets or in the laboratory. (From Neuweiler 1983.)

ing they emit pulses of 2 to 10 ms duration, repeated at intervals of 50 to 200 ms. These intervals allow time for echoes from targets as far away as 8 to 34 m to be received before the next pulse is produced. After detecting prey, the bat turns its head toward the target and begins the approach phase, characterized by shorter, more frequent pulses. The terminal phase, which starts at 30 to 50 cm from the target, includes pulses as short as 0.5 ms at intervals as short as 5 ms (see Figure 12-10 for an example). The capturing of the prey usually occurs less than a second after it is detected.

The high frequency sound bats employ has a main frequency that varies from 9 to 212 kHz. Most species employ a frequency in the 30 to 60 kHz range (Surlykke 1988). The high frequencies have the advantages of increasing the resolution of approach velocity (Equation 12-13), making narrower beams, and providing increased reflection from small targets. However, air is highly absorbing to high-frequency sound (Griffin 1971; Pye 1983). At 20° C and 50 percent relative humidity, a 100-kHz sound has an attenuation length of only about one meter (see Table 9-3). Clearly, absorption places limits on how high a frequency can be used. On the other hand, absorption can be advantageous in reducing reverberation from uninteresting targets and reducing the range at which prey can detect the approaching bat (see earlier Figure 12-6).

In fact, bats that hunt near foliage and usually detect prey at close range employ higher frequencies than those that hunt in the open and detect prey at long range (Figure 12-11). In addition, they emit sounds that are modified according to the prevailing circumstances. The short, narrow-band, constant-frequency pulses are generally used by bats searching in the open. In proximity to prey or obstacles, the broad-band components with a frequency sweep are added (Simmons and Kick 1983).

The maximum range at which bats can detect moths can be estimated as follows. The sound emitted by many bats has a power as high as one milliwatt (see Table 9-5), and bats' threshold intensity is on the order of 10^{-12} W/m² (see Table 9-4), with moths having a target strength of -42 dB (Miller 1983), equivalent to an effective radius for reflection of 1.6 cm (Table 12-2). Thus, by using equation 12-2,

$$r_{max} = 0.376 \left(\frac{10^{-3} \text{ W}}{10^{-12} \text{ W/m}^2} \right)^{1/4} (0.016 \text{ m})^{1/2} = 8.4 \text{ m}$$

The estimated range at which bats can detect moths is about 8 m, ignoring attenuation. For a typical degree of attenuation at 100 kHz, $\alpha = 1$ m^{-1} and the estimated range is reduced to 2.5 m (Figure 12-6). Over

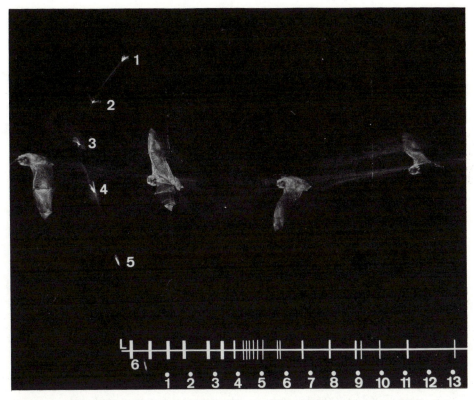

Figure 12-10 A bat echolocating a flying insect. This stroboscopic photograph taken at 50 ms flash intervals shows a bat (*Pipistrellus pipistrellus*) approaching a green lacewing (*Chroysopa carnea*). The scale at the bottom shows a record of the bat's ultrasonic cries, with the numbers giving the corresponding times of the flashes. These cries are typical of those for a bat cruising until just before flash 4, when a single cry typical of the approach phase was emitted. Then a short buzz typical of the terminal phase occurred, between flashes 4 and 6. The numbers 1–6 along the path of the insect show its position at the times of these flashes. After flash 4 the green lacewing has folded its wings and begun falling faster than it had been flying. (Photograph by Lee Miller, Odense Universitet.)

the whole range of conditions encountered by bats, attenuation can range from 0.2 to 15 dB/m (Griffin 1971), or $1/\alpha = 0.2$ to 20m. It is clear that a quantitative understanding of how bats use echolocation must be specific to the atmospheric conditions in which they hunt.

Experimental studies have indicated that bats are able to determine the direction to a target within about 2° and the distance to it to within about 1 cm (Simmons and Grinnell 1988).

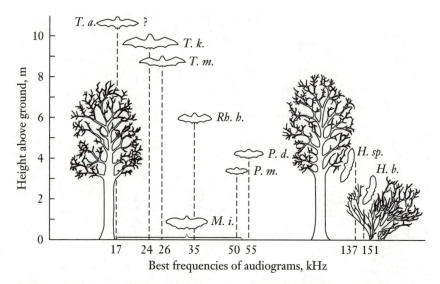

Figure 12-11   Bat habitats and frequencies: some correspondences between a particular species' preferred hunting environment and the frequency at which it is most sensitive. (After Neuweiler and Fenton 1988.)

Bats are thought to perceive information about the nature of a target in terms of "glints" (high intensity reflections) that occur at slightly different times from various distances (Simmons 1989; Simmons and Chen 1989; Simmons and Moss 1990) or from changes in the frequency spectrum of an echo created by different interference effects at different frequencies from a broadband emission (Neuweiler 1983). Fine discriminations about the texture of a target are possible (Neuweiler 1983). Bats apparently recognize flying insects by changes in their glints or in the Doppler shift during the wing-beat cycle (Schnitzler et al. 1983; Kober 1988). Insects that are not flapping their wings are often ignored. It is clearly possible to distinguish one insect type from another by the information in the echo.

Some of bats' insect prey have developed simple detectors for ultrasound and can detect bat vocalizations at distances where the bat cannot detect the echo (Surlykke 1988). When the ultrasonic pulses are infrequent and weak, indicating the existence of a searching bat at some distance, moths use this information by flying away from the sound source, and green lacewings fold their wings and drop, in fact much faster than they can fly, as was seen in Figure 12-10 (Miller 1983). When the bats' pulses are intense and frequent, indicating a nearby bat that is approaching a target, erratic flight is generally initiated. Studies have shown that these evasive maneuvers are successful about half the time.

Bats seem to have countered insects' development of "bat detectors" with various strategies of their own. In the tropics, where prey are present in high density, the so-called whispering bats employ sound emissions that are a thousand times lower than those of most other bats, which reduces the advantage of the insects (see Figure 12-7) (Surlykke 1988). With low-power emission, the range of detection is necessarily shorter so that attenuation has less effect. This frees whispering bats to shift to higher frequencies without suffering a penalty, and as a matter of fact low-power bats tend to be high-frequency bats as well.

WHALES   Apparently all Cetacea—the whales, dolphins, and porpoises—vocalize (Würsig 1989). More to the present point, the toothed whales (the odontocetes), including dolphins and porpoises, are thought generally to employ echolocation. At least one (*Platanista indi*) is blind (Purves and Pilleri 1983; Evans 1988). However, there remain doubts as to whether some toothed whales, like the sperm whale (Watkins et al. 1988), do in fact use echolocation. Detailed studies have been limited to a few species of dolphins, porpoises, and beluga whales whose hearing capabilities seem similar to those of humans except for being shifted to tenfold higher frequencies and perhaps for having greater frequency discrimination (Johnson 1986).

Most of the animals studied use clicks—brief pulses, of minimal duration and bandwidth—with a predominant frequency in the neighborhood of 100 kHz and a lower frequency peak of 20 to 30 Hz (Kamminga 1988; Wiersma 1988). Their sounds are projected in a beam about 10° wide (Au 1988) that can be aimed in different directions (Johnson 1986). The sounds they emit can be extremely intense, up to 220 dB (re 1 μPa at 1 m) (Johnson 1986); it has in fact been suggested that such intense sounds might be used to stun prey (Würsig 1989).

The maximum range at which dolphin might echolocate fish can be estimated from equation 12-2. A 10-cm-long fish or squid has a target strength of −45 dB (Urick 1983, 316–17), corresponding to an effective radius of $r_T = 0.01$ m. Then, using data from Tables 9-4 and 9-5, we can see that

$$r_{max} = 0.376 \left(\frac{10^5\,W}{10^{-14}\,W/m^2} \right)^{1/4} (0.01\,m)^{1/2} = 2{,}000\,m$$

At high frequencies attenuation is significant over such distances, as seen in Table 9-3, and reverberation from other objects such as the surface would hamper detection, as does noise. The maximum actual range of dolphin echolocation for an 8-cm metal sphere has been measured as 100 m (Au 1988). It has also been demonstrated that a single small fish 14 cm long can be detected at a range of at least 9 m, a school of fish at 100 m or more (Würsig 1989).

The distance to a target is probably determined by arrival time. The accuracy of distance estimation is a few centimeters to a distance of 10 m.

Dolphins are capable of identifying the nature of their targets with great accuracy (Roitblat et al. 1990), but little is known of the cues they utilize (Au and Martin 1989). In comparison experiments, humans listening to slowed-down recordings of echoes from dolphinlike sounds reflected off of targets could make similar discriminations. Size differences on the order of 10 percent could be discriminated (Würsig 1989).

HUMANS   Blind people frequently turn to acoustical sounding to acquire information about their environment. In many cases they may unconsciously make use of the sound of their footsteps and be unaware of how they perceive objects (Supa et al. 1944). In other cases the blind deliberately produced sounds, as for example by tapping a cane.

Another situation in which humans learned to use echolocation was reported by the World Soundscape Project:

> [The boat captains] used to get their position by echo whistling. They'd give a short whistle and estimate distance from the shoreline by the returning echo. If the echo came back from both sides at the same time they'd know that they were in the middle of the channel. They could recognize different shorelines by the different echoes—a rocky cliff, for example, would give a clear distinctive echo, whereas a sandy beach would give a more prolonged echo. They could even pick up an echo from logs. Nowadays, if the radar breaks down, they have to put out an anchor. Their ears aren't trained to listen their way through the fog. (Truax 1984, 18.)

12-3 MECHANICAL SOUNDING

In addition to sound waves, other types of mechanical stimuli may be used for sounding. The example of sounding that is perhaps most familiar to humans is the use of a stick to poke at something. Spiders make a similar use of tools for sounding. The webs of some spiders are not sticky and act as extensions of the spider's legs to provide information about mechanical disturbances (Riechert and Gillespie 1986). Some trapdoor spiders employ twigs for the same purpose (Barth 1982). However, animals' use of tools for mechanical sounding is limited. They make much more extensive use of appendages and vibrations.

SOUNDING WITH APPENDAGES

Many animals employ appendages (such as a finger) for sounding, and some even have appendages that have sounding as their primary function. The **antennae** of arthropods are the most dramatic example. Although the antennae usually carry chemoreceptors, they are also used as mechanical probes of the environment. Unfortunately, this behavior is difficult to study, and thus there is not much information about it. Nonetheless, it is probably very important to these animals.

Another type of mechanical probe is **vibrissae** such as the sensory hairs or whiskers of a cat (See Figure 8-17), which are carried by most mammals (Ahl 1986). These highly specialized hairs stick out from the body so that contact with a solid object causes them to bend, which is detected by receptor cells in the follicle at the base of the vibrissa. Each such follicle is highly innervated, and the nerves from each one project to a specific anatomical structure in the sensory cortex of the brain, suggesting that a considerable amount of information is obtained from vibrissae. Experiments have demonstrated that this information includes the amplitude, direction, velocity, duration, and frequency of the deflection (Carvell and Simmons 1990). Deflections of as little as half a degree can be detected.

Vibrissae can be moved by activating striated muscles that are under voluntary control, as distinct from ordinary hairs, which are erected by smooth muscles under autonomic control. In some species the vibrissae move continuously, whereas in others their movement is limited.

Vibrissae are surprisingly important to many mammals. For example, experiments in mice, rats, cats, and seals have demonstrated that vibrissae aid in the final stages of capturing mobile prey. Furthermore, it has been observed in rats running mazes that they suffered more from deprivation of their vibrissae than from deprivation of their senses of vision, hearing, or olfaction. In at least some mammals the vibrissae aid in swimming, protecting of the eye from impacts, and probably in detecting currents.

It has recently been demonstrated (Carvell and Simmons 1990) that rats can use their vibrissae to discriminate surface roughness with a sensitivity similar to that of humans using their fingertips. Like humans, rats continuously move their probing appendages across the surface (Figure 12-12). This is done in an asymmetrical back-and-forth motion having a frequency of about 8 Hz. Some rats can reliably distinguish a smooth

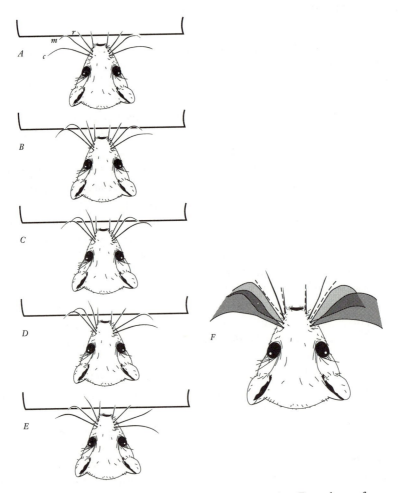

Figure 12-12 Whisker usage by discriminating rats. *A* to *E* are drawn from camera fields taken at 16.7 ms intervals, with only some of the whiskers shown. *F* shows the range of movement of the whiskers. (From Carvell and Simmons 1990.)

surface from a surface with 30 μm-deep grooves 90 μm apart. However, they could not discriminate a surface with grooves half as wide and deep from the smooth surface.

VIBRATIONAL SOUNDING

Another possible strategy for using mechanical stimuli for sounding is to generate a vibration that propagates away and to detect reflections from objects having different mechanical properties. This technique is closely related to echolocation, but in solid media more complex types of waves may be generated (see pages 254–56). This strategy is much used by geologists to gain information about what lies below the surface of the earth, but its use in nature is not so clear.

The animals that gain the most information from mechanical vibrations, excluding sound, are probably spiders, which detect vibrations in the substrate through their eight legs (Barth 1982). Spiders clearly detect vibrations generated by insects caught in a web (Masters et al. 1986). When an object in a web is immobile, a sounding method may be employed in which the spider plucks pairs of adjacent radial threads in the web. If a mass is attached to the web, that will cause echoes in the radial threads adjacent to it, which the spider can apparently use to identify the direction of the mass (Barrows 1915; Barth 1982; Foelix 1982).

Another type of vibrational sounding may be that of whirligig beetles, which swim very rapidly—up to 1 m/s—on the surfaces of ponds (Tucker 1969). When they are swimming at more than 0.23 m/s, they produce capillary waves that spread out ahead of themselves by as much as 6 cm. Typically they swim in erratic patterns that produce rings of waves that propagate out in all directions. It is thought that the whirligig beetle detects objects by the reflection of these waves (Markl 1983), but little research seems to have been done on this hypothesis.

12-4 ELECTRICAL SOUNDING

Vision is frequently only of limited use in aquatic environments, particularly in many terrestrial waters where the attenuation distance for light may be less than the length of small fish. For aquatic mammals echolocation provides a solution, but fish are not sensitive to sounds of a high enough frequency for echolocation to work well. On the other hand, many fish are sensitive to electric fields, as seen in Section 10-3. Furthermore, electric fields are associated with all cells, and in fresh water significant electric fields can be established with only modest cost in terms of the necessary current flow. As a result, it is not surprising that some fish have developed on ability to employ electrical stimuli for sounding.

The best-studied electric fish are those from tropical fresh-water habitats like the Gymnotiformes (knife fishes) of South America and the Mormyriformes (elephant-nose fish) of Africa (Westby 1984). Two species—the electric eel from the Amazon Basin and the electric catfish of Africa—are in fact strongly electric, being capable of developing sufficient electric power (10^3 watts at 10^2 volts) to stun prey. All the other so-called weakly electric fish develop at most a volt or two and are thought to employ electric fields only for sounding and intraspecific communication. There are also a few marine species of electric fish, such as the electric ray (Torpedo).

Sounding fish generate trains of pulses or continuous, nearly sinusoidal, waves of electric current. The period between pulses varies between species from one minute to one millisecond (Westby 1984; Bastian 1986). The electric currents leave or enter the fish near its tail and enter or leave more diffusely near its operculum or gill cover (Figure 12-13). The fish are nearly constant-current sources, and the electrical potentials generated depend on the conductivity of the surrounding water. The electrical potential across the skin, or currents through it, are detected via the electroreceptors spread out over the body's surface in the lateral-line system (see Section 10-3). Objects in the environment that differ in conductivity from the surrounding water cause distortions in the distribution of electrical potential and current along the fish, as shown in Figure 12-13 and demonstrated in the simple model presented in Figure 12-4. Heiligenberg (1975) has made computer calculations for a more detailed model.

The maximum range for detecting targets on the order of a centimeter in size is limited to a few centimeters, never in any case more than 10 cm (Westby 1984). The fish's nervous system seems to be attuned to detect the changes that occur as a target moves along the body, and a fish exploring a target repeatedly swims back and forth along it (Bastian 1986). An exploring fish sometimes also bends its tail around the target or adopts other specific locomotor patterns (Toerring and Moller 1984). Fish that emit brief pulses containing high-frequency components may also be able

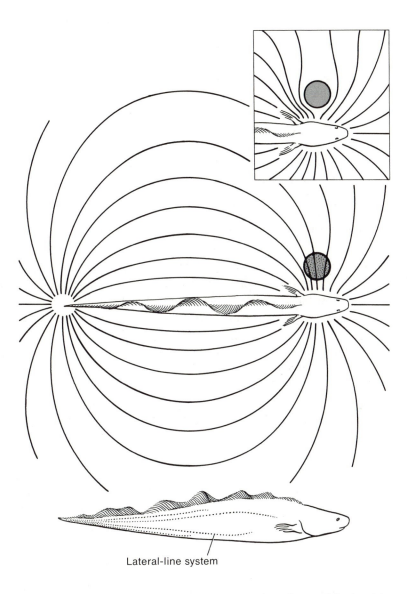

Lateral-line system

Figure 12-13 Electric current distribution around an electric fish. An object that is more conductive than water concentrates the current in its vicinity. An object with conductivity less than that of water (*inset*) diverts current away from its vicinity. (From *Electric Location by Fishes*, by H. W. Lissman. © 1963 Scientific American, Inc.)

to recognize the features of an object that affect its electrical capacitance (Meyer 1982).

In summary, it is likely that fish obtain information analogous to an image of objects close to their bodies. This capability matches their habits of being nocturnally active and hiding during daylight in confined spaces among rocks, roots, and vegetation. The short range that electrical sounding is limited to suggests that it might provide much the same kind of information as might be obtained by sounding with appendages, such as the barbels of catfish, or by sensitivity to flow, which seems important to most fish, as discussed earlier in pages 250–52.

12·5   OPTICAL SOUNDING

The majority of marine fish from a wide range of environments and phylogenetic origins have organs that produce bioluminescence (Herring and Morin 1978; Herring 1987). This light probably serves a variety of functions such as camouflage (see pages 199–201), communication, and as lures or warnings, but good evidence is lacking. One obvious potential use of bioluminescence is as a source of illumination to aid vision when sufficient sunlight is lacking.

This potential has been best demonstrated in the *flashlight fish* (*Anomalopidae*) (Herring and Morin 1978; Colin et al. 1979; Johnson and Rosenblatt 1988), one of the few groups of luminescent fish found in shallow water. There are four species; small and very dark, with a light organ under each large eye, they live around reefs in the tropics. In habits, they are strictly nocturnal, even to the extent of avoiding coming out of their hiding places on moonlit nights. Their light organ is spectacular, clearly able to illuminate the environment even for human observers. These fish seem to engage in optical sounding, but they may also employ their light organs to attract prey and communicate with conspecifics. Similar functions have been suggested for the light organs below the eyes in other groups of fish such as the midshipman (*Porichthys*).

Some fireflies increase their frequency of flashing when they are landing, presumably to help illuminate the landing spot (Buck 1978). Nonetheless, the flashing's most important function is in communication (see Chapter 13), and many species do not exhibit the increase in flashing when landing.

In summary, the use of light for sounding is not yet well documented and is probably limited in occurrence.

12-6 CHEMICAL SOUNDING

Chemicals move only by diffusion or as they are carried by currents of the medium in which they are dispersed. Since currents carry chemicals in one direction only; they offer little opportunity for chemical sounding, except for detecting the current itself. As discussed on pages 288–91, diffusion can provide a basis for sounding, but it is limited to providing a small signal with a long response time. It is thus not surprising that chemical sounding is not widely used. Nonetheless, there have been suggestions of use of chemical sounding by microorganisms, which are too small to implement other techniques.

Many microorganisms release spores from the top of a stalk elevated away from the substrate, independently of the direction of gravity (Bonner and Dodd 1962). This behavior helps the spores get through the boundary layer so that they can be dispersed by the wind (Vogel 1981, 157–62). Because many such stalks bend away from barriers, it has been suggested that the barrier is sensed by some method of chemical sounding.

Sporangiophores of the mold *Phycomyces* have been studied extensively (E. Cerdá-Olmedo and Lipson 1987; Meyer et al. 1987). In these experiments barriers have been detected to distances of several millimeters (Figure 12-14). A wide variety of materials are effective for this, including thin wires (Shropshire and Lafay 1987). The sporangiophores will also bend into a wind. These observations suggest that barriers are detected through their effect of damping air movements (Cohen et al. 1975).

The air currents are probably sensed by chemical sounding. A detailed theoretical model (Cohen et al. 1975) was developed in which a gas was released at a steady rate from the surface of a cylinder, with a fixed probability of adsorption back on the surface. Currents in the surrounding medium decrease the probability of readsorption. With an appropriate tendency for adsorption on the surface of the source, measuring any of the factors—the rate of adsorption, the gas's concentration, or the net rate of the release of the gas—can provide information about currents nearby. The presence of a barrier, which would inhibit currents, can thus be detected without having to have the gas reach the barrier, which explains the lack of dependence on the nature of the barrier. However, no chemical with the appropriate properties has thus far been identified (Shropshire and Lafay 1987).

More recently, a carefully controlled experiment (Meyer et al. 1987) has demonstrated that having air currents is not necessary to avoid barriers. The observations that the strength of a response to a large, flat

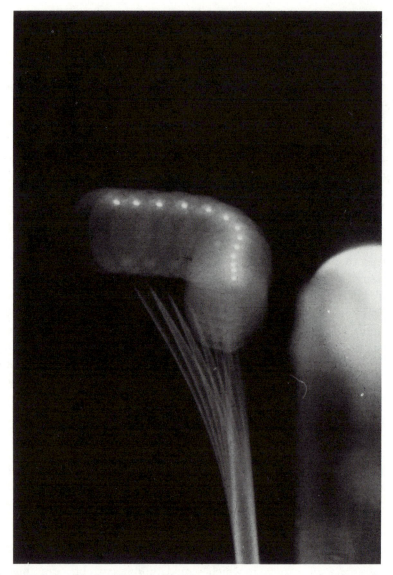

Figure 12-14   The bending of a Phycomyces sporangiophore away from a barrier. This multiple-exposure photograph was made in red light with an exposure every two minutes. The bright spots are glints of light from the light source above. After five exposures, a barrier was placed to the right of the cell. After two more exposures, the cell started bending away from the barrier, moving horizontally much faster than its upward rate of growth. The spherical sporangium, containing spores, is about 0.6 mm in diameter. (Photo courtesy David S. Dennison, Dartmouth College.)

barrier does not change rapidly with distance and that a small fiber is nearly as effective a barrier as a large, flat surface conflict with the hypothesis that the barriers simply reflect a chemical that the sporangiophores grow away from. The most promising remaining hypothesis is that the sporangiophores emit an inert precursor and the barrier surface catalyzes its transformation into a chemical that is released from the surface of the target and avoided by the sporangiophores. It has not yet been demonstrated that in the absence of air currents a wide variety of materials are avoided. Thus, it may well be that either air currents or specific chemical properties will prove sufficient to change the concentration.

With slime mold stalks it has been found that, although they avoid most surfaces, they move toward an adsorptive surface of charcoal, suggesting that the reflection of a chemical from targets made of normal materials is important (Bonner and Dodd 1962). Furthermore, there is evidence suggesting that ammonia, NH_3, is the chemical used for sounding (Bonner et al. 1986; Kosugi and Inouye 1989).

In this case pure diffusion could mediate the response, so that a rough estimate of the distance at which a target could be detected can be made from Equation 12-5 and the available data. The rate of production (J) of NH_3 is about 10^{-14} moles/s, and the observed threshold gradient is about 4×10^{-14} mole cm^{-4} (Kosugi and Inouye 1989). Assuming a typical diffusion coefficient for a small molecule in air from Table 4-3, the maximum range of detection is estimated to be 0.1 mm. This is at least in the same ball park as the threshold distance at which fruiting bodies respond to a barrier, 0.09 to 0.90 mm (Bonner and Dodd 1962).

12-7 THERMAL SOUNDING

Heat is also subject to transport by diffusion and currents. The rate at which metabolic heat is lost from the surface of an organism is influenced by the strength of the surrounding currents and the nature of the material in contact with the organism. The rate of heat loss can be sensed by its effect on the temperature of the surface of the organism. Because of these relationships modifying the flow of heat, important information can be obtained by temperature sensors about the nature of the external environment. For example, humans sense a draft of air by its cooling effects on the skin.

From Table 12-4, one can see that thermal conductivity varies significantly among materials. Thus, when different materials are pressed

Table 12-4 The Thermal Conductivity of Common Materials

Material	Thermal Conductivity $(10^{-5}\ \mathrm{cal\ sec^{-1}\ cm^{-1}\ °C^{-1}})$
Air	6
Wood	40
Water	144
Rock	400

Data from Carslaw (1959, Appendix VI).

against the skin, distinct differences can be determined from their varying effects on temperature sensors. Think of walking in bare feet on a stone floor as compared to a wood floor.

A particularly dramatic example of using temperature to identify the surrounding medium occurs in the **diving reflex** of mammals, in which contact with either cold or heat-conducting materials on the face causes a cessation of breathing, slowing of the heartbeat, and peripheral vaso-constriction (Wittmers et al. 1987). A cooling of the skin of only 5° C is sufficient to initiate such responses (Heath and Downey 1990). This reflex functions to protect the subject from drowning by reducing the chance of inhaling water and reducing oxygen demand in response to stimuli indicating that the face is submerged in water.

CHAPTER 13

Communication

■ ■

Not many sounds in life,
and I include all urban and rural sounds,
exceed in interest *a knock at the door*.
Charles Lamb, *Valentine's Day*, 1819

At present there is no precise, agreed-upon definition of communication
(Myrberg 1981; Hopkins 1988), although it is one of those things we
normally recognize. A common conception is that communication is a
transfer of information in which both the transmitter and the receiver
cooperate and agree as to the meaning of the message transmitted (Hawkins and Myrberg 1983).

Let us consider this description more precisely. Most fundamentally,
biological communication is an interaction between organisms that is
mediated by signals (that is, sensory information) rather than physical or
trophic factors or genetic information (see Table 1-3). Most biologists
would probably agree with this much. However, disagreement occurs in
two situations, because signals mediate an interaction but the transmitting
individual does not cooperate. One such situation is an interaction between organisms in which information is transmitted only incidentally, as
the result of behavior serving some other function, as when one individual
observes feeding activity in another and thus obtains information that
there is food in the vicinity of the other one. The other confusing situation
occurs when misinformation is deliberately transmitted, as in cases of
mimicry or lure. These distinctions were depicted in Table 3-1.

As examples of differences in conceptions of communication, consider
the following much-quoted definition by E. O. Wilson (1970): "Biological
communication can be defined as action on the part of one organism (or

cell) that alters the probability pattern of behavior in another organism (or cell) in an adaptive fashion." Presumably Wilson means to exclude cases in which physical forces or trophic factors mediate the interaction, but to include cases of incidental transmission. The requirement that the interaction be adaptive for the recipient excludes the transmission of misinformation. In contrast, Otte (1974) has gone to great pains to argue that incidental transmission should be excluded from communication, but he included cases of transmission of misinformation. Biologists typically use a definition that excludes the case of incidental transmission but includes the transmission of misinformation (Hawkins and Myrberg 1983; Hopkins 1988).

This book uses a narrower definition that excludes both incidental transmission and the transmission of misinformation (see Section 3-1). This narrow definition is described as **true communication** when it is important to indicate that the transfer of misinformation is to be excluded. The interactions excluded are distinct from true communication because in the excluded cases the signal is constrained to some specific stimulus pattern. For example, in incidental transmission it is constrained to behavior serving other functions, and in mimicry and lures it is constrained to imitate a model (see Figure 2-4). In true communication the signal is simply a *symbol* for which different signals could be substituted, if only both parties to the communication were to cooperate in the substitution. Thus, true communication can be defined simply as an interaction between individuals that is mediated by one or more symbols. Because of the requirement that there be cooperation by both parties, true communication probably occurs only between individuals of the same species, or individuals of different species that have symbiotic or mutualistic relationships (Haynes and Birch 1985) including cases of honest warning signals between different species. (An honest warning is one backed up by real danger, which must involve a physical or trophic interaction.) Because of this requirement, true communication mediated by a chemical would involve a pheromone or synomone, but not the other classes defined in Table 7-1. These two chemical classes could also mediate deception.

To identify true communication among real organisms presents a challenge. It is usually necessary to determine that (1) a particular behavior in the transmitting organism has generated a signal that influences the behavior of the receiving organisms; (2) the behavior is adaptive to the transmitting organism only because of its signal-generating effects; and (3) the response of the receiving organism is adaptive to it.

In the most sophisticated forms of communication a system of symbols is used to create a **language,** which can communicate a wide variety of information. All human societies use languages, which employ many common features, although the sets of symbols vary. As far as is known, no other species uses anything close to a real language. Several attempts have been made to teach languages to chimpanzees, but with little success (Gibbons 1991). The use of language is probably the one characteristic that most clearly separates humans from other animals.

13-1 GENERAL PROBLEMS OF COMMUNICATION

Because communication involves signals that are merely symbols, the parties involved have a choice in principle of what signals to use. Each choice will have its own advantages and disadvantages relative to the alternatives, making it important to consider the general problems to be solved by the choice. This section discusses these problems, along with general strategies for their solution. The problems to be discussed are those of signal recognition, specificity, range, the position of the transmitter, veracity, and privacy in communication.

RECOGNITION

One of the most basic objectives in choosing a signal for communication is selecting one that is easily recognizable as distinct from noise. For example, chemical stimuli are easily recognizable if they have chemical properties that promote specific binding to protein receptors. These needs are met by hydrocarbon skeletons with six to twenty carbon atoms and one or two polar groups, usually including oxygen, at specific locations on the skeleton (see Table 7-5). Peptides containing seven or more amino acids are also effective (see pages 135–40) for communication through water.

The type of visual communication in which the most emphasis is placed on ease of recognition is probably aposematic or warning signals. To function well, warning signals must be quickly recognizable. In addition, it is often important that they be recognized upon their first being experienced and that information for recognizing them be obtainable genetically, a limitation that provides less flexibility than their recognition by learning. These considerations suggest that warning signals should be both simple and distinctive. In fact warning color patterns generally employ only a few colors, in bold patterns that can be easily recognized (Cott 1941). Such colors are often black, with one or two highly saturated colors that are normally uncommon in the environment. Yellow and red are often used; but green is avoided. In dim light color perception and visual resolution are reduced, and visual warnings rely on broad black and white markings, as on the skunk (Lythgoe 1979, 177). It is worth observing that these same features are employed in the design of highway signs that give warnings. The relevant features of auditory warnings are discussed on page 329.

SPECIFICITY

In some communication situations there is pressure to make signals more specific so that a greater number of distinctive messages can occur, as can be seen in the wide variety of sound patterns in human speech. In other animals a similar pressure occurs in situations where many species employ similar signals at the same time and place. Bird songs and moth sex attractants are two well-studied examples.

One of the most significant advances in understanding chemical signals has been the recent appreciation that they often consist of blends of different kinds of molecules (see pages 135–40). Although this fact greatly complicates research, it provides an easy way for organisms to increase the number of potential signals available to them. Even greater specificity occurs in ants, which use tactile cues in combination with pheromones to generate specific signals (Hölldobler and Wilson 1990, 252).

In using sound there are two independent parameters, intensity and frequency, that can be varied to produce specific signals. Variations in intensity—frequently called *amplitude modulation*, or *AM*—are easily produced and perceived, but they suffer from the fact that scattering and reverberation in the environment generate intensity fluctuations that obscure the signal's pattern at long ranges. Another consideration is that abrupt changes in intensity make it easier to localize the source (see "The Position of the Transmitter," below), which may be either advantageous or detrimental.

Variations in frequency—*frequency modulation*, or *FM*—are less subject to degradation by environmental effects and are thus the most suitable for long-range communication (Wiley and Richards 1978). Concentrating the energy in one frequency band improves the signal-to-noise ratio for receivers that have tuned or frequency-analyzing receptors, such as ter-restrial vertebrates (but not insects).

Problems of signal recognition can be ameliorated by repeating the signal or its components. Acoustic signals that are used for long-distance communication in fact do include frequency modulation or amplitude modulation with repetition (Wiley and Richards 1978). A particular tem-poral pattern of some complexity that is often repeated is called a **song.** The many species of birds known as songbirds are most conspicuous to humans by their songs, but it has also recently been discovered that humpback whales have a very long, slow song (Winn et al. 1981). Songs are often composed of repeated elements, each with a particular frequency pattern. An element may consist of a particular frequency produced for a certain length of time or, more commonly, of a specific frequency sweep. The elements may be repeated, with or without modification, a certain number of times to produce a phrase. Often several phrases with distinctly different elements will be combined in the complete song.

RANGE

Another feature of communication signals that is often under pressure to increase is its range for maximum possible separation between sender and receiver. As can be seen in Figure 13-1, there is a strong correlation between the distance, over which communication is effective and the size of animals' territories. An ability to send signals to greater distances could thus help an animal defend a larger territory. An increase in the transmis-sion range can generally be accomplished by increasing the energy of the signal transmitted, but the energy available to do so is limited. For ex-ample, the energy cost of singing in crickets is about ten times their resting rate of metabolism (Prestwich and Walker 1981). This raises the question of which stimulus modalities require the least energy to transmit a signal a given distance in a particular environment. Some relevant calculations are provided in Table 13-1.

For efficient communication the attenuation distance for the stimulus in a particular environment must be equal or greater than the range of communication, if the energy required is not to become astronomical. For example, light is a practical means of communicating in the dark ocean at ranges of less than 10 m but is totally impractical at a range of one km (see Table 13-1). Other considerations include the energy levels that are re-quired to reach the sensitivity of the receptor or overcome noise in the

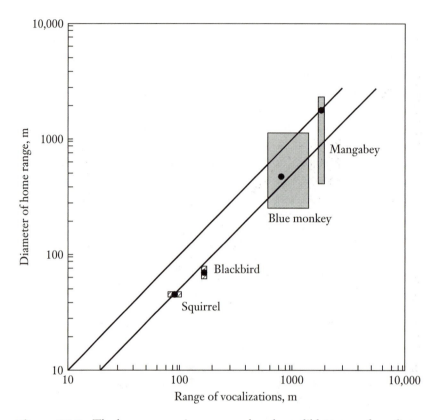

Figure 13-1 The home range size compared to the audible range of vocalizations. The gray areas cover the range of estimates; the black circles indicate
median values. Where the home range was reported by area, the diameter was calculated by assuming a circular home range. The upper diagonal line indicates
positions where the home range's diameter is equal to the range of vocalization,
or approximately sufficient range for neighboring animals to be audible to each
other from average positions in their home ranges. The lower diagonal line shows
positions where the home range's diameter is twice the range of vocalization.
Here there is sufficient range for neighbors to be audible to each other from all
the positions in their home ranges. (Even assuming the territories to be packed in
a square checkerboard arrangement, the calculations are changed very little, compared to the model having adjacent circular territories.) The data here are for thirteen-lined ground squirrels (Brown 1984) red-winged blackbirds, blue monkeys,
and gray-cheeked mangabey monkeys (Wolfheim 1983, 347, 403; Brown 1989).
These same data were used in Brown (1989b, Fig. 4) and Brown (1989a, Fig.
7.18), but his calculations there apparently contain an error, which makes all the
data fit the lower diagonal.

environment. For instance, in sunlight an organism would have to expend enormous amounts of energy to produce enough light to overcome the sunlight, and it would be much more efficient to make use of sunlight by reflecting it off the body's surface in some way that would produce a

Table 13-1 The Energy Required to Send Information by Various Modalities

Stimulus	Assumed Threshold	Energy (J) for Communication Over		
		1 mm	1 m	1 km
Heat/temperature in soil (500 kcal m^{-3} °C^{-1})	10^{-3} °C/cm [6B]	10^{-6}	10^{6}	10^{18}
Chemical (1,000 kcal/mole)				
1 μm receptor in water	10^{9} mol/cm^{3} [7C]	10^{-10}	10^{-1}	10^{8}
1 mm receptor in air	10^{2} mol/cm^{3} [7C]	10^{-17}	10^{-8}	10^{2}
Light (nondirectional receptor)				
In sunlight and atmosphere (L_A = 10 km [8F])	10^{20} photons m^{-2} s^{-1} [8D]	10^{-3}	10^{3}	10^{9}
In starlight and atmosphere (L_A = 10 km [8F])	10^{12} photons m^{-2} s^{-1} [8D]	10^{-11}	10^{-5}	10^{1}
In darkness and ocean (L_A = 10 m [8F])	10^{9} photons m^{-2} s^{-1} [8D]	10^{-14}	10^{-8}	10^{41}
Sound (1 kHz)				
Air (L_A = 1 km [9B])	10^{-12} W/m^{2} [9B]	10^{-18}	10^{-12}	10^{-5}
Water (L_A = 70 km [9B])	10^{-14} W/m^{2} [9B]	10^{-20}	10^{-14}	10^{-8}
Vibration (10 Hz)				
Water surface (L_A = 2.6 cm [10A])	10^{-8} m [10A]	10^{-17}	10^{5}	10^{16698}
Electric dipole				
Fresh water (10^{4} ohm cm)	500 nV/cm [Knudsen 75]	——	10^{-6}	10^{12}
Seawater (20 ohm cm)	5 nV/cm [10C]	——	10^{-7}	10^{11}

The three right-hand columns contain estimates of the energy (in joules, = 0.24 cal) required in the stimulus to send one bit of information over the different distances given and reach the threshold. The brackets [ ] enclose references to earlier sections where these values were obtained.

The model used in these calculations was a pulse release into three dimensions with a quantity of stimulus just sufficient to reach the threshold at the ranges indicated. An integration time of one second was assumed. The temperature and chemicals were assumed to spread purely by diffusion with no flow. For mechanical vibrations the equation in Box 4-1 was used to calculate the energy. The electric dipole was modeled as two 1-cm conductors spaced 10 cm apart, using the equations in Box 13-1, and was thus not appropriate for ranges of less than one meter.

For comparison, typical rates of metabolism are about 0.01 J/s per kg of tissue. These calculations assume 100 percent efficiency in converting metabolic energy into energy in the stimulus. In reality, these processes will have significant inefficiencies, but the calculations illustrate the constraints imposed by the environment.

distinctive signal. This is the strategy that has resulted in our having colorful birds, butterflies, and fish.

Another consideration, and one not included in Table 13-1, is the efficiency of producing a stimulus from metabolic energy. Only heat can be produced with 100 percent efficiency; all other stimuli have some degree of reduced efficiency, according to the second law of thermodynamics. To produce sound in air is a relatively inefficient process, because of the impedance mismatch between air and biological tissues (Bailey 1991, 48–50). For example, the efficiencies of converting muscle energy to sound energy have been estimated as about 1 percent in singing insects and 1 to 5 percent in a frog and a bird (Brackenbury 1982), compared to 5 percent in converting electrical energy by ordinary radio speakers. Many singing insects employ a variety of special structures, in their bodies or the environment, to improve this efficiency (see pages 219–20).

The energy required by humans to generate an acoustic signal can be crudely estimated as follows. The sound power radiated during normal speech is about 25 μW (Meyer and Neumann 1972, 269). At least this much energy must be expended in generating the sound, but much more could be required, depending on the efficiency of converting chemical energy in muscles to sound energy. The work that is done in expelling air can be estimated from the product of the pressure and volume of air expelled during speech. During normal speech the volume of air exhaled ranges from 40 to 200 cm^3/s (Meyer and Neumann 1972, 257). Normal pressures in the lungs correspond to about 10 cm of water or 100 Pa. The product of this pressure and a volume of 100 cm^3/s is equivalent to 0.01 W. Consequently the mechanical efficiency of human sound production is about 10^{-3}. To determine the efficiency of the production of sound from chemical energy (ATP), this efficiency would be multiplied by the efficiency of muscle contraction, which would probably lower it another order of magnitude (Lighton 1987).

The communication range appears to have a close link to the size of the home ranges of a variety of animals (Figure 13-1). For red-winged blackbirds, ground squirrels, and two species of monkeys with audible distances ranging from 100 to 1,800 m the estimated home range diameter was very close to half the audible distance for each species. This correspondence has been explained as enabling contact between adjacent neighbors no matter where they may be on their home ranges (Brenowitz 1982; Brown 1989a; Brown 1989b).

THE POSITION OF THE TRANSMITTER

Another factor likely to affect the relative advantages that different stimuli have for communication is the ease with which the position of the transmitting organism can be determined. In some cases, such as territorial displays or sex attractants, it is beneficial for the transmitter's position to be communicated over great distances.

Visual stimuli are inherently easy to locate, chemical stimuli difficult. Sound sources are relatively hard to localize in aquatic environments (Hawkins and Myrberg 1983), and the ease with which they can be localized in air depends on their temporal characteristics (see pages 241–44). Electric dipole sources are difficult to locate directly, but such a source can be approached simply by following the lines of current flow (Section 10-3).

In other situations it may be important not to employ a signal that can easily be localized, especially if the signals might be used by parasites or predators to locate the transmitting individual (Bailey 1991, 162–63). For example, it has been demonstrated that the songs of crickets attract both parasitic flies and predatory cats (Cade 1975).

A good example of selecting communication signals based on the degree to which they would reveal the position of the transmitting animal occurs in the warning calls of birds (Marler 1959). After seeing a hawk fly overhead, many species utter a warning call that lacks strong transients and contains only a narrow range of frequencies near 7 kHz (Figure 13-2), sounding like a high, thin whistle sometimes described by the word *seet*. These characteristics make the source hard to localize because its frequency is too high (its period 0.1 ms) for the phase differences to be resolved and too low ($\lambda \approx 5$ cm) for diffraction around a predator's head to provide binaural intensity differences (Marler 1955) (see pages 237–38). In contrast, after spotting a perched owl in daylight many small birds produce a mobbing call that attracts other birds. These calls have wide frequency ranges and sharp on–off transients (Figure 13-2).

VERACITY

Another factor influencing the choice of a signal for communication is the extent to which it can be depended upon to be truthful. If a signal can easily be produced, then competitors, predators, or parasites might be able

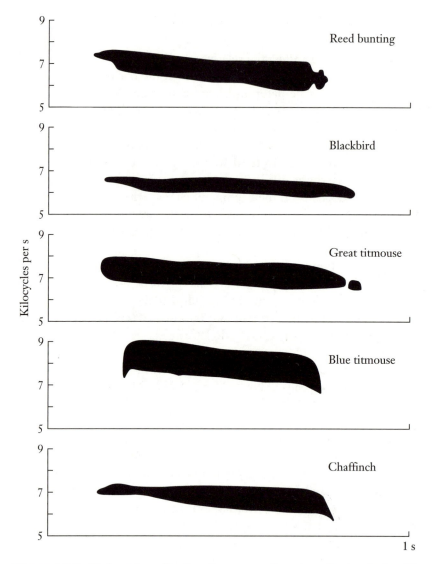

Figure 13-2 Bird warning calls. *Above*: Sonograms of warning calls given by five different species when a hawk flies overhead. Each has a frequency range of only about 2 kHz, and no sharp on–off transients. *Facing page*: Sonograms of the calls of seven species from several families given while mobbing an owl. All contain a wider range of frequencies, from 3 to 8 kHz, and sharp on–off transients. (After Marler 1959.)

to exploit the signal, to the detriment of one or the other of the parties engaged in true communication. For example, moths attracted to a pre-

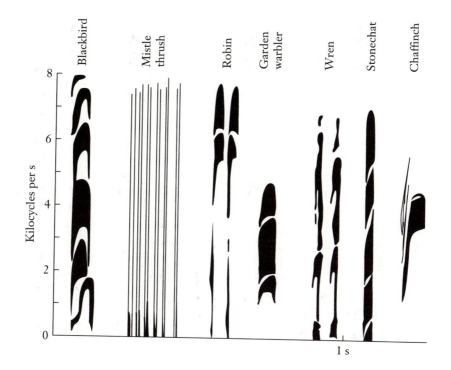

datory spider because it releases a stimulus chemical that acts like the sex-attractant pheromone of the moth (see Section 2-6) might benefit from adopting a pheromone signal more difficult to produce. The common use of specific blends of chemical constituents for communication may, in part, be a strategy to reduce counterfeiting.

Even in situations involving only parties engaged in true communication there may be an advantage to the transmitting organism to exaggerate the signal (Dawkins and Krebs 1978). For example, females usually make a larger investment in each of their offspring than do males, and they benefit from selecting the best available male as a mate. In many species the males compete with one another for access to the females, in a competition that often involves true communication. In some of these species the males communicate their fitness to the females by displays that seem dramatic even to humans. The peacock is an extreme example. In other species males communicate with one another in contests to determine among themselves which will have access to the females. Sometimes these contests involve physical fighting, but often they are settled simply by the displaying of certain specific signals, like antlers or vocalizations.

Whether the receiver of a communication is another male or a female, it is advantageous to the transmitting male to exaggerate any information about his fitness. Such signals are more likely to provide a true indication about the fitness of the transmitting individual if they are closely related to his actual fitness and are difficult to counterfeit (Dawkins and Krebs 1978). Antlers, for instance, may be good signals because weak animals cannot carry around a large set and heavy antlers may be more effective weapons if a physical contest with other males occurs. Similarly, the peacock's outrageous tail may be a burden that weak individuals cannot support. Some displays, such as the roaring of red deer stags, are highly energetic and may become endurance contests. In other vocal signals the volume may be what is important, but in many cases it is low frequency that is most effective (Dawkins and Krebs 1978). This makes sense in that their physical dimensions allow larger animals to produce lower frequencies (see Section 9-1) and the larger males are the more likely to be more fit.

Even where an interaction is purely informational, it will often not be clear to what extent the competition involves pure communication as opposed to an incidental transmission of information. In other words, females or male competitors may obtain direct information about the size and strength of others without that information being converted to a symbol. However, specialized displays such as those of the peacock appear to be arbitrarily chosen symbols of male fitness.

As we have seen, the cooperation between individuals that is required for true communication restricts its occurrence in nature. It most commonly occurs between individuals of the same species and usually has to do with sexual reproduction.

PRIVACY

Another consideration in selecting a communication mechanism may be privacy. Because communication signals are subject to exploitation by competitors, predators, and parasites, there may be pressure to use stimuli and symbols or codes that make communication signals less easy to exploit. For example, insects that employ bioluminescence for communication may often use light in pulses, which make it more difficult than a steady signal to track down the source and also provide additional specificity

(Lloyd 1983). This factor may explain why the New World has few firefly species that glow constantly, whereas the Old World has many—the New World has a large genus of fireflies that are predatory on other fireflies but are absent from the Old World. Even the flashing fireflies may employ complicated codes, to make exploitation more difficult (Lloyd 1983).

13·2 SPECIFIC EXAMPLES OF COMMUNICATION

The remainder of this chapter considers some specific examples of communication.

MICROORGANISM PHEROMONES

Even single-celled organisms have the occasion and ability to communicate. As with higher organisms, the occasion most often involves mating for sexual reproduction. It is reasonable to expect mating to be especially likely to involve communication because mating always requires the cooperation of two individuals and most organisms have no other need to cooperate with another individual. A less common function of communication in microorganisms is to aid the cooperation necessary to form a fruiting stalk that helps disperse spores. One general feature of communication among microorganisms is that the signal is chemical. There do not seem to be any known exceptions.

The general question of the use of pheromones by bacteria has been reviewed by Stephens (1986). Myxobacteria live in groups, or swarms, that secrete hydrolytic enzymes for the external digestion of macromolecules associated with wood or other bacteria. In addition, myxobacteria are unique among bacteria in developing a fruiting body composed of thousands of cells. This process has been considered a simple model system for understanding the developmental processes of multicellular organisms (Dworkin and Kaiser 1985) and has stimulated much research on myxobacteria. Because of this highly organized activity of the individual cells, myxobacteria seem to be the bacteria most likely to employ pheromones. However, none has yet been clearly identified (Stephens 1986).

Other groups of bacteria clearly employ pheromones, some of which have been chemically identified (Stephens 1986). In most cases the information conveyed probably concerns the density of the cells of a particular species in the local environment. The responses controlled by pheromones include production of antibiotics, sporulation, and biolumines-cence. However, it remains unclear whether these are cases of true com-munication or merely of incidental transmission, because the function of releasing the pheromone is not clear.

The best example of pheromones in bacteria is in plasmid transfer in *Enterococcus faecalis*, a common inhabitant of the human gut (Clewell and Weaver 1989). Cells lacking certain plasmids (recipients) produce peptide pheromones, which act on cells that contain a plasmid (donors) to induce the formation of a fibrillar surface material that facilitates the binding of one cell to another. The pheromones also activate machinery required for plasmid transfer. After a recipient cell has acquired a plasmid, synthesis of the pheromone is inhibited. This appears to be a clear example of an informational interaction, but is it true communication?

One question that must be addressed is whether the interaction is adaptive to the receiving organism. Here we face a special difficulty in that what is adaptive for the genes in the plasmid may not be adaptive for the cell's main genome. In many respects the plasmid can be viewed as an infectious agent. Taking this view, we could consider the plasmid to be the receiver in the communication, with the interaction clearly being adaptive to it. This does not therefore appear to be a case of deception.

The other question that needs to be addressed is whether the function of the peptides is to provide a signal. There is not enough known about them to provide a clear answer, but what is known suggests that they serve other functions (Clewell and Weaver 1989). Assuming this to be accurate, the interaction is best classified as one of incidental transmission rather than true communication. It appears that there are no clear cases of true communication among bacteria, perhaps because they do not engage in true sexual reproduction and thus have no strong need for highly specific communication.

In contrast, yeasts do engage in sexual reproduction. In many species, haploid cells exist as one of two mating types or sexes, each releasing a distinctive peptide pheromone (Yoshida et al. 1989). Each peptide acts on the other mating type to inhibit vegetative reproduction and induce the synthesis of a particular sex-specific glycoprotein that contributes to ag-gregation of cells of opposite mating type. Once aggregated, the cells fuse to form diploid heterozygotes, which can undergo meiosis to complete their sexual reproduction. This interaction is a clear-cut example of true communication, because both mating types benefit by sexual reproduc-tion. In some species the haploid cells are not even capable of asexual reproduction.

Another example of true communication among microorganisms occurs in the aggregation of slime mold cells to form fruiting bodies, a behavior discussed on page 272. It is a rather complex interaction in which signaling occurs in both directions over the same channel (cyclic AMP concentration). Since the excretion of cyclic AMP is not likely to serve any other function and the cells involved are in the same situation and engaged in the same behavior, it seems clear that this behavior is one that is adaptive to all parties and thus represents a case of true communication.

It might be well to mention here that there is intriguing evidence that plants sometimes use pheromones to communicate the fact of injury to their neighbors, which can thus devote more resources to producing defensive chemicals (Baldwin and Schultz 1983; Rhoades 1985). Because the chemicals have not been identified, there is no way of determining whether this is an example of true communication or of incidental transmission.

ANTS

Social organisms require extensive communication between individuals of the social group. This dependence on communication reaches an extreme in such social insects as ants. In many respects an ant colony may be regarded as a single organism, with the communication between individual ants seen as analogous to the communication between the different cells and organs of a metazoan organism. Wenner and Wells (1990, 296–98) provide a discussion of the history of this concept.

The subject of communication in ants has recently been reviewed in a spectacular book by Hölldobler and Wilson (1990, 227–97). A dozen types of responses have been identified. An estimated ten to twenty different kinds of signals occur in the typical ant colony. Most of the signals are pheromones, released from at least ten different exocrine glands, which vary greatly between species. Many of the glands produce a blend of chemicals, and the significance of each chemical has not yet been worked out (Bradshaw and Howse 1984). In different species the same chemical will often serve different functions.

In some cases, chemical mixtures can be explained on the basis of individual components eliciting different behaviors (e.g., alerting → attraction → biting) at appropriate ranges controlled by V_T (see page 65) (Bradshaw and Howse 1984). The range of action generally runs from 1 to 10 cm; the chemicals are typically 6–10 carbon alcohols or ketones. Further complexity of the chemical communication within an ant colony derives from the fact that there are frequently differences between different casts in chemical secretion and responses to chemicals.

In many cases it is not even clear whether a particular chemical mediates true communication or a trophic interaction. For instance, many of the chemicals that communicate alarm also have defensive functions (Bradshaw and Howse 1984; Hölldobler and Wilson 1990, 260); that is, they mediate a trophic interaction as well as an informational one. About fifty different chemicals have been identified as communicating alarm in at least one ant species. Ants' known propensity to aggregate is at least partly the result of a tendency to move up gradients of carbon dioxide concentration. Here we clearly have a case of incidental transmission rather than true communication, because carbon dioxide is excreted for other reasons. Trail pheromones are similar in composition to other pheromones but are normally discharged as dilute aqueous solutions, which presumably slows their evaporation. As noted earlier, such trails may persist from a few minutes to a matter of days (Bradshaw and Howse 1984).

Ants often supplement pheromones with vibratory signals (Hölldobler and Wilson 1990, 255–57). For example, ants that nest in wood, such as carpenter ants, respond to a disturbance of their nest by a violent rocking motion in which their bodies strike the substrate, causing it to vibrate. These vibrations can be sensed up to several centimeters away, and alert recipients to protect the nest. Other ants stridulate by rubbing two body parts together. These vibrations permit communication through several centimeters of soil to allow the rescuing of ants buried by a cave-in. In all these cases the vibrations appear to transmit only a single signal, and only one pattern of vibration is recognized.

Tactile communication in ants also occurs, but the amount of information transmitted by it is much less than cursory observation of the complex behavior might suggest (Lenoir 1982; Hölldobler and Wilson 1990, 258–59). Although some ants have large eyes, the existence of visual communication has not been clearly demonstrated in any ant species.

THE HONEYBEE DANCE

Honeybees are among the most convenient animals with which to study behavior: they spend most of their time in a nest or beehive, which can be made with transparent walls, illuminated, and located indoors for ease of observation (von Frisch 1967, 7–11). In addition, they live in a complex society requiring a high degree of communication (Seeley 1985, 1).

It has long been known by people interested in honey that, as with ants, once an individual bee discovers a rich resource like honey other bees will quickly appear (von Frisch 1967, 3; Seeley and Visscher 1988). Apparently, others in the nest somehow acquire information about the location of the resource and are recruited to it by the first bee. One description of this process is often attributed to Aristotle (ca. 330 B.C.), but the meaning of his single sentence is ambiguous (Wenner and Wells 1990, 270–93). More clearly there are old traditions of locating honeybee trees by attracting bees to a piece of honey and observing the direction in which they fly away. John Burroughs described this practice in 1875 (Wenner and Wells 1990, 36).

There seem here to be two distinct items of information involved: that a rich resource is available, and where it is located. Ernst Spitzner described his view of the mechanism in 1788:

> When a bee has come upon a good supply of honey anywhere, on her return home she makes this known in a peculiar way to the others. Full of joy she twirls in circles about those in the hive, from above downward and from below upward, so that they shall surely notice the smell of honey on her; for many of them soon follow when she goes out again. (von Frisch 1967, 6.)

This description suggests two hypotheses: that the existence of the resource is communicated by specific locomotor movements, called the **dance,** and that the location of the resource is communicated simply by following the dancing bee when she leaves the hive again. The latter hypothesis is similar to what is known of ant behavior and applies also to colonies of cliff swallows (Brown 1986). What stimuli bees might use to follow a rapidly flying bee is not clear, but vision was probably what was assumed. However, the previously noted Burroughs description of 1875 mentioned that the recruited bees always arrived many seconds behind the leader. At the speed bees fly, 8 m/s (Wenner and Wells 1990, 304), one bee would not be visible to another if they were even one second apart. In 1924, Bruce Lineburg concluded that the recruits followed a scent trail left in the air when the dancing bee returned to the resource (Wenner and Wells 1990, 55).

The famous German zoologist Karl von Frisch conducted extensive experiments on this question beginning in 1919 (von Frisch 1967, 4) that were finally summarized in a 1965 book translated into English in 1967. Von Frisch believed that scent was the prime indicator of position until 1944 when he discovered a correlation between the orientation of the dance and the direction to the resource (von Frisch 1967, 222; Wenner and Wells 1990, 58).

There are actually two kinds of bee dances. In the simpler, the **round dance,** a bee returning from a successful foraging trip to a source near the hive first gives up her harvest to other bees, then runs in a tight circular path, alternating clockwise with counterclockwise directions, after completing one-half to two circles (Figure 13-3) (von Frisch 1967, 29). The dance may last from a few seconds to a few minutes. It is always performed on a region of the comb that is crowded with bees, and several of these bees usually follow along behind the dancer. The followers seem to be stimulated by the dance to leave the hive and search for the resource, which they can identify from chemical stimuli carried by the dancer, but not by color or shape (von Frisch 1967, 46–50). The existence of the chemical stimuli may be a case of incidental transmission, but the dance itself appears to be a symbol that there is a nearby resource currently available.

The second type of dance, called the **waggle dance,** is produced when the resource is more than about 100 m away (von Frisch 1967, 57). In this case the dancing bee runs in a straight path for a short distance while wagging her body at about 13–15 Hz and producing a buzzing sound, then circles around to the starting point, repeats the straight run, and finally circles around to the other side. The sound has a 250 Hz carrier frequency and is pulsed at 30 Hz. Although the bodily waggles and the sounds that occur are intermixed during the straight run, the sound pulses do not have a fixed relationship to the waggles. Sometimes the curved part of the dance is S shaped and the next straight run starts at a different location, but even so the straight run occurs in the same direction. For resources at inter-mediate distances of 0.1 to 1 km from the hive, the runs alternate between two different directions a few tens of degrees apart. As the distance increases, the angle between the two directions decreases (Figure 13-3) (von Frisch 1967, 61). The dance is sometimes interrupted for a short time, when the dancer shares a sample of her harvest with a following bee.

The length and duration of the straight run have a direct relationship with the distance to the resource (Table 13-2) (von Frisch 1967, 100–101). These and other features of the dance contain information about the distance to the resource. A variety of experiments have suggested that the distance is measured in terms of the amount of energy expended in making the trip (von Frisch 1967, 109–16).

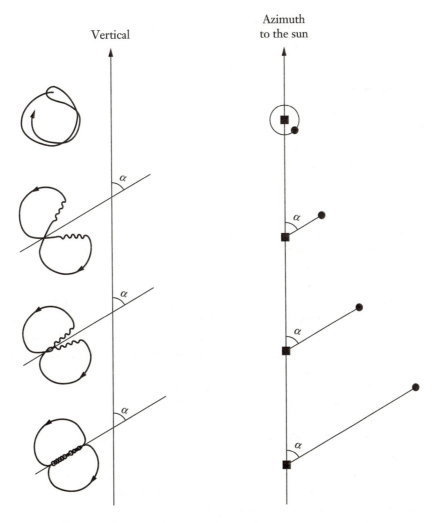

Vertical

Azimuth
to the sun

Dance on honeycomb
(viewed from the side)

Relation of hive, sun, and resource
(viewed from above)

Figure 13-3 The honeybee dance and its correlation with the distance and direction of the resource from the hive. *Left*: The resource at four different distances from the hive, showing the dance pattern. *Right*: The arrangement of the hive (filled square), the sun, and the resource (filled circle). When the resource is close by (*top*), the bees perform a round dance, which contains no information about direction. At greater distances the waggle dance is performed. As the distance increases, the waggle component increases in duration and is more consistent in direction. The average direction of the waggle run makes an angle (α) with the vertical that is the same as the angle between the direction to the resource and the azimuth of the sun.

Table 13-2 Correlations Between the Honeybee Dance and
              Distance to the Resource

Dance Parameter	Distance of Resource from Nest (km)					
	0.2	0.5	1.0	2.0	3.5	4.5
Length of run (comb cells)	—	2–3	3	3–4	4	4–5
Number of waggles per run*	6	12	17	27	38	51
Duration of run (s)*	0.45	0.95	1.3	2.1	3.1	4.0

* Data from von Frisch 1967, 100–101.

In addition, the direction of the straight run has a relationship to the direction of the resource from the nest. Rarely dances may be performed on a horizontal surface out in the open, and, on average, the run is made in the direction of the resource, using the position of the sun or skylight polarization as a directional reference. Dances are usually performed in the dark on the vertical surface of a comb, and the angle between the straight run and the vertical is on average within a few degrees of the angle between the path to the resource and the direction of the sun (von Frisch 1967, 142). The vertical is determined by gravity. The waggle dance can thus be seen to contain information about the direction of the resource as well as its distance (von Frisch 1967, 144–64).

Although this behavior is often referred to as the dance language of bees, it is not sufficiently flexible to meet the usual meaning of language. In fact, it can be argued that this behavior is not even symbolic in the strictest sense, since the waggle run can be viewed as a miniature flight approximating the length and direction to the resource. On the other hand, this dance seems unique among the known nonhuman communication systems in its broadcasting of quantitative information rather precisely. This uniqueness no doubt accounts for the tremendous interest von Frisch's discovery aroused.

The next major question is how much of the information presented in the dance is obtained and used by the recruited bees. Von Frisch obtained data indicating that most recruits search for sugar water within about 100 m, or 25 percent of the distance, of a feeding site at distances of 300–450 m from the hive, increasing to within about 500 m, or 10 percent of the distance, at distances of 4,400 m. With regard to direction, most recruits appeared at stations within about 15° of the direction of the feeding site (von Frisch 1967, 156–61). These amazing results were initially greeted with great skepticism but became rapidly accepted as established scientific dogma (Gould and Gould 1988, 69), and von Frisch shared a Nobel Prize in 1973 with Konrad Lorenz and Nikolaas Tinbergen (von Frisch 1974).

One reason for the ready acceptance of von Frisch's work was that his experiments were successfully repeated by many others (Wenner and Wells 1990, 63). The other major reason was that it was hard to believe that bees would perform dances containing valuable information—and with no other apparent function—but not make use of the information.

In analyzing this dance language it is important to keep in mind that the following bees do not have the bird's-eye view of the dance that human observers have. The two essential questions about the dance are, Do the following bees in fact acquire information about the location of the resource? And what are the signals the following bees receive?

Honeybees usually nest in dark cavities, so vision is not necessary to follow the dance. Perhaps they use touch to determine the movement pattern of the dancing bee. Another possibility is that they may use the near-field vibrations of air discussed on pages 250–52, associated with the sounds, which are produced by the dancer vibrating its wings (Michelsen et al. 1987). The flow of air around the wings approaches velocities of 1 m/s, which would be produced in a far-field sound intensity of 120 to 140 dB SPL. The maximum velocity falls off within a few millimeters of the rear edge of the wings, but the following bees are usually positioned so that their antennae are in the zone of maximum flow velocity. Those to one side of the dancing bee would experience a 12 to 13 Hz modulation of the "sound" intensity caused by the waggles of the dancer's body.

Despite the general acceptance of von Frisch's work, his simple view that bee recruits make use of the scout's information and fly straight from the hive in the direction and distance indicated (at least to human observers) by the dance, is not found in practice. In fact, on average the new recruits take about twenty times longer to appear at a feeding station than the time taken by regular foragers in their "beeline" flight (Wenner and Wells 1990, 147, 177, 224).

Frustrated by unsuccessful attempts to determine the details of what stimuli were actually used by the following bees, Adrian Wenner and colleagues came to question whether the dance communicates anything but the existence of the resource and suggested instead that the resource's location is transmitted only by chemicals acquired by the dancing bee at the resource (Wenner and Wells 1990, 9, 367–69). One problem of interpretation is that bees at a rich resource may help other individuals find it by releasing a pheromone into the air (von Frisch 1967, 50–55; Wenner and Wells 1990, 312–18) or by dispersing hive odor (Gould et al. 1970). Another problem is that several researchers have discovered that a purely chemical stimulus can cause bees that have experienced that specific chemical in association with a particular resource to fly out to check the location of that resource (Wenner and Wells 1990, 125–28). It appears that von Frisch was not aware of this effect, which casts doubt on the interpretation of his experiments. The bees he recorded as new recruits may instead have located the site previously and simply been stimulated to check it again.

The basic problem is that no one knows what the recruits do between the time they attend a dance and when they appear at an observation station in the field. During this interval they could be searching more or less at random until they locate the scent of the resource or of the pheromone released by other bees at the resource. Wenner and colleagues argued that the use of scent should lead the bees to the center of an array of scented observation stations, and von Frisch had usually arranged his stations in a symmetrical array around the resource. (It might be argued that the use of scent should lead bees to the downwind edge of the array, a bias that in fact is often observed [Towne and Gould 1988].) In any case, it is clear that the bees do search and do make use of chemical stimuli. What needs to be demonstrated is that the following bees have more information than a pure search-and-smell strategy could provide. Definitive experiments might be designed to create conditions in which olfactory information and dance information are in conflict, then determine to which the following bees respond.

Does attending a dance make any difference in where a bee searches for a resource? This simple question is difficult to test directly, because individual bees must be marked and dances in the hive must be observed during the same time period recruits are recorded at stations in the field, but a few such experiments have been carried out. Esch and Bastian (1970) obtained the data shown in Table 13-3. This experiment, with only one feeding station, indicated that attending a dance at least alerts a bee to the presence of a resource. Around the same time Gould, Henerey, and MacLeod (three undergraduates at Caltech) conducted a series of experiments with two observation stations at an equal distance but in opposite directions from the hive and took elaborate precautions to reduce odor cues (Gould et al. 1970). Some of their results are shown in Figure 13-4. These observations indicate that the dance does in fact impart some directional information. Recruits arriving at a station within fifteen minutes of having been observed attending a dance arrived preferentially

Table 13-3 The Effect of Dance Attendance on Searching*

Found Station	Attended Dance	
	Yes	No
Yes	14	0
No	20	36

* The success of marked bees in finding a station with a familiar scent but in a new location. The station was frequented by only a single foraging bee. When the foraging bee did not dance, no recruits were observed at the station. From the data shown, bees attending dances by the forager were more likely to appear later at the observation station. The flight time of the recruits from the hive to the station was three to six times that of the forager. Data from Esch and Bastian (1970).

at the station indicated by the dance by a 3:1 ratio (21/7). (The chance of achieving a result this good from random choices is only 0.006.) The few recruits arriving later, from 23 to 75 minutes after attending a dance, showed no preference. This seems like clear evidence that a few recruits flew more or less directly to the station indicated by the dance. However, the researchers observed 277 marked bees attending dances, so only a few (1 to 5) percent of the attendees performed as well as implied by von Frisch's descriptions. Also it is desirable to have more observations; since this data is selected for its clear implications, a probability of 0.006 of occurrence by chance is not very comfortable. An additional problem is that these observations leave open the possibility that the dancing bee might have picked up odors from its surroundings at the foraging site that then provided information to the following bees about where the dancer had been.

A still more definitive experiment is to alter the dance experimentally so that its information would lead the followers to search in a location different from where the dancer had actually been. Such "misdirection" experiments have been attempted by forcing the dancing bees to walk to the food station, which causes them to dance in a way that indicates that the location is much farther away than if they flew. Unfortunately, technical problems cloud the interpretation of this experiment (Gould et al. 1970). A more recent attempt made use of observations that the bees would accept a bright light as the sun and orient their dances to it, and that painting over the ocelli of individual bees would cause them to require a brighter light before they would orient their dances to it. In the presence of an artificial sun of an appropriate intensity painted bees would thus use gravity as their directional reference, whereas unpainted bees would use the light. This should cause the painted

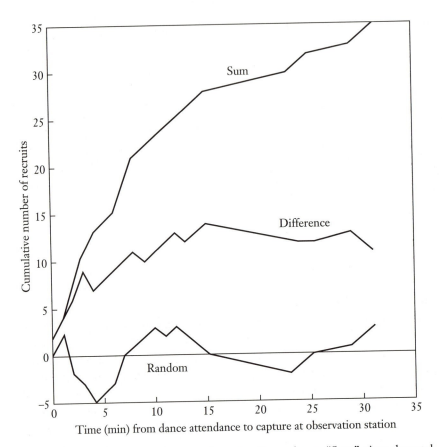

Figure 13-4 Specific recruitment after attending a dance. "Sum" gives the total number of recruits captured at both of two observation stations, at equal distances but in opposite directions from the hive. "Difference" is the number of recruits captured at the station indicated by the dance attended minus the number captured at the other station. "Random" is an example of what the difference would be like if there were no preference shown for one station over the other, as drawn from a table of random numbers. For perfect performance the difference would coincide with the sum; if no preference, the difference would fluctuate around zero, as do the random data. For the first minute or two after the recruits attended a dance their performance was perfect (4 vs. 0), and for the first 15 min there was an excess of recruits at the station indicated by the dances (21 vs. 7). At longer intervals the recruits were equally likely to arrive at either station, as indicated by the zero average slope of the difference line. The rate at which recruits arrived at either station also dropped after 15 min. (Data from Gould et al. 1970.)

bees to misinterpret the dances of the unpainted bees and vice versa. These experiments demonstrated dramatically that moving the position of a light near the hive could alter which station in a fixed array obtained the most recruits in a predictable manner (Figure 13-5). This outcome clearly demonstrates, at least in principle, that the dance communicates positional information aside from any chemical stimuli. However, critics see technical flaws in the technique (Wenner and Wells 1990, 242–43, 250, 252–53).

The ultimate in rigor for testing the dance-language hypothesis is to build a physical model of a dancing bee and test whether it can cause following bees to search in the appropriate location around the hive. Several attempts have been made to do this, with some success recently reported (Michelsen et al. 1989). Although the mechanical model was relatively crude, it seemed to direct the bees in the expected direction and influenced the distance of their search for food in the proper sense. Eliminating wing vibrations or wagging caused a reduction in their searching and a loss of preference for the predicted direction. These observations are good controls against a purely olfactory mechanism. Further refinement of this model and experiments with it should go a long way to elucidating the dance language of honeybees.

In spite of this evidence, many people who have extensively reviewed the conflicting research (Ohtani, Rosin, Taber, Wells, Wenner) remain unconvinced that the dance language conveys direction and distance information (Wenner and Wells 1990, 235–52). As an alternative, Wenner and Wells (1990) have proposed a detailed model in which the dance communicates only the existence of the resource and chemical stimuli provide the positional information. Their model includes two features that push beyond the known limits of insect abilities. For one, they propose that dancing bees pick up chemical stimuli that are characteristic of the neighborhood around the resource, in addition to the chemical stimuli associated with the food. And they also propose that having numerous foragers visit a particular feeding station establishes a chemically defined trail in the air, which new recruits can use to help find the resource.

In summary, we are left with a choice between two surprising hypotheses: (1) that bees exhibit complex behavior in a dance that contains useful positional information but the information is not used by the bees, and recruits find new (to them) resources on the basis of random search and an amazing ability to make use of chemical stimuli; or (2) that bees have a remarkable dance language that is without known parallel among

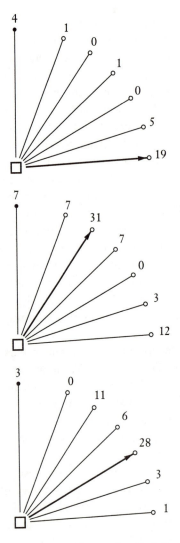
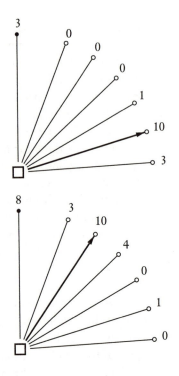

Figure 13-5 The results of misdirection experiments. Foraging bees were trained to a feeding station (filled circles) 150 m from the hive (squares) and their three ocelli were painted over so that they oriented their dances to gravity rather than toward an artificial light. In contrast, the recruited bees used the light, representing the sun, as their point of reference. The light's position was adjusted so that if the recruits used the direction of the waggle run to determine the direction in which to search for a feeding site, they would be directed to one (arrows) of the six observation stations (open circles). The numbers show how many recruits were captured at each station as reported in Gould (1975). The three experiments on the left were carried out on a different day and with a different scent than the two on the right.

other animals. A good guess is that the truth is some combination of these two hypotheses, but the question certainly deserves more study. Further progress probably depends on developing new techniques. For example, it would be revealing to be able to track the searching bees in the field; perhaps new electronic technology will make this possible.

Whatever the mechanism bees use to locate resources, it can be argued that their accuracy is as high as is useful to them (Towne and Gould 1988). The idea here is that most natural resources are spread out, so that it is not adaptive for the recruited bees to go to exactly the same spot the dancing bee visited. If the recruitment system is to be fine tuned to this consideration, the following bees should be scattered at equal distances in all directions from the resource visited by the dancing bee. This scattering should cover the same area, no matter how far the resource is from the hive. When Towne and Gould (1988) tested this prediction they obtained data suggesting that errors in distance and direction in fact produced scattering that was spread over the same extents along a radius from the hive (distance errors) as along a circumference concentric with the hive (direction errors), and found that the extent of the spread did not change much with the distance from the hive. If a resource's position is indeed communicated by the dance, this constancy would require increased precision in how distance is communicated as distance to the resource increases, which in fact occurs as the transition from round dance to the waggle dance (Figure 13-3).

We might now ask how much information is actually communicated to the following bees. In the experiment described above, Towne and Gould (1988) used many more stations than had von Frisch and found that the standard deviation of the distance from the resource to the station that was entered by the following bees was 50 to 100 m for distances from the hive of 100 to 700 m. Unfortunately, their array of stations was not extensive enough to clearly define the whole distribution, and von Frisch's stations were too far apart. Thus, there are still not good data on the accuracy of recruitment, but it appears that a follower bee is directed to an area that is roughly 100 m in diameter. Within 1,000 m of a hive, there are 400 such areas. Consequently, it would take 9 ($2^9 = 512$) bits to specify this number of choices. The actual number could be much higher, however, because the dance can direct bees out to 10,000 m (von Frisch 1967, 70), which includes 100 times more area than within 1,000 m. If we assume the same level of absolute precision, there would be 40,000 areas to be specified, which would require about 15 ($2^{15} = 32,768$) bits. Thus, with the present state of knowledge we can say that the recruitment process, whatever its mechanism, probably involves the communication of ten to fifteen bits of information—about twice the amount of information estimated by Wilson (1962) using older data. By comparison, he estimated that there were two to eleven bits communicated for trail following in fire ants.

FISH

The problem of communication in an aquatic environment has been well reviewed by Hopkins (1988). In water the diffusion of chemicals is ten thousand times slower than it is in air (see Table 4-3). As a result, chemical communication in aquatic environments can occur only over short distances, unless currents are employed. Accordingly, fast-swimming fish probably have little use for chemical communication. Similarly, the attenuation of light is about a thousand times greater in water than in air (see Table 8-6). Even the clearest water has a transmission range for light that is equivalent to that of a dense fog. Hence, aquatic communication by vision is limited to ranges of a few meters and in turbid water is nearly useless.

In contrast, sound travels farther in water than in air, by a hundredfold at 1,000 Hz (compare Table 9-3). Thus, sound is probably of greater importance underwater than in air, a condition that has only recently been appreciated.

Aquatic invertebrates with hard body parts produce a great deal of sound (Hawkins and Myrberg 1983). Much is probably generated incidental to other activities, but some of these animals, like spiny lobsters and snapping shrimp, are specialized in ways that appear to have a sound-producing function. Nonetheless, it is unclear whether these invertebrates can detect sounds, and little is known about the function of their sound production.

In contrast, many species of fish are known to detect sounds with great sensitivity, and some species also produce sounds (Hawkins and Myrberg 1983). The sound-producing mechanisms vary widely and have presumably evolved recently. The sounds usually have low frequencies, below 1,000 Hz, matching the sound sensitivity of fish. The sound signals made by different species are usually distinguished on the basis of their duration and intervals between pulses. Sound frequency is, however, of relatively little importance, and fish have poorer frequency discrimination than do terrestrial vertebrates.

Some species of fish clearly employ sound for true communication between the individuals of a mating pair (Hawkins and Myrberg 1983). Usually the male produces a sound to attract a female, but males also employ sounds to communicate with other males, to help defend a territory. Most aquatic mammals probably make extensive use of sound for communication, but it is difficult to study them, and therefore relatively little is known. This subject has been reviewed by Hawkins and Myrberg (1983).

The use of sound by fish for purposes of deception has thus far not been observed (Hawkins and Myrberg 1983). In contrast to the common occurrence of visual lures and mimicry, it may be that appropriate sound-producing structures are simply too difficult to develop.

As described earlier (page 219), sound does not propagate well in shallow water, which limits its usefulness for species that live in shallow-water environments. In very shallow turbid water the most effective means of communication is to use electrical signals. Although it was discovered only a few decades ago, electrical communication has attracted much research interest, because such signals are relatively easy to record and reproduce, which facilitates experimentation (Hopkins 1988a).

Although electrical sounding is limited to ranges of only a few centimeters (see Section 12-4), electrical communication occurs over distances of about a meter (Hopkins 1986). At typical communication distances the electric field around a discharging fish resembles that of a dipole source. Under these conditions the electric field falls off with the third power of the distance and the required energy increases with the sixth power of the distance (Box 13-1). It is thus very costly in energy to increase the range of communication.

An electrical signal has two principal features that convey information (Hopkins 1988a). The basic unit is the electric organ discharge, which has a waveform and duration that are determined by details of the anatomy and physiology of the electric organ. The nervous system controls only the occurrence of the discharge. The intervals between discharges thus become prime features that can be modulated for communication. However, some features of such a discharge change during development, and they often differ between the sexes. In at least some species these differences are under the control of sex hormones and seem best explained as signals that convey information about the transmitter. It has also been demonstrated that fish can in fact discriminate different discharge waveforms.

In some species of fish the interval between discharges (a few milliseconds) is constant over long time intervals, and information is encoded in the frequency of the discharges (Hopkins 1988a). It has even been demonstrated that a sinusoidal electric field of the proper frequency can substitute for the complex pattern of a particular fish discharge. The information

---

Box 13-1 The Cost of Generating an Electric Dipole

What is the cost in energy of generating an electric dipole strong enough to be detected at a given distance? In order to make the calculations practical, let us model the transmitting organism as two conducting spheres of diameter d separated by distance s, with electric current I between them. The receiver, at distance r, has a threshold electric field sensitivity of $\mathcal{E}_{Th}$, and the surrounding medium has resistivity ρ. The required electric dipole, $I s$, is calculated as

$$I s \approx \frac{2\pi}{\rho}\,\mathcal{E}_{Th}\, r^3 \qquad r \gg s$$

on the axis and twice as large perpendicular to the axis (Knudsen 1975, Eq. 9).

The energy, E, expended in an electric current is

$$E = I^2 R$$

The electrical resistance, R, between two small conducting spheres ($d \ll s$) is (Jeans 1941, 352)

$$R = \frac{\rho}{2\pi}\left(\frac{2}{d} - \frac{1}{s}\right) \approx \frac{\rho}{\pi d} \qquad d \ll s$$

Combining these equations, then, we have

$$E \approx \frac{2\pi\,\mathcal{E}_{Th}^{\,2}\, r^6}{\rho\, s^2 d}$$

This sixth-power dependence on the range of communication, which would occur in any model having a fixed R, means that a doubling of the range would require a sixty-fourfold increase in energy.

---

communicated includes that of species and sex recognition and dominance status. Courtship often involves more complex signals such as warbling or even songs composed of patterns of frequency changes.

In species that provide more variable intervals, information is encoded in changes in the frequency of the discharges. For instance, during aggressive encounters between conspecifics, warnings are communicated by brief silent periods in some species, brief increases in frequency in others. The fact that opposite changes are used to communicate similar information in different species is an indication that the signal is in fact a symbol and shows that true communication rather than incidental transmission is occurring.

BIRD SONGS

The example of animal communication most readily apparent to most humans is bird songs particularly where small songbirds nest. A species-specific song pattern is often produced as a display showing that a territory is occupied by a particular species. Species that have dramatic songs in fact tend to be without a dramatic appearance, and vice versa (Baylis 1982). This type of vocal communication is particularly easy to study, because territorial males will usually respond to a recording of a song appropriate to their species, which facilitates experimentation (Becker 1982).

A number of general principles have emerged from studying some two dozen bird species (Becker 1982); a sample is shown in Figure 13-6. The intensity pattern of songs is unimportant, which is understandable because intensity fluctuations are common as a result of changing diffraction (see pages 231–34) when one of the parties or some vegetation or air moves. Similarly, removing either high or low frequencies usually has little effect, because in any given situation some frequencies are likely to be transmitted better than others.

In most species only certain features of a song are important for its recognition, and they vary relatively little between repetitions and individuals. This specificity permits other features to be used to encode information about an individual's identity or motivational state. The critical features in species identity often involve frequency contrasts or sweeps, or specific intervals between elements. Sometimes birds discriminate between songs that sound quite similar to humans, or appear similar on sound spectrographs.

Fifteen to 20 percent of bird species are considered to mimic the songs of other species, leading to much speculation about what functions this behavior might serve (Baylis 1982). A few species mimic aggressive or predatory species, probably as an aid in defense. More commonly, mimicry may aid in excluding the model species from a territory. This process may involve deception, but when the singing bird belongs to an aggressive

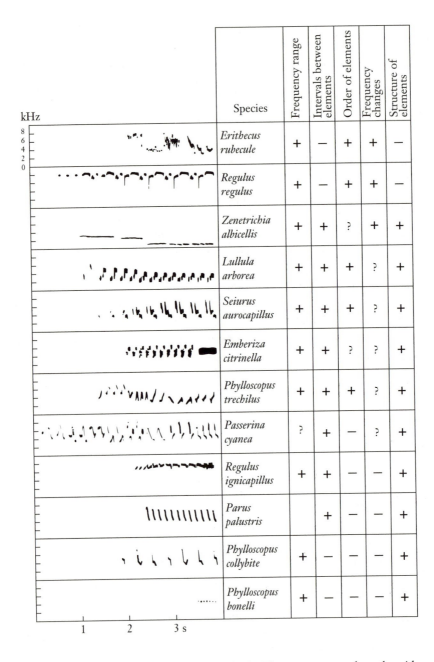

	Species	Frequency range	Intervals between elements	Order of elements	Frequency changes	Structure of elements
	Erithecus rubecule	+	−	+	+	−
	Regulus regulus	+	−	+	+	−
	Zenetrichia albicellis	+	+	?	+	+
	Lullula arborea	+	+	+	?	+
	Seiurus aurocapillus	+	+	+	?	+
	Emberiza citrinella	+	+	?	?	+
	Phylloscopus trechilus	+	+	+	?	+
	Passerina cyanea	?	+	−	?	+
	Regulus ignicapillus	+	+	−	−	+
	Parus palustris		+	−	−	+
	Phylloscopus collybite	+	−	−	−	+
	Phylloscopus bonelli	+	−	−	−	+

Figure 13-6 A variety of song patterns in birds. These sonograms show the wide variety of song patterns among birds, with a classification of the characteristics that have been found important in pattern discrimination. (From Becker 1982.)

species such as the mockingbird even a crude mimicking of the receiver's song may serve to gain the receiver's attention for a true warning.

Another interesting possibility for how mimicry may function is to provide information about a singing male that is valuable in mate selection. For example, the size of an individual's repertoire may correlate with its age and that in turn with its fitness. Alternatively, in migrating species identifying the songs mimicked might provide information about where an individual had wintered, which might be important in mate selection (Baylis 1982).

For communication over maximal distances certain features of auditory displays by birds can be recognized. Territorial singing, for instance, is most common in the early morning when the atmosphere is the most stable, which reduces the scattering and reverberation of sound (see pages 231–34). In many situations low frequencies are desirable because they may suffer less attenuation (see Figure 9-15). However, the small size of birds limits their ability to produce low-frequency sound, so that some have adopted special mechanisms for producing sound. For example, grouse use their wings to generate very low frequencies (Wiley and Richards 1982), and woodpeckers often drum on hollow trees or other resonant structures.

Finally, it should be mentioned that about twenty years ago it was discovered that humpback whales produce a reproducible pattern of sounds resembling bird songs, except they are lower in frequency and on a slower time scale (Winn et al. 1981). A single song may last from 6 to 30 minutes. The same song is produced by all the individuals at a given location—usually the calving grounds—but the song pattern changes from year to year and differs between ocean basins. The function of the songs is not clear, but they presumably serve in some way to maintain social contact.

Exploiting Spatial Goals

■ ■ ■

The most common function of information obtained from the environment is to help an organism move to the location of a resource. Such a location is conveniently referred to as a **spatial goal,** but without intending any implication concerning mental processes, as is sometimes ascribed to the term *goal.* Part 4 concentrates on strategies for traveling to spatial goals. The first chapter discusses terminology and the classification of the basic behaviors. The succeeding chapters describe strategies and examples of traveling to spatial goals, then conclude with a chapter on migration.

CHAPTER 14

Classifications of Behaviors Related to Spatial Goals

■ ■

He that thinks with more subtlty
will seek for terms of more nice discrimination.
—Samuel Johnson, *The Idler*, 1759

Organisms often benefit from obtaining information about the location of some resource such as food, a mate, or a favorable place for reproduction. To achieve that benefit it is usually necessary to move to the location of the resource. Jander (1975) calls this behavior "object orientation" and points out that it is the most universal type of behavior. Here such behavior is described as one directed toward a spatial goal.

14-1 PATHS TO SPATIAL GOALS

Behaviors involving movement toward spatial goals are frequently described with terms such as *dispersal, homing,* or *migration.* These terms are not very specific, being sometimes defined on the basis of patterns of movement (that is, tracks), sometimes by the function of the movement. This section discusses the basic patterns of movement and attempts to make clear distinctions between them, but it should be understood from the beginning that we will not arrive at a rigorous classification scheme.

One of the most common occasions for an organism to move toward a spatial goal is for it to find a place to live apart from its parents and siblings. This problem exists for microorganisms and plants as well as animals—for all, the fundamental need is to get away so as not to compete with relatives. There may thus be little need to direct the spatial movement, which is likely to occur in any direction. This is the concept of **dispersal.** In its pure form, information is not necessary for guidance during dispersal, but it may be used to gain a free ride with a current or some other agent, or to maintain a straight course so as to maximize the distance moved for a given effort, through what is known as a collimating stimulus. Dispersal occurs when the tracks of organisms depart in arbitrary directions from their origins (Figure 14-1). Most organisms have some means of dispersal, but their behaviors leading to long-range dispersal are often dependent on their receiving information about the quality of the present environment. In particular, information indicating a high density of conspecifics often promotes behavior leading to dispersal to great distances.

The next most common occasion for an organism's moving to a spatial goal is to make use of a specific resource such as food or a mate. This type of behavior I call **hunting.** In this case the goal consists of one of a set of specific objects, which will be of interest to other, similar individuals. The tracks of different individuals may therefore converge on the same location, as seen in Figure 14-1.

Probably the next most common occasion for moving to a spatial goal is to return to a personal goal such as a nest, burrow, or roosting site. This type of behavior, called **homing,** may be recognized by observing that the tracks of different individuals converge on different locations (see Figure 14-1). This behavior has been much studied, because the homing drive is often quite strong and the goal is well defined; both factors aid in experimentation.

The last type of movement is **migration.** Unfortunately, biologists use this term with a variety of meanings (Johnson 1974; Baker 1978; Aidley 1981). The most consistent element in all the uses of the term is as a reference to situations in which many individuals—often all the individuals of certain stages in a population or deme—move in the same direction, not necessarily together but apparently toward a common spatial goal, which may be very diffuse such as a geographical area. This is the meaning illustrated in Figure 14-1. Migration thus applies to a herd of caribou moving toward their calving grounds, a flock of geese flying south in the fall, the movement of locust swarms, the seasonal movement of independent monarch butterflies, and even to the chemotaxis of a population of bacteria in a chemical gradient.

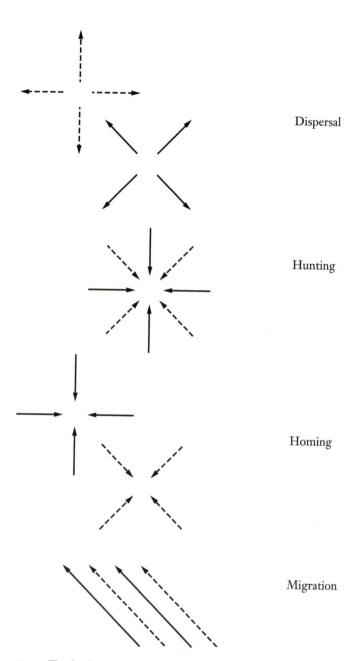

Figure 14-1   Tracks directed toward spatial goals. The solid and dashed arrows represent tracks followed by two different individuals. Different arrows of the same type indicate tracks from different starting positions or equally likely tracks.

Most behaviors related to spatial goals, with the exception of dispersal, are aimed at specific locations, and certain strategies may commonly be involved whether the path is one of hunting, homing, or migration. In order to analyze these strategies it is useful to define the common phases in the process of exploiting specific spatial goals, which is the subject of the next section.

14-2   PHASES OF EXPLOITING SPECIFIC SPATIAL GOALS

The overall process of finding and using a specific resource can be divided into several distinct phases, the first being **search,** in which the animal has no information about the location of any specific goal. The type of search conducted when goals of equivalent value are equally likely to be found anywhere within the relevant area is here called **ranging.** This is an idealization. In many real situations animals have information, whether learned or inherited, that a goal is more likely to be found in certain stimulus situations, leading them to concentrate their search in those areas. Nonetheless, this idealization remains a useful concept with which to start analyzing actual behavior.

In other situations an animal may have information that a goal is nearby, or a resource found nearby may be more valuable than one farther away. The latter occurs when the searching animal must return home. In either case, it would be appropriate for searchers to concentrate their efforts locally, a behavior called **local search.** Some strategies for both ranging and local search are discussed in Chapter 16.

If it is successful, the ranging or local search phase leads to the **detection** of a signal providing information about a specific goal. Such a goal might either be sighted or have its scent picked up. The range at which detection can occur is of particular interest. Some estimates for particular stimuli were presented in the appropriate chapters in Part 2.

After a stimulus providing information about the location of a specific goal has been detected, the **approach** phase begins. This stage might be as simple as bacterial chemotaxis, as complex as a dog following a scent trail, or as indirect as a bee using the sun as a compass. Approach might be complicated by the need to remain undetected, as in stalking prey. Because approach strategies are a main focus of interest in sensory ecology, they are classified in the next section and discussed in detail in Chapters 17, on guiding, and 18, on navigation.

After a goal has been contacted, there may be a need for **confirmation** of its identity. Tasting fills this role when the goal is food, by providing information that would otherwise not be available about whether the goal is indeed edible. For example, insects frequently use their antennae and mouth parts to confirm a potential food resource (Miller and Strickler 1984).

When the goal is finding a mate, specific behavioral sequences involving both individuals (mating rituals) are common. These behaviors may serve several functions, such as physiological priming of the pair for mating, or providing nutrients to the female, but surely one purpose is to ensure that two individuals of the same species but opposite sex are involved. Mating rituals take a wide variety of forms, but they invariably involve an exchange of signals between the parties. In insects these signals may be olfactory, visual, mechanical, or auditory. In birds, vision predominates. In humans vision and touch seem most important. And although this subject has been less well studied, insects' choice of egg-laying sites often involves a complicated series of behaviors that probably serve to confirm that the site is appropriate (Miller and Strickler 1984).

Finally, behavior directed at a specific spatial goal is completed by **consummation,** which may involve eating, mating, egg laying, and so on. In any case, the behavior is usually complicated and specific so that little of a general nature can be said about it. Because of this specificity the confirmation and consummation phases are not discussed further in this book.

The various phases of attaining a specific spatial goal are summarized in Table 14-1, along with an indication of the most important considerations relating to sensory ecology.

These phases may be seen in Tinbergen's classic study of the wasp called the bee wolf (Tinbergen 1972, 130ff). After opening its nest and

Table 14-1 The Phases of Exploiting Specific Spatial Goals

Phase	Principal Consideration
Ranging	Pattern of movement
Local Search	Pattern of movement
Detection	Range of detection
Approach	Stimuli used, efficiency
Confirmation	Stimuli used
Consummation	Stimuli used

starting to retrieve a previously killed bee, the wasp flies in loops in a local search. Upon its flying downwind of the bee (within ≈ 1 m), detection occurs and it then flies upwind on the approach. Upon contact confirmation occurs and it grabs the bee and carries it to the nest in the consummation phase.

Somewhat different terms and ways of classifying phases of exploiting resources have also been proposed. Miller and Strickler (1984) provide a description of the phases employed in the optimal foraging literature and propose an improvement for use with insect–plant interactions. In Table 14-2 these two classifications are compared to the one proposed above.

It may be seen that the classification proposed here corresponds to the most detailed phases of the other schemes and also subdivides the search phase into local search and ranging, depending on the amount of information available to the organism. Some of the terms proposed under Sensory Ecology correspond in their meaning to different terms in the other schemes. Reasons for introducing the new terms include the following. In the optimal foraging literature *encounter* is used in ambiguous ways (Miller and Strickler 1984), but *detection* specifically indicates the appropriate sense of obtaining new information. On the other hand, *pursuit* in its ordinary meaning is too specific, so that *approach* is preferred. Similarly,

Table 14-2 A Comparison of Terms Used to Describe Phases of Exploiting Resources

Sensory Ecology*	Optimal Foraging†	Miller and Strickler‡
Ranging	Search	Finding
Local Search		
Detection	Encounter	
Approach	Pursuit	
Confirmation	Handling	Examining
Consummation		Consuming

* Terms proposed in this book.

† Terms used in the optimal foraging literature.

‡ Terms suggested by Miller and Strickler [1984].

examining is more suggestive of a specific mechanism than the other terms and is not appropriate for mating rituals; thus, *confirmation* is preferred. In the same vein, *consuming* suggests the destruction of the resource, which is inappropriate for mating or egg laying, making *consummation* the preferred term.

In reality, the process of exploiting resources is likely to be more complex than the sequence of phases presented here. Often, more than one type of stimulus will aid in the approach phase, and there is likely to be a series of detections of stimuli, providing ever more precise information. For example, when the bee wolf hunts bees it initially responds to moving objects by using its vision, but hovers 5 to 20 cm downwind to test for the appropriate scent, then pounces on the bee, guided again by visual stimuli (Tinbergen 1972, 139ff). The bee wolf apparently acquires limited information for identification, however, because it often makes mistakes and pounces on something other than a bee. Similarly, a hunting fox may catch the scent of a rabbit and approach it only with difficulty as it follows the scent trail alone. Then at some point it may hear the prey and thus be able to approach more directly. Finally it will catch sight of the prey and can then approach at full speed.

The importance of the various phases described above varies with the particular situation (O'Brien 1979). For example, lions often live within sight of large numbers of prey, making the search and detection phases inconsequential. In contrast, search and detection are the predominant problems of predators in pelagic environments, to the degree that environmental factors that influence these phases modify the structure of the whole community (see pages 198–99).

Of all the behavior phases involved in exploiting spatial goals, the approach phase deserves special attention here. This phase may make use of any of several general strategies, depending on how the stimulus is distributed in the environment and on the capabilities of the organism. A classification of these strategies is presented in the next section.

14-3 APPROACH STRATEGIES

The strategies organisms use to travel to a goal almost always employ information to make their approach more efficient. These techniques are discussed in detail in later chapters, but first it is useful to develop a classification scheme for them and establish terms by which to refer to the different strategies. Unfortunately, the literature contains a confusing diversity of terminology in this area. The fundamental problem is that researchers working on specific types of organisms often define terms based on concepts that are relevant only to their own area of interest. Another problem is that terms used to classify different types of behavior have not been defined so as to meet the elementary principles of classification in that the categories are not exclusive and exhaustive. Wim J. van der Steen (1984) has suggested a classification based on logical combinations of some of the criteria that have been traditionally used. However, this results in having forty-eight possible classes—too many to be useful. In addition, many of the combinations are not likely to be efficient strategies—and some even seem to be physically impossible. The emphasis on information promoted throughout this book should help clarify these matters. In what follows I propose terminology and a classification scheme that is intended to apply to all organisms and meets the criteria of being both exclusive and exhaustive. A summary is provided in Table 14-3.

Each criterion in the table is used to distinguish two classes. As a result, this classification scheme is analogous to a dichotomous key commonly used for identifying organisms. I have attempted where appropriate to use terms that have been commonly used by others, but in a few cases it has proven desirable to invent new ones. Furthermore, I have avoided using the term *orientation*, because it has been used with many meanings

Table 14-3 A Classification of Approach Strategies

Passive Movement (*Hitching*)*

Self-propelled Movement

    No stimuli that can be used to move closer to the goal is available (*Search*)*

        Information suggesting a nearby goal (*Local Search*)

        No information favoring a nearby goal (*Ranging*)

    A stimulus that can be used to move closer to the goal is available

        The stimulus has a simple relationship to the goal (*Guiding*)*

            Information is obtained for direction to the goal (*Direct Guiding*)

            Information is not obtained for direction to the goal (*Indirect Guiding*)

        The stimulus has a complex relationship to the goal (*Navigation*)*

            The stimuli are specific to the goal (*Piloting*)

            The stimuli are general (*True Navigation*)

* These strategies are the subjects of later chapters.

and has led to much confusion, principally because it has quite different connotations in German and in English (Kennedy 1986).

The first distinction is one based on the mode of transport. Many organisms employ agents such as wind or water currents to provide them with transport. This behavior has often been called drifting and might also be described as passive approach. However, these two descriptions suggest a state of uncontrolled transport, and control is what is of most interest in this book. Control is exerted by choosing conditions in which to initiate or terminate drifting. In order to emphasize the active, and controlled nature of this strategy, it will be called **hitching,** as in hitching a wagon to a horse or "hitching a ride." Another type of transport agent is a larger mobile animal; this behavior is called *phoresy.* Hitching will be considered a general strategy including phoresy.

For self-propelled transport a further distinction can be made based on whether or not stimuli are available to provide information that can be used to move the organism closer to the goal. In the absence of such information the only strategy to use in finding the goal is to search for a clue. A further distinction can be based on whether the organism has information that the goal is more likely to be near its own present position than at an arbitrary position, or would be more valuable if found close by. If the searcher has such information, it becomes efficient to concentrate the search nearby, which is thus called **local search.** Otherwise the search is opened up to all areas, a process called **ranging.**

If stimuli are available that provide information about how to move closer to the goal, a distinction can be made as to whether or not the stimuli have a simple relationship to the goal and can thus be used to move toward it without needing additional information. The strategies for making use of such simple stimuli are called **guiding.** In this case the stimulus usually emanates from the goal and forms a stimulus field around the goal having predictable features. The goal can often be approached by simply moving up or down an intensity gradient or against the direction of stimulus movement. In other cases the stimulus may simply trigger orientation to another stimulus, such as when the odor of a sex pheromone triggers the upwind flight of a male moth. In any case, guiding can be carried out by simple reflex mechanisms.

The strategies of guiding are further divided into two subcategories, depending on whether or not the organism obtains information about the direction of the stimulus field, as for instance by simultaneously sampling a chemical gradient with spatially separate receptors, and turns in the appropriate direction. The method of **direct guiding** is based on this ability. If sufficient information for turning in the appropriate direction is not obtained, a strategy of **indirect guiding** can be employed to move gradually along a gradient by using a biased random walk. Further possible subdivisions of these different strategies are discussed in Chapter 17.

If the stimuli do not have a simple relationship to a goal, their use is called **navigation.** In order to use such stimuli, information from them must be compared to information about the goal, which is usually learned. In particular, there must be information about the intensity of the stimulus, its distribution, or its quality around the goal. This category is subdivided on the basis of whether the stimuli are general or are specific to particular locations. Strategies of **true navigation** employ general cues like celestial stimuli or the earth's magnetic field that vary in predictable ways over the area of movement. Strategies of **piloting** employ specific cues such as landmarks that have no predictable relationship to different locations. Consequently, piloting can work only in familiar areas where specific relationships have been learned, whereas true navigation can work in unfamiliar regions. The term *true navigation* has previously been used by Able (1980) with an even more restrictive meaning requiring a map sense (see pages 469–74).

More able animals are likely to employ a variety of approach strategies. For example, a homing pigeon might employ true navigation to reach a familiar area around its loft, followed by piloting to move within sight of the loft, then finishing the approach phase by employing direct guiding using visual stimuli emanating from the loft. What is generally of the greatest interest is to determine the most sophisticated strategy an organism is capable of exploiting.

All of these approach strategies are discussed in detail in the chapters that follow on hitching, searching, guiding, and navigation. The last chapter discusses some specific examples of migration.

CHAPTER 15

Hitching

Catching a Ride to a New Location

■ ■

> There is a tide in the affairs of men,
> which, taken at the flood, leads on to fortune.
> —William Shakespeare, *Julius Caesar*

For both migration and dispersal, one strategy organisms use for cost-effective movement to a new location is to hitch a ride on an agent of some sort. In fact, this is the only practical strategy for small organisms to move large distances. From a sensory ecology point of view the interesting questions concerning this behavior are what information is used to decide when to hitch to the agent, to orient to the agent, and finally to decide when to terminate the ride. These questions are addressed in this chapter.

15-1 TRANSPORT AGENTS

The hitching strategy may employ any of several transport agents like currents of air or water or mobile animals. The important features of these agents and some examples of how they are used are presented in this section.

In order to transport itself by wind or water currents, an organism hitches onto it by getting far enough from the substrate so that it is no longer in the boundary layer of the fluid flow (see pages 75–86). The thickness of the boundary layer is not well defined, because a current's speed increases continuously with the distance from the substrate. A convenient definition of an organism hitching to a current is to say it is doing so when it is in a current that is flowing faster with respect to the substrate than its rate of locomotion with respect to the fluid (Gibo 1986).

WIND

Currents of air, or winds, are used by many animals for transport. The **prevailing winds** vary with their position on the earth's surface (Figure 15-1). At the equator, heating causes air to rise, which makes the prevailing winds light and creates what sailors call the doldrums. Over continents the land's temperature varies with the season. The region of maximum heating is called the intertropical convergence zone, because the surface winds converge on it to replace the warm, rising air (Rainey 1976). Air tends to sink near latitudes of 30° in what are known as the horse latitudes, where again the winds are light. However, between the equator and the horse latitudes there is a strong return flow at the surface, called the trade winds, which blow toward the equator and are deflected to the west by a phenomenon known as the Coriolis force, which results from the rotation of the earth. At latitudes between about 30° and 60°, surface winds blow toward the poles and are deflected to the east. These winds provide the prevailing winds out of the west that are familiar to the residents of temperate zones.

Given this pattern of winds, it is often economical to make seasonal migrations in a loop, which many birds seem in fact to do. For instance, sandpipers and plovers that nest far north in North America and winter in South America migrate south over the Atlantic Ocean and north through the Mississippi Valley in a clockwise loop (Richardson 1976; McNeil and Burton 1977; Williams et al. 1977). Note in this regard the migration circuit in Figure 15-2.

The wind patterns must provide a return circuit of air flow. Consequently, above the earth's surface there is usually a counterflow. For instance, the antitrade winds blow to the east above the westerly trade winds, with the boundary between these flows having little wind. In the fall over Puerto Rico this boundary is at an altitude of 5 to 8 km, and migrating birds frequently fly at altitudes of 4 to 5 km, where they experience much weaker headwinds than they would closer to the surface (Richardson 1976).

In addition to the prevailing winds, more local and temporary patterns of air circulation also occur. The strong but infrequent winds of hurricanes and tornadoes are not known to be actively exploited by any organism.

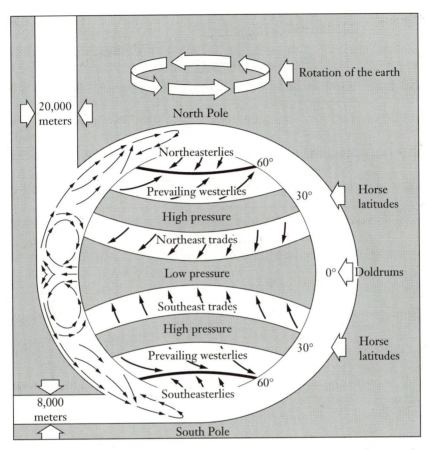

Figure 15-1 The prevailing winds in the earth's atmosphere. The sun heats up the oceans, with the large heat capacity of water, maximally at the equator, whereas land masses, which have a lower heat capacity, are heated maximally in a region called the intertropical convergence zone that varies with the season. Air over these regions of maximum heating rises, producing little wind in the region and resulting in what are called the doldrums by sailors. On both sides of these regions the surface winds blow toward them to replace the risen air. These flows are then deflected to the west by the rotation of the earth, thus producing the trade winds. The risen air flows north and south, cooling until it sinks to the surface at latitudes of about 30°, where it produces another zone with little prevailing wind, known as the horse latitudes by sailors. Some of the sunken air moves toward the poles but is deflected to the east by the rotation of the earth, producing the prevailing westerly winds of the temperate latitudes. The air at the cold poles primarily sinks and flows along the surface away from the poles and is deflected to the west, thus forming the prevailing winds at latitudes higher than 60°.

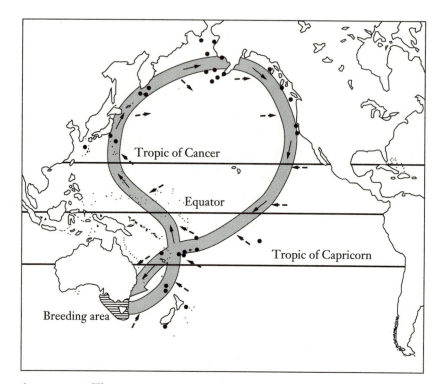

Figure 15-2 The migration loop of the slender-billed shearwater, that takes advantage of the prevailing winds. (After Baker 1978.)

However, in temperate latitudes the atmosphere contains high- and low-pressure systems that drift from west to east (Richardson 1978). As they move, the accompanying **weather fronts** sweep across a given position, causing predictable changes in the weather conditions (Figure 15-3). Winds in the Northern Hemisphere blow clockwise around highs and counterclockwise around lows. As a result, winds are warm and from the south at positions west of a high or east of a low, but they are cold and from the north at positions east of a high or west of a low. Because of the eastward drift of such systems, warm winds from the south tend to occur with falling pressure and increasing cloudiness, and cold winds from the north with increasing pressure and clearing skies.

The autumn migration of monarch butterflies occurs predominantly in winds out of the north, which often follow the passage of a cold front

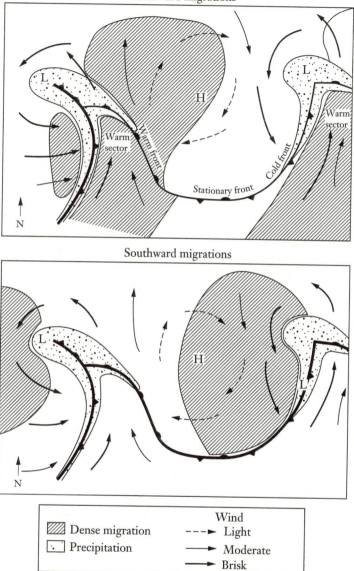

Figure 15-3 Bird migration patterns as related to weather fronts. These diagrams show weather fronts with the typical patterns of low (L) and high (H) pressure, wind (*arrows*), and precipitation (*stippling*) for the north temperate latitudes. The ross hatching shows the regions where migration tends to occur in the spring (*above*) and the fall (*below*). Migration is most likely to occur in regions with tailwinds and a lack of precipitation. (From Richardson 1978.)

(Urquhart 1960, 258–59). There exists at least one observation suggesting that monarchs can predict approaching bad weather (Urquhart 1960, 271).

Much research has been aimed at determining whether birds choose particular conditions for migration and this has been critically analyzed by Richardson (1978). The general conclusion is that some birds are aloft in almost any weather conditions, but the greatest numbers are aloft in fair weather when winds are blowing in their direction of migration, as in Figure 15-3. Land birds that fly over the Atlantic, from Northeastern America to the West Indies or South America, or over the Gulf of Mexico usually depart in following winds behind powerful cold fronts. A number of weather factors correlate highly with the numbers of birds migrating, but common weather patterns make all these factors highly correlated with one another. It is not now possible to determine which cues birds use to decide to begin a migratory flight. The presence of following winds was the most highly correlated factor and seems likely to be the most important. It remains unclear whether birds are in effect weather forecasters, using temperature and pressure or other such information to predict favorable weather or whether they simply select currently favorable wind and sky conditions to initiate their migratory flights.

Local winds, of any origin, are often employed for shorter distances of travel. In a behavior called **ballooning** many species of spiders hitch rides on the wind by letting out a strand of silk until it has sufficient air resistance to carry the spider (Duffey 1956). They usually choose warm days, likely to generate thermal updrafts able to carry them high enough to be transported by persistent winds. Indeed, spiders have been observed some 5 km above the earth's surface (Southwood 1962) using this mechanism, which can provide transport over long distances. Two years of sampling in the sterile blast zone around Mount Saint Helens in Washington state produced forty-three species of spiders that must have arrived in this way (Edwards 1986). Of these, thirty-six species must have traveled at least 30 km.

Massive studies have been focused on the adult moths of the spruce budworm, which sometimes defoliate trees over wide areas (Greenbank et al. 1980). The budworm flies at an airspeed of about 2 m/s. However, on summer evenings adults often migrate by flying several hundred meters above the canopy, where the winds average 8 m/s, and they fly downwind, generating an overall ground speed of 10 m/s. They can thus move on the order of 100 km during a single night's migration. Other flying insects are known that also average about 100 km/day (Mikkola 1986).

Under certain conditions, organism's rates of movement can be much higher. For example, an outbreak of the fall armyworm in Sault Ste. Marie,

Canada, was thought to have resulted from a weather front that carried adults some 1,600 km from Mississippi in thirty hours (Mitchell 1979).

This outbreak was probably enhanced by a concentration of the insects brought together by converging winds. Rising air causes winds to converge at the surface of the earth. If insects are not carried upward in the rising air, they will over time become concentrated by the converging winds. Patterns of air flow appropriate for concentrating insects in this way occur at all size scales, from the global intertropical convergence zone to storm cells down to small eddies (Greenbank et al. 1980). This effect has often been observed in locusts (Rainey 1976) as well as in the spruce budworm (Greenbank et al. 1980). In some instances spruce budworms have suddenly appeared in concentrations that are hazardous to automobile traffic and piled up ankle deep on the ground, apparently as a result of converging winds. These atmospheric effects indicate that observing concentrations of flying insects does not necessarily mean that the insects are using information to aggregate or that the aggregation is of adaptive value.

WATER CURRENTS

Water currents are probably also quite important transport agents, although we know much less about the behavior of organisms below the surface than of those above it.

Rivers provide reliable currents to use in transport—but only in one direction. Consequently, a complete migration circuit must include one leg that is either self-propelled or uses some other agent. For instance, the larvae of many aquatic insects drift downstream in water currents and later the adults fly upstream, completing a migration circuit during their life cycle. Similarly, adult salmon actively swim upstream after having moved down to the ocean as juveniles.

Winds that are sustained for twelve hours or more over the surface of the ocean generate significant water currents (Bowditch 1984, 815). Thus, the prevailing winds generate sustained ocean currents. In addition to their being deflected by the Coriolis force, ocean currents are deflected by land masses. As a result, each ocean basin has its own flow pattern (Figure 15-4). The strongest currents, like those that make up the Gulf Stream, flow at about 5 km/hr. Currents tend to flow in closed patterns, called *gyres*. Many fish seem to use these circular flows to make a migration loop (Jones 1981) (Figure 15-5).

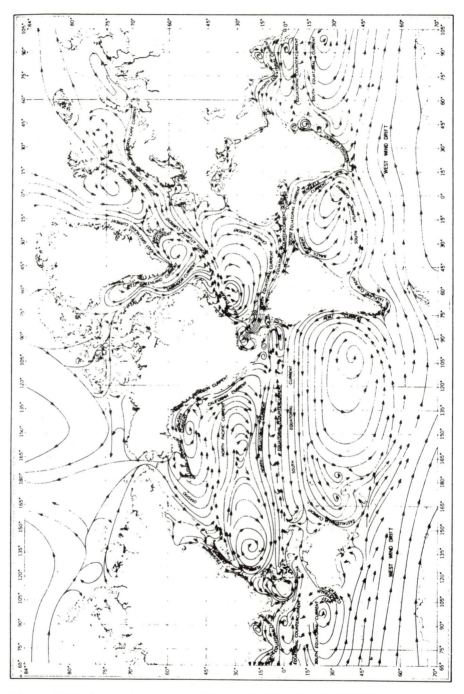

Figure 15-4 Ocean currents around the world. (After Bowditch 1984.)

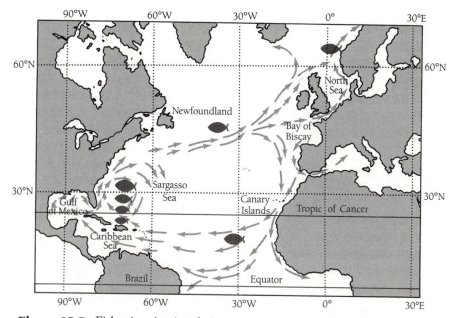

Figure 15-5 Fish migration in relation to ocean currents. The giant bluefin tuna annually follow a transoceanic feeding migration that circles around the North Atlantic, generally in the direction of its clockwise surface currents. (From Waterman 1989.)

Tides cause currents near shore that are often used for transport. Their most characteristic feature is to reverse direction twice a day, providing an opportunity for organisms to catch a ride in either of two opposite directions. Many aquatic species seem to employ this opportunity (Leggett 1984). An example of the pattern of tidal currents is shown for the English Channel in Figure 15-6. An interesting point to be made is that there are some locations that become dead ends, where the tidal currents die out, and others where the currents converge or diverge, complicating any attempt to travel through them.

The flatfish called plaice have been shown to make use of tidal currents during seasonal migration (Walker et al. 1978; Arnold 1981). They hitch to the current by swimming up from the bottom into the water column when the current is flowing in the appropriate direction; otherwise they stay down where the currents are weak (Figure 15-7). They use this mechanism to travel between their spawning and feeding grounds.

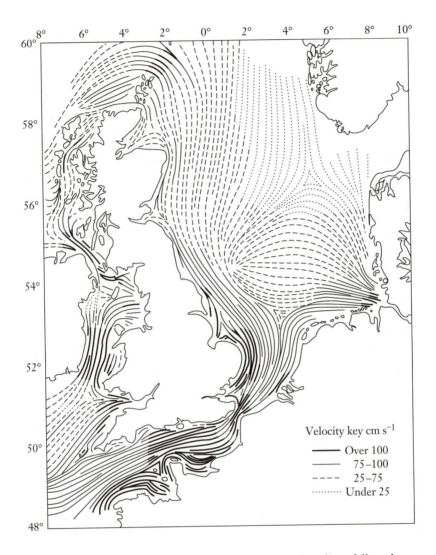

Figure 15-6 Tidal currents in the North Sea. The various lines follow the stream paths of different tidal currents, with the type of line indicating the speed of the current. Observe that some locations have no tidal current. Organisms can move between two locations by hitching only if both of the locations are on the same stream paths. (After Harden Jones et al. 1978.)

Computer modeling of various strategies of hitching in the English Channel by Arnold and Cook (1984) indicates that an optimum strategy of riding a particular tide for the maximum of six hours can lead to migration

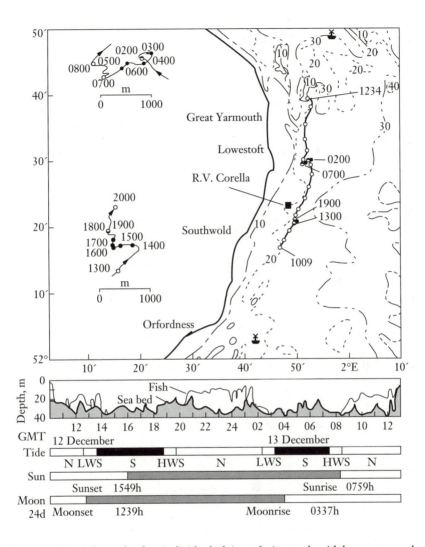

Figure 15-7 The path of an individual plaice relative to the tidal current, tracked with an attached sonar responder. *Top*: The open circles show positions during northgoing tides, the filled circles during southgoing; the half-filled circles show its positions during slack tides. The 10 and 40 m depth contours are shown as dashes, the shore as solid. *Bottom*: The depth of the fish as a function of time, along with the timing of the tides, sun, and moon. When the tide is flowing against the plaice's direction of travel, it spends most of its time on the bottom and moves very little. In contrast, when the tide is flowing in its direction of travel it spends most of its time in midwater and makes a great deal of progress. (From Walker et al. 1978.)

averaging over 10 km/day, without requiring any horizontal swimming. By this mechanism alone migration can occur only between certain sets of locations, which may help explain the distribution of the feeding and spawning areas of fish. The crucial nature of the timing of the takeoff was demonstrated by models in which the hitching occurred at a fixed time of day. With such a strategy net movement over a period of many days was restricted to about 10 km.

ANIMALS

The remaining common class of transport agents is mobile animals. The relationship in which one animal obtains transport on another is often called **phoresy.**

Many plants benefit from getting animals to carry their seeds or pollen, a process that usually involves elements of advertisement and payment. The payment usually takes the form of nectar or fruit mediating a trophic interaction between the two species. The advertisement involves signals such as colored flowers or fruit or a fragrance that inform the animal about the presence of the payment in an informational interaction relevant to sensory ecology. Since these signals are produced for the benefit of both the sender and the receiver they are a form of true communication, which is discussed more fully in Chapter 13.

Another strategy for dispersing seed for the plant is to produce a seed that will stick to the coat of a passing animal. Such an interaction does not seem to involve information and thus need not concern us here. A marginal situation is one in which plants discharge seeds in an explosive fashion when the seed pods are mechanically disturbed. With such a behavior the seeds are often discharged when an animal comes close, so that some of the seeds may land on it and be carried away. In this case it is not clear whether the mechanical disturbance provides information or it is simply accidental that the coiled spring is released sooner than it would otherwise be because of the disturbance.

Nematodes move slowly through soil, but many species have developed specialized survival/dispersal stages called *dauer larvae* that do not feed and often obtain transport by attaching themselves to larger animals (Poinar 1983, 108–114, 162). For example, *Rhabditis coarctata* feeds on bacteria in cattle dung. When the dung pat dries out, juveniles develop into dauer larvae, which attach themselves to the surface of dung beetles.

When a beetle arrives at a fresh dung pat, the nematodes release themselves from the beetle and begin further development in the fresh dung. Similar relationships occur between other nematodes and bark beetles and burrowing rodents (Poinar 1983, 176, 114).

The stimuli that the dauer larvae use to attach themselves to animals are not known. However, some of the stimuli that cause formation of dauer larvae in the widely studied model species *Caenorhabditis elegans* have been elucidated (Golden and Riddle 1984). A fatty-acidlike pheromone provides information about the population density, and higher levels of the pheromone increase their tendency to develop into the dauer larval stage and stay in it. Other chemicals from cultures of their bacterial food have the opposite effect. Thus, the decision to become a dauer larva appears to be based on assessing the density of competitors relative to the food supply.

Another example of transport by animals is that many insects that are parasitic on the eggs of other insects attach themselves to adult females of the host species (Clausen 1976). Doing so automatically provides them transportation to a suitable site for egg laying. Little is known about the behavior controlling the attachment to the host, but it must be very specific, as attachment occurs only to the appropriate species. A few other arthropods also seem to obtain rides on animals for dispersal (Southwood 1962).

15-2 ORIENTATION TO A TRANSPORT AGENT

In order to hitch a ride on a transport agent, an organism often must employ some kind of information, to orient appropriately to the agent. There is not enough information available for a systematic treatment of this process, but a few specific examples will suggest what is known.

Because fluids are generally less dense than solids, air or water currents are usually above the substrate, so that transport by them is initiated by flying or swimming upward and is terminated by moving downward. Thus, hitching a ride on a current frequently requires distinguishing up from down. In slow-moving animals this orientation is often accomplished by gravity detection (see Section 10-2). However, in flying animals rapid acceleration distorts the force of gravity, and light is therefore frequently employed as a reference.

Among marine invertebrates the larvae of 80 percent of the species are photopositive and swim toward the surface early in life, where they join the plankton, but later most become photonegative, when they metamorphose to bottom-dwelling adults (Dingle 1980). Similarly, insects are often photopositive when migrating but become photonegative at the end of the migration (Johnson 1974).

Aphids are small enough that movements of more than a few meters must rely on wind currents. When the host plant of the black-winged aphid becomes too crowded or senescent, winged forms of the aphid are produced that climb to the top of the plant and hitch to winds by flying toward blue light (Kennedy et al. 1961; Dingle 1980). After a period of flying, they become attracted to yellow light and descend to vegetation to search for a new host.

Animals with greater sensory capabilities may employ more sophisticated mechanisms for determining the vertical direction. Many flying insects, as well as some other animals, reflexly orient their dorsal side to the brightest diffuse source of blue or ultraviolet light (Wehner 1981), a response called a **dorsal light reaction.** Another point of reference is a visual horizon, which is a sharp, straight, boundary that is brighter above, a device probably used commonly by animals that fly above the canopy.

In order to take full advantage of a current it is necessary for an organism to get out of the boundary layer next to the substrate (see pages 75–86). In the atmosphere this layer may be as much as several hundred meters thick. Animals must fly through it if they are to take full advantage of wind as a transport method (Riley and Reynolds 1986). During the day, as the sun heats the ground *thermals* may be produced that help carry animals aloft. Locusts, for instance, are sometimes carried upward without flying at all, with their maximum altitude being determined by the height of convection (Riley and Reynolds 1986). Under the right conditions, soaring birds can also climb high with little effort. At night, however, the atmosphere is more stable and there are few vertical currents, so night-flying insects must actively fly through the boundary layer. For example, grasshoppers migrating at night have been observed to climb at rates of 0.3 m/s to heights of more than one kilometer (Rainey 1976; Rainey 1979). Likewise, the spruce budworm climbs at about 0.6 m/s to a height of a few hundred meters (Greenbank et al. 1980), then orients itself downwind; what information it uses to determine the wind's direction remains a mystery.

With insects, hitching to air currents is often confined to a particular developmental stage that is adapted to flying (Southwood 1962). This stage is a regular part of the reproductive cycle in some species and is facultative in others. Individuals that are ready to travel are often seen to walk up to the tips of vegetation, launch themselves, and fly upward, frequently in spirals. Clearly, these insects are using their flying ability primarily to get up to where the winds are strong.

Even animals that are more able actively to propel themselves may be concerned with similar problems, to save energy. Plaice, for example, seem to be able to maintain a midwater position for many hours (Walker et al. 1978). Many flying insects and birds fly close to the ground when the wind is blowing against their movement but higher when the wind is from a favorable direction (Alerstam 1981, 81; Baker 1982, 179, 182). Furthermore, birds seem to be able to select an altitude where the most favorable winds are blowing (Figure 15-8) (Alerstam 1981). In some cases they may fly at enormous altitudes. For example, on their autumn migration from Northeastern America to South America passerines pass over Puerto Rico (Richardson 1976) and Antigua (Williams et al. 1977) at altitudes of 4 to 5 km, where the crosswinds are not as strong. Migrating flocks have even been seen as high as 10 km. How these birds determine the optimum altitude is a mystery, but one hypothesis is that, at least for mass migrations, birds at different altitudes that are in visual or acoustic contact can compare their relative speeds. This might explain why geese "honk" while flying.

15-3 TIMING INFORMATION

The optimal use of the strategy of hitching requires appropriate timing for *initiation* and *termination* of the ride. Some organisms may hitch themselves to a transport agent whenever they reach an appropriate stage of development, but many will find it advantageous to control the time of initiation to improve their chances of hitching to an optimal agent. For example, if the goal is known to be in a particular direction, it would be beneficial to wait for a transport current flowing in that direction.

In most insects, hitching is influenced by assessing the quality of the present environment and of the conditions for travel, specifically the present temperature, wind, and light intensity (Johnson 1974; Greenbank et al. 1980). Temperature is particularly influential in the morning, probably because cold prevents flying. Wind blowing directly on insects generally inhibits their take-off, perhaps by cooling them. The spruce budworm appears to fly 10 m up above the canopy to sample the wind, then it returns to the canopy if the air is still (Greenbank et al. 1980).

The decision by aphids to hitch a ride is based on the quality of the present location, taken as the age of the leaf and the competition from other aphids, and in part on factors related to wind (Walters and Dixon

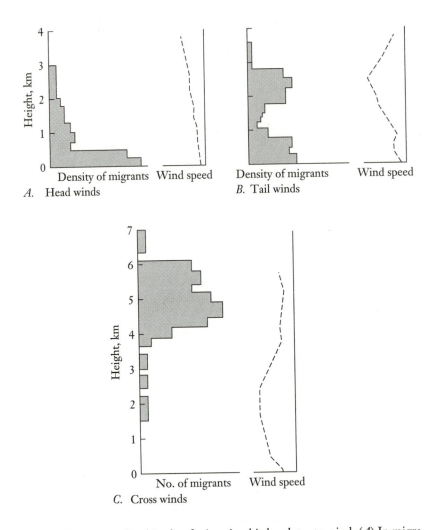

Figure 15-8 How the altitude of migrating birds relates to wind. (*A*) In migration during a period of headwinds the birds flew low where the winds were less strong. (*B*) In migration during a period of tailwinds the birds flew high; many seemed able to select the altitude with the maximum wind speed. (*C*) In migration over the ocean approaching Puerto Rico the birds flew very high where the trade winds were light. (After Alerstam 1981.)

1982; Dixon and Mercer 1983). Dixon and Mercer (1983) found that those most inclined to move tended to take off near noon, when thermals were likely to lift the aphids high in the air for long-distance transport.

Among ballooning spiders, at least some species initiate additional rounds of hitching if they arrive at an unfavorable location the first time (Riechert and Gillespie 1986).

Once an organism is riding with a transport agent, it must make the important decision of when to terminate the ride. One obvious approach is to use some sort of cue associated with a desirable goal. Spruce budworms appear to decide when to land while they are still quite high (100–200 m) in the air (Greenbank et al. 1980). The information they employ is unknown, but they do seem to be able to select for the presence of host trees. Once the decision is made, they seem to simply fold their wings and fall into the canopy.

Many aquatic invertebrates are bottom dwelling or sessile as adults but have free-swimming planktonic larvae that provide for their dispersal. Their decision to settle seems to be based primarily on chemical signals. Some of the identified chemicals are peptides or proteins attached to the substrate (Carr 1988) or fatty acids released by conspecifics (Pawlik et al. 1991). Some larvae appear also to select for appropriate current flows for settling (Pawlik et al. 1991).

If information about the proximity of a goal is not available, the best strategy may simply be to terminate the ride after a fixed time interval, independent of any stimulus. Aphids seem also to use this strategy (Dingle 1980). Most insects seem to have a minimal time interval before they respond to cues terminating the ride (Southwood 1962). However, other insects simply fly until their fuel runs out, which may allow for more than twenty hours of flight (Johnson 1974).

When animals use back-and-forth tidal currents to move in a particular direction, it is especially important to make appropriate decisions about when to initiate and terminate hitching. It has been suggested that, when they are resting on the bottom, current flow in the appropriate direction might be detectable by changes in chemical composition such as salinity (Hughes 1972; Arnold 1981). This change could be used as a cue for initiating a ride. However, for terminating no chemical change is expected, since the animal is moving with the water mass in its vicinity. In this situation an indigenous timing mechanism is likely to be useful for making termination decisions. There is evidence for the existence of such mechanisms in shrimp (Hughes 1972) and crab larvae (Cronin and Forward 1979).

Plaice might have access to compass information and hitch to the current when it is flowing in the right direction, or they might use some information to indirectly determine the appropriate phase of the tide such as changes in hydrostatic pressure, the phase of the light–dark cycle, or changes in chemical composition. In tracking studies (Walker et al. 1978), plaice seem to often make exploratory excursions between the bottom and the water column. In addition, the initiation of hitching seemed to be more closely tied to slack water than was termination, which is consistent with the idea that information about the phase of the tide is difficult to obtain in midwater.

From this brief review it should be clear that the strategy of hitching is important to a wide variety of organisms. As seen, many of them employ information about the transport agent to determine when to initiate a ride and to orient themselves appropriately toward the agent.

CHAPTER 16

Searching

Looking for Relevant Stimuli

■ ■

It is advantageous for an animal propelling itself to a spatial goal to do so efficiently. In particular, the path traveled to the goal should be as direct as possible. The advantage of directing locomotion is such that all motile organisms and cells studied have been found to employ sensory information for this purpose. If appropriate sensory stimuli are not available, an organism may adopt a pattern of locomotion efficient for gaining contact with such stimuli. This behavior is described by saying that the organism is searching for a stimulus. Using the term *search* is not meant to imply consciousness on the part of the organism; the term is used merely for its convenience in describing behavior. Unfortunately, little data are available about the travel patterns of searching organisms. This chapter thus focuses on theoretical strategies for conducting an efficient search.

16-1 RANGING

What is the optimal pattern of movement when an animal is in the first phase of moving to a spatial goal and has no information about the location of a specific goal? Do organisms in fact move in an optimal pattern? There is surprisingly little data on these matters. This section therefore deals primarily with theoretical models that set limits on what is possible and also identify the most important factors in an efficient search. These concepts lead to hypotheses that should be tested in the future.

A theory of optimal search strategies has been developed, primarily for locating ships at sea (Koopman 1980), which is referred to here as the **standard search theory.** Unfortunately, this theory is more limited in scope than the variety of circumstances of interest in biology. Standard search theory is restricted to a two-dimensional area either of the surface or the bottom of the sea and, in comparison to their range of detection, the searchers maintain a straight course over large distances and search large areas. Standard search theory is described in the following section, then modified to make it more appropriate for biological situations (Dusenbery 1989d).

BASIC SEARCH THEORY

The parameters used in standard search theory are presented in Figure 16-1. In this standard theory the searcher is assumed to have effective sweep width W. If the searcher is assumed to detect all the goals within a range of detection r_d and none beyond this range, $W = 2r_d$. If the searcher travels distance L in a straight line, it detects all the goals in an area $W L$ so that area $s = W L$ has been searched. A **clean sweep search** of area A is one in which the area is completely covered by the searcher with a 100 percent probability of finding all the goals but there is no overlapping of

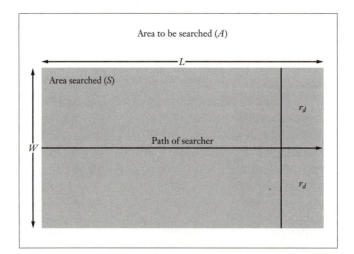

Figure 16-1 The parameters used in standard search theory. L is the length of the path of the searcher, W the width of the area searched, as determined by the range of detection, r_d.

searched areas, meaning no wasted efforts. If a goal is in area A, the probability of finding it in a clean sweep search of length L is

$$p(L) = \frac{WL}{A} \qquad L < A/W \tag{16-1}$$

and the mean search distance to detection is thus

$$\langle L_d \rangle = \frac{A}{2W} \tag{16-2}$$

A **random search** is one in which goals are uniformly distributed over A and the search path is random, in the sense that one segment is unrelated to the others but the path is straight on the scale of the range of detection. In this case it is not necessary to assume a definite detection range, so W can represent an average sweep width. The probability of detection of each goal is then

$$p(L) = 1 - \exp\left(-\frac{WL}{A}\right) \tag{16-3}$$

and the mean search distance to first detection is

$$\langle L_d \rangle = \frac{A}{W} \tag{16-4}$$

On average a random search takes twice as long as a clean sweep.

The clean sweep and the random search are compared in Figure 16-2. The clean sweep search is more efficient because in it there is no overlapping of effort. The difference is greatest when the search has gone on long enough so that the area to be searched has been well covered. In the initial stages there is no significant difference, because there is then little probability of overlapping in a random search.

Neither of these models is very realistic in detail. For one thing, searchers rarely have an absolute range of detection, and even if they did, errors in traveling in an appropriate pattern would still make a clean sweep difficult to conduct. Furthermore, the assumptions just made about the properties of a random search track do not apply to real searchers (Stone 1975, 24). Strictly speaking, using a random search model requires that the search path be straight, with any changes of direction coinciding with

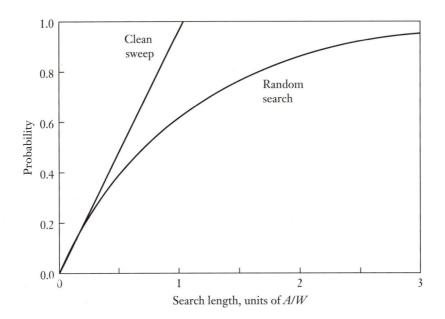

Figure 16-2 The probability of finding a target in area A as a function of the search length. In the initial stages of a search there is no difference in efficiency between a clean sweep and a random search, but as the area becomes more thoroughly searched the random search decreases in efficiency because some areas will be searched more than once.

discontinuities, whereas real searchers usually move in a continuous pattern, making changes of direction that cause overlapping in the searched areas. However, the random search model is useful both for its mathematical simplicity and for providing a worst-case approximation for an attempt at a clean sweep (Stone 1975, 24). In biology the random search model has been used with success to describe the behavior of female butterflies searching for host plants on which to lay their eggs (Jones et al. 1980).

MODIFIED SEARCH THEORY

Standard search theory can easily be generalized from two to three dimensions. In doing so the sweep width, W, is replaced by the cross section, σ, for detecting the goal. The **cross section** is a much-used concept in physics that measures the probability of two objects that are both moving in straight lines coming sufficiently close together for some interaction to occur (Figure 16-3). In the present application of this concept the interaction is detection. The basic idea is that the probability of the intraction occurring is proportional to the two objects' joint cross-sectional areas in the plane perpendicular to the direction of movement and containing both objects. Thus, for two spheres of radii r_1 and r_2 the cross section for collision is

$$\sigma = \pi \, (r_1 + r_2)^2 \qquad\qquad (16\text{-}5)$$

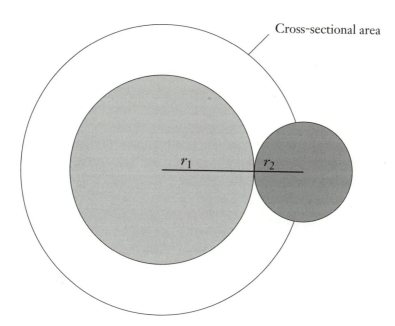

Figure 16-3 A cross-sectional area. Here two objects have ranges of interaction r_1 and r_2. The cross-sectional area for their interaction is the circle that has a radius equal to the sum of their ranges. If the smaller object is considered as moving perpendicular to the page, interaction of the objects will occur if the smaller one's center is within the cross-sectional area centered on the larger one.

In many cases one of the objects will have a much larger effective size than the other, so that it becomes appropriate to treat the latter as a point with zero radius. In standard search theory the goal is assumed to be a point, and the cross section is determined by the range of detection of the searcher. When searching for chemical stimuli emanating from a goal it is useful to think of the cross section as being determined by the volume in which the stimulus concentration is above the detection threshold of the searcher, a volume referred to as the **active space** (see sections 4-2 and 4-3). Since the searcher must make contact with this volume for detection to occur, the searcher can be treated as a point. Observe that it is only the relative motion of the searcher and the goal that is important and one can treat as stationary whichever is the most convenient.

Applying this cross-section concept to standard search theory in three dimensions then $\sigma = r_d 2$ and in two dimensions $\sigma = 2\ r_d$. Let ρ be the density of the goals per unit area or per volume as may be appropriate, and v be the velocity of the searcher. Then for a clean sweep search the probability of detection is

$$p = \rho\ \sigma\ v\ t \qquad \rho\ \sigma\ v\ t \leq 1 \tag{16-6}$$

For the random search,

$$p = 1 - \exp\left(-\rho\ \sigma\ v\ t\right) \tag{16-7}$$

Note that, strictly speaking, in three dimensions it is not possible to do a clean sweep while assuming an equal range of detection in all directions. The reason for this is that, as the searcher moves along, cylindrical volumes are searched, which cannot fit together to fill up all of a volume without overlapping. The equation of the random search model applies also to a straight-line search path through a random distribution of goals, which we might call simply a **straight search.** This interpretation would seem more applicable to the search patterns of animals.

Another type of search is a **discontinuous search,** in which the searcher moves a certain distance, then stops to test for detection of the goal. Such a model might represent, say, a robin flying between locations where it alights on the ground to listen for worms, or a situation in which animals probe randomly in the ground for a goal. If the animal repeats the search at frequency f, the product $\sigma \cdot v$ *in the above equations is replaced by* $s \cdot f$, *where s* is the area or volume searched in one test. For a two-dimensional search the area searched is

$$s = \pi\, r_d^2 \qquad\qquad (16\text{-}8)$$

and for a three-dimensional search the volume searched is

$$s = (4/3)\, \pi\, r_d^3 \qquad\qquad (16\text{-}9)$$

These values would apply accurately to a random search, but not to a clean sweep, because again the circular patterns do not fill up space in a precise pattern.

RANDOM-WALK SEARCH

Neither the clean-sweep nor the random-search models apply very well to organisms, which rarely move in patterns that are as precisely controlled as is assumed in the clean-sweep model. Furthermore, the assumption made in the random-search model that the search path is straight over a distance that is large compared to the range of detection is inappropriate in many biological situations. For example, bacteria can swim in a straight line for only about 10 μm, because of rotational diffusion (Berg and Purcell 1977), but they can respond to chemical gradients a millimeter from the source (Adler 1969).

Consequently, what is called a **random walk** model becomes more appropriate in many biological situations. This model has been applied to a variety of problems from molecular movements to population genetics. Its basic assumption is that in each of a series of steps a parameter such as position, r, changes by a fixed amount, L_s, in a random direction—either in any direction, or restricted to steps on a square lattice. There is a simple expression for the root-mean-square average of the change in the parameter after N steps:

$$r_{rms} \equiv \langle r^2 \rangle^{1/2} = L_s\, N^{1/2} \qquad\qquad (16\text{-}10)$$

If steps are taken at frequency f for time t,

$$r_{rms} = L_s\, (ft)^{1/2} \qquad\qquad (16\text{-}11)$$

When this model is applied to molecular motion it is consistent with diffusion, if the diffusion constant is given by

$$D = L_s^2 f / 2d \qquad (16\text{-}12)$$

where d is the number of dimensions in which diffusion is occurring (one, two, or three as appropriate). In terms of the diffusion coefficient D,

$$r_{rms} = (2d \, D \, t)^{\frac{1}{2}} \qquad (16\text{-}13)$$

These are the standard results for the random-walk model (Berg 1983).

MEAN TIME TO CAPTURE

These standard results of random walk theory do not apply directly to the problem at hand. We would like to know how long it would take an organism moving by a random walk to come within a certain distance of a goal. This problem is much more difficult to solve, but some recently derived results provide an approximation (Dusenbery 1989). The results are based on calculations for what is referred to as the **mean time to capture**, t_c, by an absorbing surface of radius r_d upon starting at random positions between it and a reflecting boundary at radius r_g. For a one-dimensional movement (Adam and Delbürck 1968; Berg 1983, 44) and substituting for D,

$$\langle t_c \rangle = \frac{2 \, (r_g - r_d)^2}{3 \, f L_s^2} \qquad (16\text{-}14)$$

For two dimensions (Berg and Purcell 1977),

$$\langle t_c \rangle = \frac{2 \, r_g^2}{f L_s^2} \left(\ln \frac{r_g}{r_d} - \frac{3}{4} \right) \qquad r_g \gg r_d \qquad (16\text{-}15)$$

Finally, for three dimensions (Berg and Purcell 1977),

$$\langle t_c \rangle = \frac{2 \, r_g^3}{f L_s^2 \, r_d} \qquad r_g \gg r_d \qquad (16\text{-}16)$$

These mean times to capture are applied to the problem of ranging by assuming that the searching animal moves in a pattern describable by the random-walk model and that identical goals are distributed uniformly in space (Dusenbery 1989d). The way to interpret the parameters is that r_d is the distance from a goal at which detection occurs and r_g is half the spacing between goals. Upon reaching r_g the reflection back toward the original goal is equivalent to continuing on into the domain of an adjacent goal. This interpretation necessarily involves some small errors for two and three dimensions, because circular geometries are assumed in the derivations but these shapes cannot fill space in a regular array. If the goals were randomly instead of uniformly distributed the formulas would have to bee modified, but they are probably still accurate to a factor of two, which is sufficient for many applications.

COMPARATIVE EFFICIENCY

In order to compare the effectiveness of the different types of search patterns, the random-walk equations are converted to a form that includes the velocity of locomotion, $v = f L_s$ (Dusenbery 1989d). Thus,

$$D = \frac{L_s^2 f}{2d} = \frac{L_s v}{2d} \qquad (16\text{-}17)$$

If the goals are assumed to be located in a square array at density ρ with half-spacing r_g and dimensionality d, then

$$\rho = (2\, r_g)^{-d} \qquad (16\text{-}18)$$

This leads to the formulas given in Table 16-1.

From this table, it can be seen that for a random walk search the mean time to detection is always inversely related to the velocity of travel. Detection time is nearly inversely proportional to the square (in two dimensions) or the third power (in three) of the distance between the goals. This strong dependence on the density of goals provides a powerful selective force against the clumping of organisms that are prey to others, as has been pointed out by Tingergen (1967).

The range of detection is iversely related to the time for detection. In a straight search the dependence on the range of detection is one power lower than the dependence on the distance between goals. And in a random-walk search the dependence on the range of detection is even less. This dependence can be explained by the fact that during random walk or diffusion, if any region is reached it is usually explored thoroughly (Berg

Table 16-1 The Time to Detection During Ranging

Dimension of Space Being Searched	Mean Time to Detection		
	Random-Walk Search	Random or Straight Search	Ratio of Random Walk to Straight
1	$\dfrac{2\,r_g^2}{3\,L_s v}$	$\dfrac{r_g}{v}$	$\dfrac{2\,r_g}{3\,L_s}$
2	$\dfrac{2\,r_g^2}{L_s v}\left(\ln\dfrac{r_g}{r_d}-\dfrac{3}{4}\right)$	$\dfrac{2\,r_g^2}{r_d v}$	$\dfrac{r_d}{L_s}\left(\ln\dfrac{r_g}{r_d}-\dfrac{3}{4}\right)$
3	$\dfrac{2\,r_g^3}{r_d L_s v}$	$\dfrac{8\,r_g^3}{\pi r_d^2 v}$	$\dfrac{\pi r_d}{4\,L_s}$

The parameter r_g is half the spacing between goals; r_d is the distance from a goal at which detection occurs; L_s is the step size of the random walk; v is the velocity of locomotion. The formulas are approximations that assume $r_g \gg r_d \gg L_s$. The formulas for the random-walk search neglect certain geometric factors that would require an increase of r_g by less than $\sqrt{}$ twofold in two dimensions and twofold in three dimensions.

1983, 12) (Figure 16-4). Dependence on the range of detection increases slowly with dimensionality, from independence, in one dimension; to logarithmic dependence, in two-dimensional random-walk search; to proportionality, in two-dimensional straight search or three-dimensional random walk search. The strongest dependence on the range of detection (r_d^{-2}) occurs in a three-dimensional straight search. This helps explain why the range of visual detection has such a dominant effect on three-dimensional pelagic communities (see pages 198–99).

In the random walk, the time to detection is inversely proportional to step size. In other words it will take longer than a straight search for all dimensions, as long as the step size is less than the range of detection. In two- and three-dimensional conditions, which are the ones of most interest, the relative disadvantage of the random-walk search (see the ratio column in Table 16-1) is proportional to the ratio of the step size to the range of detection (exactly in three and approximately in two dimensions). This distinction points up the advantage of maintaining a straight course when ranging, as depicted in Figure 16-4. Measurements on birds and bees foraging on flower nectar indicate that they indeed ten to move in straight lines, especially when they have not found much nectar in flowers recently visited. (Waddington 1983).

The models described probably encompass all real behaviors in terms

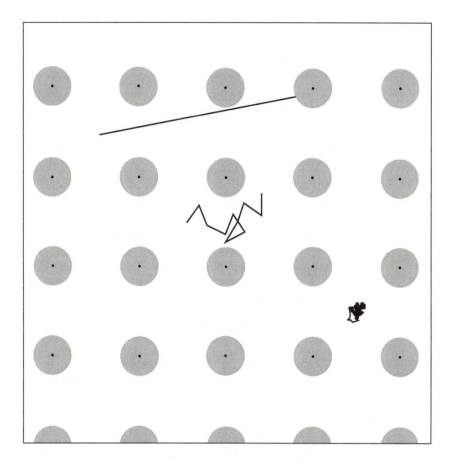

Figure 16-4 Search paths of various step lengths. The black dots represent goals, the surrounding gray areas the range of detection. The three paths all have a total length of ten times the detection range, equal to the radius of the circles. The step lengths of the three paths are > 10, 1, and 0.1 times the detection range. The straighter paths with their longer step lengths have a greater chance of coming within detection range of a goal.

of their efficiency. The clean sweep is the most efficient possible, the random or straight search probably close to the most efficient kind of practical search, and the random walk, with appropriate step sizes, has the lowest efficiency likely to occur.

If the goal is moving in addition to the searcher's movement, the average time to detection for movement in three dimensions can be taken from the kinetic theory of gases (Kauzmann 1966). The relationship is the same as for a straight search in three dimensions except that the effective speed, v, is replaced by

$$v = v_{fast} + \frac{v_{slow}^2}{3\, v_{fast}} \tag{16-18}$$

where v_{fast} is the speed of the faster of the searcher and the goal and v_{slow} is for the slower of the two (Kauzmann 1966, 284).

This model has been employed to analyze the mechanism by which male copepods find females (Katona 1973). This analysis indicates that a purely random search leads to a female within a few seconds at high natural population densities (25,000 m^{-3}) but takes hours at low densities (25 m^{-3}). This study nicely illustrates the important role that the density of goals has on the success of a search.

MAINTAINING A STRAIGHT COURSE

As just pointed out, there is often a significant advantage to maintaining a straight course while ranging, because it avoids searching the same area twice, but this is not easy to carry out. In the absence of external signals, all animals (and machines) can maintain a straight course only for limited times or distances. On a microscopic scale the limitation is due to Brownian motion; swimming bacteria can in fact maintain a straight course for only a few seconds (Berg and Purcell 1977). For large animals, random perturbations are less important than the inevitable biases that occur in locomotion.

When Schaeffer (1928) studied the tracks of people deprived of visual and auditory information, he found that they indeed tended to move in circles, as reported by folklore. Whether the human subjects were walking, swimming, or driving a car, they traced irregular paths containing many circular elements, with a radius of curvature typically on the order of 10 m.

Some aquatic organisms reduce the effect of bias in their direction of swimming by constantly rotating around an axis parallel to their direction of locomotion. This movement causes the bias to be sequentially aimed in opposite directions so that the bias is largely canceled out. Such spiral swimming can be observed in many protozoa, rotifers, and motile algae (Jennings 1904; Schaeffer 1928; Foster and Smyth 1980). It undoubtedly helps them maintain a straight path over larger distances than would be possible otherwise. This strategy is also used in the design of unguided rockets (Molitz 1966).

Another strategy for maintaining a straight course is to use environmental stimuli. In this respect, following a stimulus may be useful even if the stimulus is unrelated to a particular goal. As suggested earlier, a stimulus that serves the purpose of maintaining a straight course can be called a **collimating stimulus** (Pline and Dusenbery 1987), to distinguish it from stimuli that serve the function of leading to a goal. An organism may make use of a collimating stimulus by a variety of different behaviors like klinotaxis, tropotaxis, or menotaxis (see Section 17-1), but the stimulus serves the same function.

For example, Robin Baker (1969) observed that certain species of butterflies and moths—which do *not* exhibit long-range migration—frequently fly in straight lines as if they were in fact going somewhere, but different individuals fly in different directions. They seem often to use celestial cues but not to compensate for rotation of the earth. This method would seem to provide sufficient guidance to let them avoid backtracking, but not enough accuracy for migration to a specific place.

The concept of collimating stimuli also provides one explanation of why night-flying moths are attracted to lights, which they presumably mistake for the moon (Janzen 1984). Nielsen (1961) described cases where butterflies followed leading lines (see pages 456–60) but broke away toward a goal when downwind of it. Presumably the wind carried a chemical cue, but following the leading line produced a straight search path until the chemical stimulus was detected. Similarly, fruit flies fly straight when illuminated from above with polarized light, but not when the light is unpolarized (Wehner 1984). Jander (1975) provides other examples of this type of behavior.

The concept of collimating stimuli predicts some otherwise unexpected interactions between stimuli. For instance, it suggests that microscopic, motile algae, which swim toward or away from light sources (Foster and Smyth 1980), can follow shallower chemical gradients in a beam of light rather than in diffuse light, because they can use the light to maintain a straight course. This prediction apparently remains to be tested.

CROSS-CURRENT RANGING

In the discussion of ranging at the beginning of this chapter, it was tacitly assumed that the range of detection was the same in all directions from the goal. There are, however, several situations in which this assumption is significantly violated. One of these occurs when a chemical stimulus is carried by a current of air or water. In such a situation the stimulus spreads symmetrically across the current, by diffusion and perhaps turbulent mixing. However, along its flow it is retarded in its movement up the flow but is carried down the flow farther than it spreads in other directions. The result is that the stimulus is distributed in an elongated shape called a **plume,** described in Section 4-3.

If an animal is searching for a stimulus plume can determine the direction of current flow, it may make sense to search by moving in a particular direction. The basic argument here is that the plume presents a larger target size when it is approached across the current than when approached parallel to the current.

The overall dimensions of a plume (width W_p, length L_p, and height H_p) for uniform flow can be estimated by

$$L_p \approx J/4\pi \, D \, I_{Th}$$
$$W_p = H_p \approx (J/8.54 \, v_C \, I_{Th})^{\frac{1}{2}} \qquad (16\text{-}20)$$

where I_{Th} is the threshold intensity, J the rate of release of chemical stimulus, D its diffusion constant, and v_C the velocity of the current. Calculations of cross sections from different angles have been made for rectangular and ellipsoidal shapes (Dusenbery 1989c). If θ is the angle between the course direction and the downstream direction (Figure 16-5) then the cross section for a rectangular plume is given by

$$\sigma(\theta) = (L_p \sin \theta + W_p \cos \theta) \, H_p \qquad (16\text{-}21)$$

For a two-dimensional search $H_p = 1$. A cross-current search can thus be seen to be more efficient than a parallel search, by the ratio L_p/W_p. The average cross section for random directions in two dimesions is

$$\langle \sigma(\text{random}\theta) \rangle = (2/\pi) \, (L_p + W_p) \, H_p \qquad (16\text{-}22)$$

On average, searching cross current is more efficient than searching in a random direction, by about 50 percent, since $\pi/2 \approx 1.57$.

This calculation assumes that the animal can move with the same velocity or cost in any direction with respect to the goal, a reasonable assumption for an animal walking on a solid surface on which the goal is

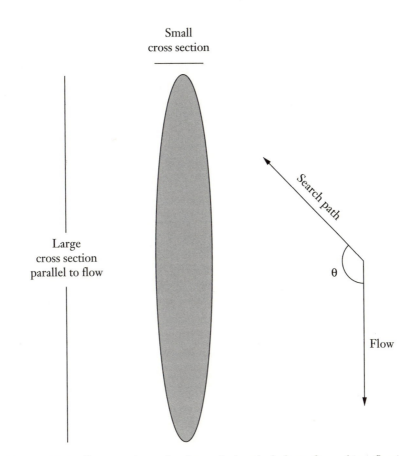

Figure 16-5 Cross sections of a plume. A chemical plume formed in a flowing medium has a large cross section along the direction of flow but only a small cross section across the flow. Consequently, the probability of finding the plume depends on the direction of the search path.

stationary. However, with an animal flying or swimming in the same medium carrying the stimulus the situation becomes more complicated, because the animal's net velocity with respect to a stationary goal depends on its heading. If v_a is the velocity of the animal with respect to the medium (air speed) and v_c the velocity of the current with respect to the goal, then breaking down the motion into components that are parallel and perpendicular to a rectangular plume indicates that the search rate (that is, the volume or area searched per unit of time) is

$$s/t = [(L_p \sin \theta + W_p \cos \theta)\, v_a + W_p v_c]\, H_p \qquad (16\text{-}23)$$

Figure 16-6 shows the relationships for ellipsoidal plumes, assuming that $v_a = v_c$. For a highly elongated plume the optimal course heading is nearly across the flow. However, an active space that is spherical has the same cross section in all directions, so that the optimal course direction is with the flow, which produces the highest speed with respect to the goal. Plumes that are only slightly elongated have optimal search directions that are intermediate between the downstream direction and across the flow (Dusenbery 1989c).

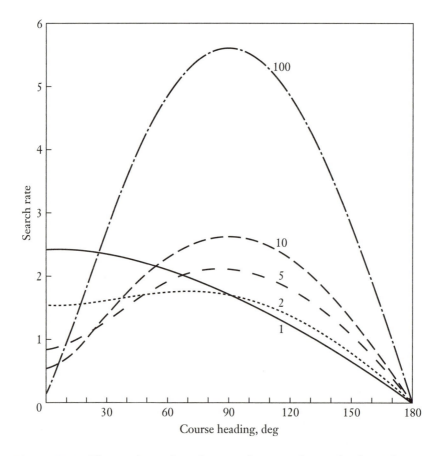

Figure 16-6 The search rate for a plume as a function of course heading. These calculations are for ellipsoidal plumes of equal volume, with the indicated axial ratios of 1, 2, 5, 10, and 100. The searcher's air speed is assumed to be the same in all directions and equal in magnitude to the speed of the current with respect to the goal.

In all these cases, the advantage of searching in the optimal direction over random directions is about 50 percent. If the active space is spherical or highly elongated, or the locomotor's speed is much greater than the speed of the current, the advantage is a factor of $\pi/2 \approx 1.57$ (Dusenbery 1989c).

Another situation in which cross-current ranging proves less than optimal occurs when the direction of flow fluctuates. If the fluctuations are rapid compared to the searcher's rate of movement, the plume can be thought of as a sector with its vertex at the goal and extending downstream with a radius equal to the range of detection and an arc equal to the range of the current's directions (Figure 16-7). If the range of current directions is more than $\pm 30°$ from the mean current direction, the plume will be wider than it is long, and the greater cross section across the current will make a search along the mean current direction more efficient than one across it (Sabelis and Schippers 1984). This analysis is more appropriate to searchers that travel by pushing against the stationary substrate (walking or crawling) than to organisms that push against the moving medium (swimming or flying). The original conclusion that downstream searching is superior to upstream is now thought to be true only if the searcher moves toward the goal relatively inefficiently within the area of the plume's influence. Otherwise, upstream searching reduces the total distance traveled to a goal (Dusenbery 1989f).

A basic question is whether the effective search rate depends on the particular shape of the active space aside from its volume. It might be anticipated that spreading a given volume of active space into a thinner (in the search direction) shape would increase the search rate, because additional depth along the search direction does not contribute to success, but extending the shape in other directions increases its cross section. Calculations and computer modeling indicate that there is a small decrease in search rate for axial ratio increases, from one to about three, and show an increase at larger ratios that becomes proportional to $(L/W)^{1/3}$ (Dusenbery 1989). Thus, an active space of a given volume is indeed easier to locate if it is elongated (see also Figure 16-6).

These concepts need to be tested, but there is unfortunately very little data on what free-flying insects or free-swimming fish actually do, or on their success rate in finding goals. It is hoped that advances in electronics will make such experiments increasingly feasible.

TRAIL FINDING

Another type of goal is a trail, which is to be distinguished from the point, spherical, and elongated goals previously discussed by being extremely

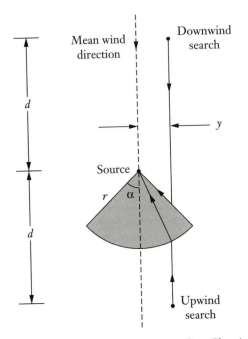

Figure 16-7 Searching in a fluctuating flow. Chemical stimuli emanating from the source are assumed to generate the plume that is sector shaped when averaged over time periods greater than those of the fluctuations in the direction of the current or wind. The search paths (the heavy lines) are shown for searchers starting distance y from the axis of the mean wind direction and distance d up- or downwind of the source. If the searcher can move efficiently to the source once it is in the plume, then an upwind search produces a shorter search path than one downwind, because it avoids backtracking.

thin in its width as compared to its length. Let us assume here that width can be entirely neglected. Trails represent an extreme case of elongation, whereas plumes are intermediate in their length.

In order to analyze trails, let us assume them to be long and straight over distances typical of those between different trails and assume also that the range of detection is small in comparison to both the trail's length and the distance between trails. Doing so allows the search to be modeled by lines where detection occurs only when a search path intersects a trail. If trails are short or very curved they may be more appropriately modeled as point goals. Assume in addition that trails are distributed at random, with random orientations.

What is the optimal search path or ranging pattern in such conditions? The search path can be represented by connected straight-line segments.

These segments may be made sufficiently short so that this is a good approximation to any realistic path. Next, what is the optimal sequence of segments? Since there is assumed to be no directional preference for the first segment, it is possible to move in an arbitrary direction. If a trail is not located, what is the optimal direction for the next segment? As the results of the first segment indicate that there are no trails crossing this path, the optimal choice for the second segment becomes a path that does not test for trails that are already known not to exist, from the results of the first segment. The only choices that meet this requirement are segments on the line passing along the first segment and not overlapping any segments previously searched, as shown in Figure 16-8. Therefore, continuing on in the same direction is an optimal strategy. In this case a discontinuous

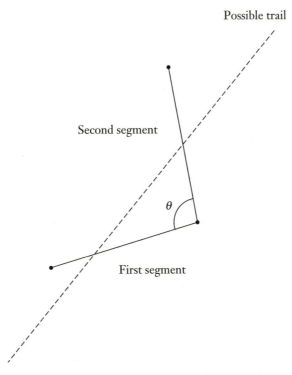

Figure 16-8   The optimal search path for trails. The second segment of the search path tests for possible trails, some of which have already been found to be absent during the first segment, unless θ = 180°. Thus, searching in a straight path is the optimal pattern.

search does not have the advantage it has for a point goal. If the trails are known to have some preferential orientation, the optimal strategy is to search in a direction perpendicular to the preferred orientation, as for an elongated plume.

SEARCHING DEFINED AREAS

In some situations, the appropriate search area can be divided up into many isolated areas that are each small in diameter in comparison to the length of the probable search path but larger than the detection range. In such cases a straight-line strategy is not appropriate, because the boundary of the search area will often be encountered. An alternative strategy is to follow this boundary, always keeping it on the same right or left side. This simple strategy allows an animal to explore an area without overlapping its previous search paths, assuming it recognizes its starting point when it returns (Figure 16-9). This strategy is apparently used by ants in searching

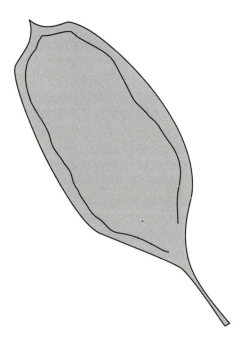

Figure 16-9 Searching by keeping a boundary to one side. The heavy line represents a search path using this strategy.

the leaves of a plant (Jander 1977, 155). In this situation the edge of the leaf is serving the function of a collimating stimulus.

16-2 LOCAL SEARCH

In ranging it was assumed that goals of equal value were distributed either uniformly or randomly in space. If, on the other hand, an animal finds itself in a situation where a goal is likely to be nearby or a nearby goal is more valuable after discovery than a distant one, it becomes expedient to search the local area more thoroughly than distant ones. The center of such a local search is here called its **focus.** Appropriate situations for employing a local search include the following:

1. Searching for a goal, such as food, near a home to which the searcher must return. This behavior is a component of *central place foraging.*
2. Searching for a goal in an area indicated by information that is insufficient to find clues leading directly to the goal, as, for example, when an ant employs celestial cues to return to the vicinity of its nest.
3. Searching for a goal typically found in a patchy distribution in an area where a similar goal was previously found.

There are a number of different search patterns an organism might adopt in a local-search situation. If the search is confined to two dimensions, as on a surface, one simple strategy is a **spiral search**. In this case the animal starts at the focus and circles around it in one continuous curve. If the animal has a definite range of detection (r_d) it can, in principle, perform an optimal clean-sweep search by maintaining a path that is always $2r_d$ from the path already traversed. Such a path forms an Archimedean spiral ($r = r_d\,\theta/\pi$, with θ in radians), as shown in Figure 16-10. In practice, it may be difficult for a searching animal to follow such a path. Furthermore, if the searcher does not have a reliable definite range of detection it may miss a nearby goal and will not have a second chance to find it with this strategy (Hoffmann 1983b). These disadvantages probably account for the fact that search pattern is rarely observed (Wehner and Srinivasan 1981). However, many people who work with honeybees seem to believe that they employ spiral search patterns (Wenner and Wells 1990, 322, 323, 326).

Another simple strategy, which works in three as well as two dimensions, is to pick a direction at random and move out along a radius, then return to the focus and repeat the process in a new direction. This strategy,

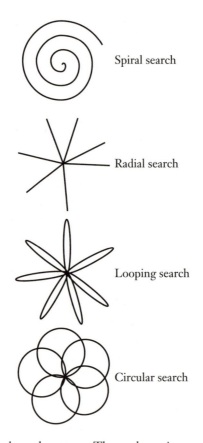

Spiral search

Radial search

Looping search

Circular search

Figure 16-10 Local search patterns. These schematics represent some idealized patterns for local search. The focus of each search is at the center of its pattern. The spiral search can, at least in principle, be a clean-sweep search; the other three cannot. The bottom three patterns can be viewed as variations on the theme of testing a randomly chosen direction out to a certain distance. They could be improved beyond what is shown here by increasing the distance to which each new search is made as the shorter excursions prove unsuccessful.

which may be termed a **radial search**, automatically generates a higher search density close to the focus. For two dimensions its search density is $1/2\,\pi r$ per trip; for three dimensions $1/4\,\pi r^2$. Harvester ants, for example, often use this type of search pattern (Hölldobler and Wilson 1990, 615). The searching efficiency of this process can be nearly doubled by returning along a different radius. Since this strategy requires moving along straight lines, it is probably most appropriate for animals that can define their direction by using celestial cues or distant landmarks.

A path that returns along a different radius generates a loop. Such **looping search** patterns (Figure 16-11) have been described for a variety of animals (Wehner and Srinivasan 1981; Bell 1985; Bell et al. 1985; Fromm and Bell 1987; Hoffmann 1983a). An idealized version of this strategy would be a circle with the focus on its circumference (**circular search**) (Figure 16-10). One lobe of this search path is generated simply by maintaining a constant tendency to turn to one side. However, if this were to be done exactly the same circle would be followed over and over. An efficient search therefore requires a certain amount of variation in direction near its focus to change the orientation of the different lobes of the search. This strategy generates a search density similar to that of a radial search close to the focus, but the density increases above it as the outer limits of the search area are approached.

Radial, looping, and circular search patterns may be further improved by starting with small excursions out from the focus and expanding the search with ever-larger excursions as the search continues unsuccessfully.

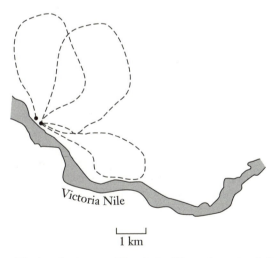

Figure 16-11 Tracks of primates. The dashed lines show the daily tracks of savanna baboons, which return to the same area (*solid dots*) for the night. The daily tracks are similar to the looping search patterns in Figure 16-10. (After Baker 1978, 386.)

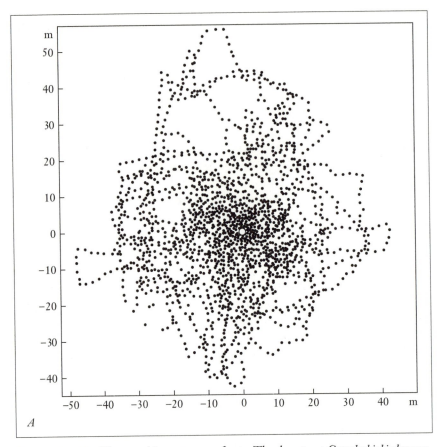

A

Figure 16-12   The searching patterns of ants. The desert ant *Cataglyphis bicolor* uses celestial cues to return to the vicinity of its nest but is not sufficiently accurate to come within range of detection. Thus, it must initiate a local search with the focus at the point where the nest is expected to be. The panel above shows the distribution of search paths for several individual ants. Notice that the density of their searching is higher near the center but falls off with distance from the focus. This relationship is shown explicitly in the second panel on page 409. The third panel on page 410 shows the distance of four individual ants from the focus as a function of time. They alternately move away from and back toward the focus, as in the looping model. In addition, the maximum distance of the loops increases with time. (From Wehner and Srinivasan 1981.)

This pattern has been observed both with desert ants (Figure 16-12) (Wehner and Srinivasan 1981) and isopods (Bell 1991, 76; Hoffmann 1933a). With circular search patterns this change corresponds to a decrease in the turning angles, and the pattern gradually approaches that of

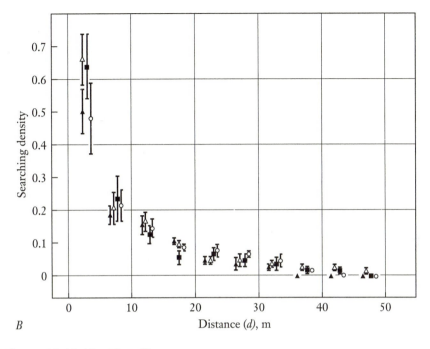

B

Figure 16-12 (*Continued*)

the straight-line search of ranging as the curvature decreases. Bell and associates (Bell 1985; Bell et al. 1985; Fromm and Bell 1987) have good evidence that walking flies reduce their rate of change of direction in this way when they are searching for food distributed in a patchy manner.

Next consider an argument for some features of an optimal strategy (Wehner and Srinivasan 1981). In most cases, the simplest assumptions about the probability distribution of a goal around the focus ($r = 0$) is that it is radially symmetric. If so, it can be described by function $p(r)$. It is reasonable to expect that this distribution will have the form of a two-dimensional Gaussian distribution, (Hoffmann 1983b) but the argument requires only that the probability decrease as r increases. It is further assumed that in searching a given part of this area that in fact contains the goal the probability of discovery is limited ($p_d < 1$), meaning that this is not a clean-sweep search. Then the search should be driven by the consideration of searching whatever area has the highest remaining probability of discovering the goal, given the previous searching in the local area. The search path should also be continuous, since movement cannot

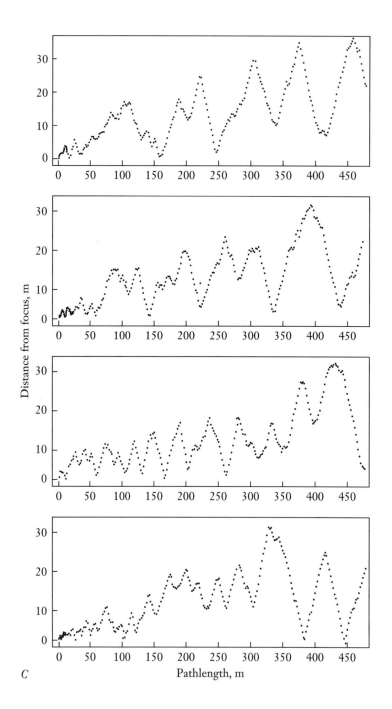

C

Pathlength, m

Figure 16-12 (Continued)

be discontinuous, and searching usually does not require much reduction in the speed or cost of movement. The optimal pattern of searching is, then, to begin at the focus ($r = 0$) where the initial probability of finding the goal is highest, then search concentric rings of increasing radii until the point r_1 is reached where the original probability becomes lower than that at the focus by the probability of discovering the goal in a searched area, or $p(r_1) = p_d\, p\,(0)$. The maximum remaining probability will then be back toward the focus and back out again to the larger radius (r_2), where $p(r_2) = p_d{}^2\, p(0)$. This process repeats indefinitely in concentric rings with an increasing radius followed by decreasing to zero, followed by an increase to a still-larger maximum radius. Concentric rings have not been observed in the search paths of real animals, but the feature of varying the distance from the focus is a characteristic of the radial, looping, and circular search patterns and can be seen in the actual search behavior of desert ants (Figure 16-12). A quantitative analysis of searching by desert isopods (Hoffmann 1983b) suggests that their search strategy has the basic pattern of search density suggested by this type of argument.

In conclusion, animals are rarely found searching in the idealized patterns described here, probably because in most cases the goal occurs in association with certain features of a nonuniform environment that provide information concerning the probability of a goal being nearby, which alters the search pattern constantly in response to the new information. In such cases a search should be guided by various cues as they are obtained, and no general prediction about the ranging patterns can be made. This situation is probably the most common one for animals that can easily be observed. Nonetheless, these basic strategies must be understood in order to make sense of more-complex behaviors. A book discussing searching behavior in detail has recently been published (Bell 1991).

CHAPTER 17

Guiding

■ ■

This way it passed: the scent lies fresh;
The ferns still lightly shake.
—Henry Augustin Beers, *Ecce in Deserto*

After an organism has detected a stimulus that can be used to move closer to a specific spatial goal, the approach phase begins. If the stimulus has a simple relationship to the goal, simple reflex behavioral mechanisms may be employed, and the behavior is termed **guiding.** In this chapter the organism is assumed to be in a stimulus field, information from which is used to guide locomotion. In most cases the stimulus field provides information by the way in which its intensity is spatially distributed. For light or sound the field has directional components as well, and often special strategies are employed for determining the position of the source. These techniques were discussed in the chapters in Part 2 devoted to the particular stimulus in question. (In addition, the visual guiding of arthropods is discussed in great detail by Wehner [1981].) This chapter concentrates on general strategies appropriate to stimulus fields that have only intensity information. It is worth pointing out that these simple strategies are of interest primarily in understanding organisms that cannot employ sophisticated ones, and this chapter does not provide much enlightenment concerning vertebrate behavior.

Depending on the characteristics of the stimulus field and the abilities of the organism, guiding may either be *direct* in the sense of taking a straight-line path to the goal or *indirect*, as in using a biased random walk to reach the vicinity of the goal. This distinction is important, because the direct approach requires that the animal have some mechanism for rapidly determining the direction to the goal, whereas the indirect approach does not. Special strategies may be employed for certain kinds of stimulus fields such as chemical plumes or trails. Each of these types of guiding is discussed in turn after a discussion first of terminology.

17-1   TERMINOLOGY

A general scheme for classifying the various strategies used by simple animals to move toward a goal was presented by Fraenkel and Gunn (1961) in an influential book. Since then a number of improvements on their scheme have been proposed (Bursell and Ewer 1950; van der Steen and ter Maat 1979; Bell and Tobin 1982; Burr 1984; Kennedy 1986; Benhamou and Bovet 1989; Dunn 1990), but conflicts in terminology remain a problem. In particular, microbiologists tend to use different terms or meanings than do zoologists. The history of the development of terminology in this area has been presented in detail by Schöne (1984). In some cases the terminology has proliferated to the point of excess (Wehner 1981, 379), but there are some important concepts involved, so it is desirable to trim the terminology to what is generally important.

Burr (1984) has provided a detailed discussion of the problem and proposed a classification scheme that attempts to reconcile the differences in use. Because his scheme has a lot of merit, some of his definitions are presented in Table 17-1.

In addition to the terms defined in Table 17-1, several others are in common use. Terms like **chemotaxis** and **phototaxis** are frequently used in the more general sense of meaning simply oriented migration in response to the indicated stimulus, in which case *taxis* does not suggest any specific mechanism, as it does in the table. The term *taxis* should at least be restricted to behaviors that involve orientation in response to stimuli carrying information and should not include behaviors in which an orientation is imposed by physical forces. Thus, it is misleading to use *magnetotaxis* to describe the forced orientation to the earth's magnetic field of bacteria containing magnets, as was discussed on page 35. Similarly, *geotaxis* is a poor term for the vertical orientation of the vinegar eelworm, which is oriented by buoyant forces because its center of gravity is behind its geometric center (Peters 1952). The term **tropism** is frequently used for behavior in which the response is oriented growth. **Menotaxis** is sometimes used as a term for the orientation at an angle to the direction of the stimulus field.

Table 17-1 Classification of the Elementary Mechanisms of
Locomotion Guided by a Stimulus Field*

Kinesis: "Behavior comprised of *undirected* responses that are dependent on the intensity or temporal change in intensity of a stimulus." Undirected responses have no bias with respect to the orientation of the stimulus or stimulus field.

  Orthokinesis: "A kinesis in which *translational* motion is affected."

  Klinokinesis: "A kinesis in which *rotational* motion is affected." *Rotation* refers to a change of the direction of locomotion.

Taxis: "Migration oriented with respect to the stimulus direction or gradient which is established and maintained by *directed* turns." Directed turns are ones that are biased in some way with respect to the orientation of the stimulus field.

  Klinotaxis: "A taxis which results from directed responses to *sequential samples* of stimulus intensity or direction."

  Tropotaxis: "A taxis which results from directed responses to two (or in three dimensions, three) *simultaneous samples* of stimulus spatial distribution or directional distribution."

  Teleotaxis: "A taxis due to directed responses to information gathered by a raster of many receptors. Directed turning occurs until the 'fixation area' of the raster is exposed to the directional stimulus." An extreme form of tropotaxis.

* The quotations are from Burr (1984). Some of italics in original and some added.

With regard to the terms in the table, orthokinesis has relatively little effect on the distribution of organisms—even single cells have more effective strategies—so it is not discussed further.

In order for an organism to determine the orientation of a stimulus field, it is necessary to measure stimulus intensity at different positions. The two fundamental alternative strategies for obtaining the information are

1. **Simultaneous sampling** by multiple receptors separated on different parts of the organism's surface. In this case the organism directly measures the spatial gradient by comparing the intensity at the different positions.
2. **Sequential sampling** by a single receptor moved from one place to another. In this situation the organism directly measures a temporal gradient, then infers the spatial gradient from information about how the receptor was moved.

The simultaneous comparison of signals requires having widely spaced receptors for sensitivity to detect intensity gradients, which makes large body size advantageous. In addition, locating receptors on antennae can provide even greater separation. Bacteria are too small for them to use simultaneous comparison effectively (Berg and Purcell 1977; DeLisi 1982). On the other hand, sequential sampling requires a coherent pattern of movement such as locomotion in a straight line. This need can also be limiting for small organisms that must contend with the random Brownian motion induced by molecular collisions. Sequential sampling also requires some form of memory so that comparisons with previous intensities can be made.

Another fundamental distinction is based on whether an organism is able to make turns in its locomotor path that will direct it toward its goal. This fundamental distinction is the basis of the difference between kinesis and taxis. However, it seems better to use the terms *direct* and *indirect guiding*, because *taxis* is used with other meanings, and few biologists now know Greek. The two fundamental alternative strategies for responding to a stimulus field are, therefore,

1. **Direct guiding,** in which the organism obtains information about the orientation of the stimulus field in relation to itself and is more apt to turn in the appropriate direction than in another one. This category includes various forms of taxis as defined in Table 17-1.
2. **Indirect guiding,** in which all turns are oriented randomly with respect to the stimulus field, because the organism does not obtain information about its direction. Indirect guiding is equivalent to klinokinesis.

The relationships between the fundamental distinctions in sampling and response and the associated terms are shown in Table 17-2.

In indirect guiding, some organisms make simple turns of various angles, whereas others turn by initiating a relatively fixed maneuver that produces a specific change of direction. Kennedy (1986) has suggested separating these cases into two classes, but this does not seem desirable at

Table 17-2 The Logical Relationships Between Elementary Mechanisms

Sampling Pattern	Turning Response	
	Indirect	Direct
Sequential	Klinokinesis	Klinotaxis
Simultaneous	(Not appropriate)	Tropotaxis

this level of analysis. Some microbiologists (Nultsch and Häder 1988) use the term **phobic response** for such fixed maneuvers.

17-2　INDIRECT GUIDING

In terms of an organism's sensory abilities, the least-demanding strategy for following a stimulus field is to change direction to an arbitrary direction more frequently if the stimulus is changing unfavorably, or less frequently if the stimulus is changing favorably. More simply, it should keep on its present course if the stimulus is improving, but otherwise try a new direction. An important feature of this strategy is that the organism never needs direct information concerning the direction of the goal, being able to make progress toward the goal by means of a biased random walk.

This strategy requires only that the organism have one receptor. In addition, its efficiency can be greatly improved by having some sort of memory to enable the organism to compare the intensity of the present stimulus with intensities it has recently experienced. This is in fact a weak requirement. Nearly all biological receptor cells exhibit some degree of sensory **adaptation,** which emphasizes information related to temporal changes in intensity at the expense of information about the absolute level of intensity (see Section 5-3).

Early researchers proposed that sensory adaptation may be necessary if indirect guiding (klinokinesis) is to result in net migration in a stimulus field (Ullyott 1936; Fraenkel and Gunn 1961), but later mathematical analyses suggested that adaptation might not be essential (Patlak 1953; Doucet and Wilschut 1987). However, a more recent analysis (Schnitzler et al. 1990) has identified flaws in the earlier mathematical treatments and demonstrated that an organism moving at a constant speed must have some kind of memory if it is to migrate in a gradient.

A computer model has demonstrated that in an ideal situation adaptation can actually degrade performance (Dusenbery 1989b). However, real organisms must deal with noise of various kinds. **Stimulus noise** is often considered, and has been previously discussed, assuming intensity is the only important aspect of the stimulus (in Section 5-1 as intensity noise). Less well recognized is that noise in the response is also important; **response noise** leads to unintended variations in the locomotor response of the organism (Dusenbery 1989b). Response noise can be further subdivided into **motor noise,** in which variations occur in locomotion from one time to another in an individual; and **developmental noise,** in which

Table 17-3 Types of Behavioral Noise

Stimulus Noise: Undesired variations in a stimulus, usually in its subjective intensity (that is intensity noise).

   Channel Noise: Noise and distortion associated with transmission and the nature of the stimulus.

   Environmental Noise: Extraneous signals from other sources.

   Receptor Noise: Thermal and molecular noise in the receptor or the associated signal-processing systems.

Response Noise: Unintended variations in the path of locomotion or some other response.

   Motor Noise: Response noise in which variations are between different time intervals for any given individual. This should average to zero over long time intervals.

   Developmental Noise: Response noise in which variations are between individuals within a population. This should average to zero over many individuals, but not for any single individual.

variations occur between individuals, as in a bias to turn toward one side. All these types of noise are compared in Table 17-3. If any of these noise types is large enough to degrade performance, adaptation can improve their efficiency (Dusenbery 1989b). The effects can be seen in the computer-generated tracks shown in Figure 17-1.

Usually there is an optimal adaptation-time scale that depends on the situation, including whatever noise the organism must contend with. For free-swimming bacteria the performance will be limited by motor noise resulting from Brownian motion (Berg and Purcell 1977). The adaptation-time scale is in fact matched to the noise level (Block et al. 1982). Bacteria that swim straighter, and thus have lower motor noise, swim for longer periods between changes of direction, meaning they have a longer time scale for adaptation (Mitchell et al. 1991).

Computer modeling indicates that performance can be further improved by having different adaptation time scales for increases in stimulus intensity than for decreases (Dusenbery 1989g). Similar asymmetries in signal processing have been observed in bacteria and nematodes. Examples of optimal adaptation rates are shown in the contour plots in Figure 17-2.

Performance is measured by **efficiency,** calculated as the ratio of the net distance moved toward a goal or along a stimulus field to the total distance moved during the same time interval (Benhamou and Bovet 1989; Dusenbery 1989b). Assuming that an organism moves at a constant speed, its efficiency is proportional to the time required for it to reach the goal. Measurements of orientation in studies of navigating animals often use the

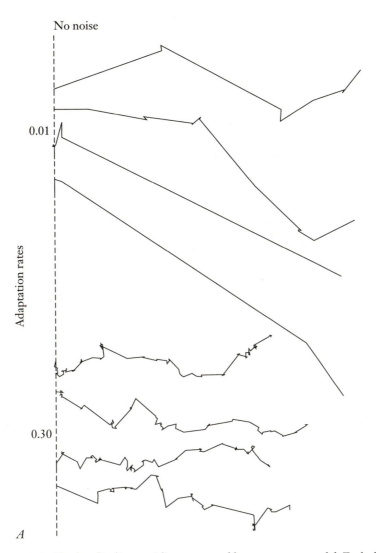

No noise

0.01

Adaptation rates

0.30

A

Figure 17-1 Tracks of indirect guiding generated by a computer model. Each thin line is the 1,000-step track of one individual, starting on the vertical dotted line. The initial step direction is vertical, perpendicular to the stimulus gradient, which increases to the right. A perfectly efficient track would appear as a horizontal line to the right of the dotted line. The top four tracks of each set were made with an adaptation rate of 0.01 per step, which is so low as to be equivalent to no adaptation. The bottom four tracks were made with an adaptation rate of 0.3 per step, close to the optimum. A set of eight tracks is shown for each of four noise conditions: none, intensity noise, motor noise, and developmental noise. This computer model and its results are described in more detail in Dusenbery (1989b). (*Figure continues*)

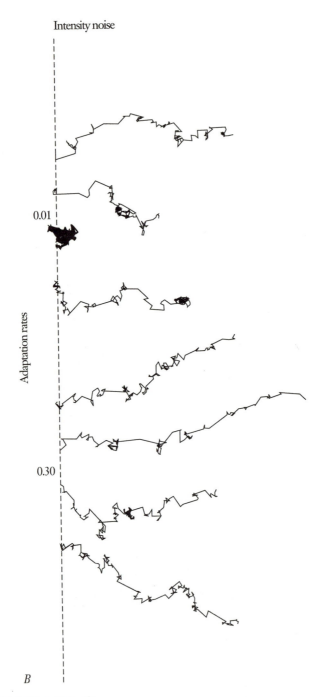

Figure 17-1 (*Continued*)

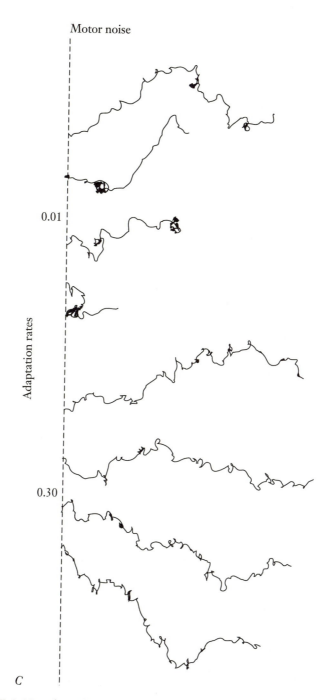

Figure 17-1 (Continued)

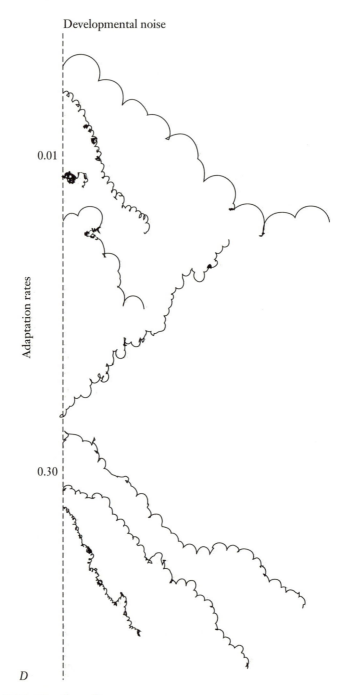

Figure 17-1 (*Continued*)

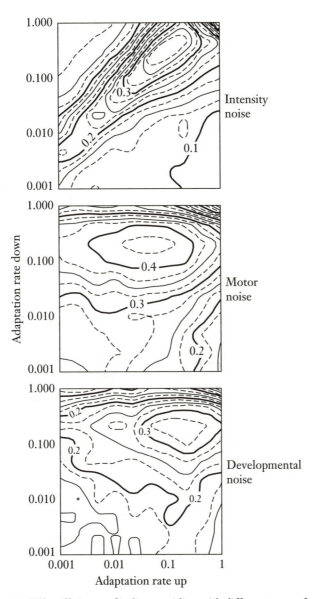

Figure 17-2 The efficiency of indirect guiding with different rates of adaptation for increases and decreases in intensity. The same computer model was used as in Figure 17-1 except that the rate of adaptation of increases in intensity was allowed to differ from the rate for decreases in intensity. This figure shows contour plots for the efficiency of guiding for each of three noise conditions. In all three cases the maximum efficiency was obtained with a higher rate of adaptation for decreases in attractant than for increases. For details see Dusenbery (1989g).

r vector devised by Batschelet (1981, 32–33). If the track of an organism is divided into segments and the vector of each segment is summed to determine r, then the length of r is the same as the efficiency of the track, assuming that r is aimed toward the goal.

In bacteria the efficiency of indirect guiding is related to the steepness of the intensity gradient (Dahlquist et al. 1976). The steeper the gradient, the greater the efficiency, at least up to a point. The maximum efficiency is surprisingly high. A noiseless computer model had efficiencies of about 90 percent for one-dimensional, 70 percent for two-dimensional, and 60 percent for three-dimensional movement in unpublished results I have obtained. Incorporating noise emulating the constraints placed on typical bacteria in this computer model produced efficiencies as high as 45 percent. Efficiencies of about 25 percent for bacteria (Dahlquist et al. 1976) and 10 percent for nematodes (Pline and Dusenbery 1987) have been observed, but probably neither measurement was a maximum. Bacteria that contain magnets are kept aligned with the earth's magnetic field (Blakemore 1982) and are thus confined to one-dimensional movement. This effectively doubles their potential efficiency of chemotaxis, again in certain of my unpublished results.

Even a small bias in the appropriate direction can greatly improve the searching efficiency over a completely random-walk search. For example, Fisher and Lauffenburger (1987) have analyzed mathematically the general problem of oriented movement in two dimensions. They decomposed movement into four cardinal directions, which can each occur with different probabilities. These directions they designate as toward, away from, and left and right of the line toward the goal. Thus, a variety of bias patterns can be analyzed. Unfortunately, the resulting equations involve integrals than can only be solved numerically. After evaluating the equations for a variety of parameter sets, an important conclusion is that a relatively modest bias—probability of movement toward the goal that is 0.2 greater than of movement away—can reduce the average time to reach the goal by as much as one hundredfold over a completely random walk (Figure 17-3). Larger biases provide relatively little improvement. These results have been used to demonstrate that oriented chemotaxis is both necessary and sufficient for alveolar macrophage to clear the lungs of bacteria at observed rates (Fisher et al. 1988).

In addition to their efficiency, the *sensitivity* of organisms using indirect guiding is important. Here sensitivity refers to how steep the intensity gradient must be in order for it to be followed. In order to be as general as possible, let us describe the stimulus gradient in terms of the

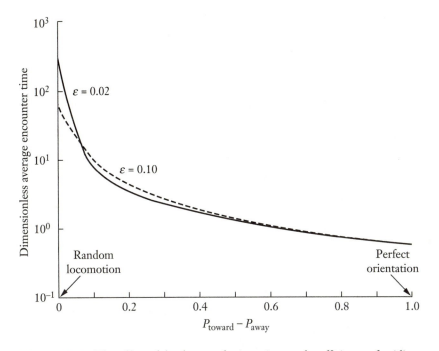

Figure 17-3 The effect of the degree of orientation on the efficiency of guiding. This graph, for a theoretical model shows the time to reach a goal as a function of the bias in locomotion toward the goal. The horizontal axis is the excess probability of moving toward the goal compared to away from it. The two curves are for different step sizes in the model. Notice that even a very small bias, producing a probability of moving toward the goal that is only 0.2 greater than of moving away, reduces the time to reach the goal by ten- to a hundredfold, and that stronger biases improve efficiency relatively little. (From Fisher and Lauffenburger 1987.)

relative gradient, $G = (1/I)(dI/dx)$, which eliminates the units in which intensity is measured and is equal to the reciprocal of the distance over which the intensity declines by $1 − 1/e ≈ 63$ percent. In this sense sensitivity is closely related to detection. Berg and Purcell (1977) have analyzed this question in great detail for chemical stimuli. Rearranging their results, one can conclude that the relative gradient (G_{Tb}) that can just be detected by an organism moving at speed v is

$$G_{Tb} ≈ (D\, C_0\, r\, v^2\, t^3)^{-1/2} \qquad (17\text{-}1)$$

where t is the time available for making the measurement, r is the effective radius of the receptor, C_0 is the mean concentration, and D is the diffusion coefficient of the stimulus (Dusenbery 1987). This relationship has been used to calculate that the gradient of carbon dioxide around roots can be used to guide bacteria only within the thin film of water on the root, but nematodes can potentially use it from a distance of a meter or two away, because of the greater time available to them to measure the signal's intensity before their orientation becomes randomized (Dusenbery 1987).

Although these indirect guiding mechanisms are very simple, other studies indicate that indirect guiding can lead to effective exploitation of a patchy environment (Benhamou and Bovet 1989).

17-3 DIRECT GUIDING

Directly orientating to a stimulus field is possible when an animal can obtain enough information to determine the orientation of the field with respect to itself. Then it can turn in the direction that brings it closer to the path to the goal. There are three basic strategies for obtaining information about the orientation of a field: klinotaxis, tropotaxis, and teleotaxis (see Table 17-1).

Klinotaxis requires only a single receptor. Information about the orientation of the stimulus field is obtained by moving the receptor or receptors from one position or orientation to another and then comparing the intensities. In simpler organisms, receptor movement is often the result of regular movements accompanying locomotion.

For example, swimming algae like Euglena and Chlamydomonas usually rotate around an axis parallel to their direction of locomotion and have photoreceptors that point to the side so that they scan the environment as the algae swim along. In response to changes in stimulation the algae turn in such a way that their direction of locomotion becomes oriented closer to the direction of illumination (Foster and Smyth 1980). Certain rotifers swim and respond to light in a similar fashion (Jennings 1904).

Another example is presented by wormlike organisms that move their heads in an undulatory motion to either side of their direction of locomotion. As a consequence, the sensory receptors located on their heads may be used to compare the stimulus intensity on both sides of the axis of the path, and small biases to turn toward the side having the greater stimulation can lead to their achieving an accurate orientation to a stimulus field.

This type of orientation is common in fly larvae guided by light (Fraenkel and Gunn 1961, 59–69) and is thought to occur in nematodes guided by chemical gradients (Dusenbery 1980; Burr 1984).

A critical test for klinotaxis is to pulse the stimulus. Pulsing at the frequency of the organism's oscillatory movements will be highly disruptive to its orientating itself by klinotaxis. Even better, if possible, is to pulse the stimulus during a particular phase of receptor movement. This procedure allows the experimenter to steer the animal in a predictable direction by using any stimulus orientation if klinotaxis is at work. Fraenkel and Gunn (1961, 65–67) described an experiment in which this was done by turning on and off a light shining down on fly larvae.

The efficiency of klinotaxis can be quite high, as the animals can quickly become oriented to the stimulus field and stay oriented to it as they move. But their sensitivity is a different matter. In many cases there is relatively little time available for a measurement to be made before the receptor is moved. In the case of chemical stimuli, the threshold gradient, G_{Th}, can be estimated (Berg and Purcell 1977) as

$$G_{Th} \approx (D\ C_0\ r\ v^2 t^3)^{-\frac{1}{2}} = (D\ C_0\ r\ d^2 t)^{-\frac{1}{2}} \tag{17-2}$$

where d is the distance between positions where measurements are made and v is the speed of movement of the receptor between them. If the receptor's speed of movement is comparable to organism's speed of locomotion, which is the usual case, this relation is the same as that presented for indirect guiding in Equation 17-1. However, the threshold gradient is likely to be much steeper (i.e., less sensitive), because the time interval available for sampling is less in klinotaxis than in indirect guiding (klinokinesis). This happens because an organism can usually maintain a straight path over a much longer interval than that in which it holds its receptors to one side of its body. One calculation for nematodes (Dusenbery 1987) predicts a thirtyfold difference in sensitivity. Thus, an animal may use indirect guiding when it is far from a goal, then switch to klinotaxis when it gets closer. The paths of several kinds of animals approaching a source are suggestive of this (Figure 17-4).

One possible modification of klinotaxis is for the organism to maintain its body orientation by ceasing locomotion, but *scanning* the receptors by moving the head or antennae. This would permit a much longer period for integrating of the signal. A nematode (Burr, personal communication) and flatworms (Taliaferro 1920) have been reported to do this.

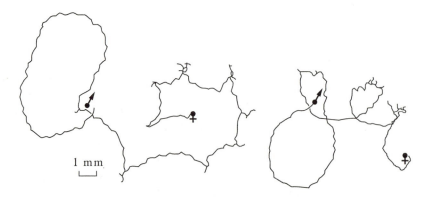

Figure 17-4   Track switching from direct guiding (klinokinesis) to klinotaxis. This plot shows the tracks of two male nematodes, *Heterodera schachtii*, as they approach females. Notice that the males make dramatic changes in direction, represented by the spurs in the tracks, at points where they may be assumed to start moving down a radial concentration gradient surrounding the females. However, once they are within a few millimeters of the female they move more directly toward her. (From Green 1966.)

These calculations assume that the searcher only compares the intensity at one extreme of motion with the intensity at the other extreme of the previous cycle of oscillatory motion. If the searcher possessed a more sophisticated signal-processing apparatus analogous to a lock-in amplifier, it might improve its sensitivity by averaging over several cycles the measurements at one extreme separately from those at the other (see page 294). This would require having at least two separate memories. Apparently, no organism has yet been shown to employ such a strategy, but it seems feasible and advantageous.

In contrast to klinotaxis, in which a single receptor is moved from one position to another, **tropotaxis** involves several receptors that simultaneously gather information about stimulus intensity in different places, to determine the orientation of a stimulus field. In many cases a bilateral pair of receptors will be involved. In the usual situation, if the receptors are stimulated unequally the animal turns in the direction of the one more strongly stimulated, assuming it is attracted to the stimulus. It then continues turning until the receptors on each side are equally stimulated. This strategy permits it to move straight to the source. Burr (1984) points out that for an animal moving in three dimensions at least three receptors are required to remove all ambiguity about its orientation, and many pelagic larvae have three or more pigment spot eyes.

A critical test for tropotaxis is to stimulate only the receptors on one side and determine whether the organism turns in the expected direction. This test is often carried out by inactivating one receptor of a bilateral pair and placing the animal in a uniform stimulus field. Tropotaxis will then often produce a continuous turning to one side, called **circus movements.** Many insects reveal tropotaxis in this way (Fraenkel and Gunn 1961; Bailey 1991).

Another type of receptor organization is found in motile cells that have thousands of receptor molecules distributed over their surface. It seems appropriate to include in tropotaxis the oriented locomotion of single cells that make simultaneous comparisons of receptor molecules distributed widely over their surface or throughout their volume. The crucial requirement is that the cell be able to determine differences in stimulation and not simply total the activity of all its receptors.

Some likely examples of tropotaxis by cells covered or filled with receptor molecules include the gliding bacteria (Häder 1987; Armitage 1988), and the slime mold responding to temperature gradients (Bonner et al. 1950) but apparently not to chemical gradients (Gerisch 1982; Vicker et al. 1984). Although leucocytes are not independent organisms, they face similar problems in locating infective agents and are reported to utilize tropotaxis with an efficiency of 85 percent (Zigmond 1974).

The sensitivity of tropotaxis is governed by the same equation given for klinotaxis except that the distance between sample positions, d, is here reinterpreted as being the distance between receptors. In many cases this distance is less than what is possible with klinotaxis, which leads to reduced sensitivity. On the other hand, a distinct advantage of tropotaxis is that the sampling time can be easily extended because spatially distinct samples are obtained in parallel, simultaneously. Some bacteria may grow in filaments in order to take advantage of this greater spatial separation and the longer integration times that occur since Brownian motion is reduced.

A quantitative comparison of the sensitivity of indirect guiding (klinokinesis) and tropotaxis has been made by Delisi et al. (1982), extending the quantitative analysis of chemotaxis developed by Berg and Purcell (1977). They include in their analysis the possibility that the rate of reaction of the stimulus with its receptor may be rate limiting, and they develop relations for the minimum integration time required to detect a gradient with indirect guiding (a temporal gradient) or tropotaxis (a spatial gradient). Tropotaxis requires relatively longer integration times for shallower gradients, smaller forward reaction rates, smaller cell sizes, greater locomotor speed, and higher stimulus concentrations. Delise and associates analyze the situation for bacterial and lymphocyte cells and conclude that tropotaxis would be too slow because it requires integration times on the order of a day with moderate concentrations, but they suggest that klinokinesis would be practical with its required integration times of only seconds to minutes. However, the ratio of the required integration times for the two strategies is a strong function of cell size, with larger sizes favoring tropotaxis. When there are rapid reaction rates such that diffusion is rate limiting and an excess of receptors over the surface, the integration time required for tropotaxis is less than that for klinokinesis when the speed of locomotion is

$$v < \pi\, r^4\, D\, C\, G^2 \tag{17-3}$$

With cell size influencing the relationship as the fourth power, a relatively small increase in the size of an organism makes tropotaxis preferable for it. For example, if an organism moved at a speed of 1 body radius per second in a stimulus concentration of 1 μM and a relative gradient of 1 cm^{-1}, and $D = 1 \times 10^{-9}$ m^2/s, then tropotaxis would require shorter integration times for organisms larger than 4 μm radius. It appears that cells do not have to be very large to make tropotaxis superior.

Because the sensitivity of tropotaxis is proportional to the distance between receptors, highly elongated animals might benefit from using head and tail receptors. However, for most, if not all, animals the head is much more highly endowed with sensory receptors than the tail, and no animal has been clearly shown to use head-to-tail comparisons. Suggestive evidence does exist for this mechanism in the terrestrial arthropod *Peripatopsis* (Bursell and Ewer 1950), and many nematodes have appropriate sense organs on their tails, but definitive tests have not been performed (Dusenbery 1980).

A large number of directional photoreceptors formed into an appropriate array can create an image of the external world with much more information than a single receptor can provide. This information allows for much more precise orientation, particularly in a complex environment. In particular, if more than one stimulus source is present an animal with an array of photoreceptors can choose one of them and move directly to it. With the previously described strategies, the best that could be done was to move along the plane of equal stimulation running between two sources. The strategy of obtaining and using this kind of information is called **teleotaxis**. This strategy really differs only in degree from tropotaxis, which has led some authors to suggest dropping the term (Burr 1984).

Many, and perhaps most, organisms are capable of employing more than one of these strategies, depending on circumstances. Thus, attempts to assign a given organism to a particular strategy often lead to contradictions. One way out of this confusion is to recognize that the various strategies represent a hierarchy of increasing sophistication, in the order presented in Table 17-1. What is of principal interest is to determine the most sophisticated strategy of which the organism is capable, for in many cases it will also be able to employ less-sophisticated ones when appropriate. A more detailed description of the organism's capabilities would include its sensitivity and efficiency for each strategy.

17-4 CHEMICAL PLUMES

A particular type of stimulus field that deserves special attention is the chemical plume produced when a chemical stimulus is carried by a current of air or water away from a stationary source. The problem of following a plume to its source is one encountered by many kinds of organisms. For a long time it was argued that many animals could determine the average concentration gradient by making simultaneous comparisons and consequently orient themselves to the plume. However, it is now recognized that turbulence causes such severe discontinuities in concentration that this strategy is of doubtful utility (see pages 75–86). In fact, many flying insects seem to expect a pulsed chemical stimulus and stop flying in an artificially formed cloud of stimulus that has a uniform concentration (Gillies 1980; Baker 1985). What makes up for this difficulty is the fact that the stimulus is carried with the current, and thus the source must be upstream. If the direction of the current is determinable, this information becomes quite valuable.

How can the direction of a current be determined? If a searching animal is walking on a substrate, the current can easily be sensed by the physical forces it exerts on the searcher and its sense organs. However, a flying or swimming animal—or machine, for that matter—experiences no such forces from a current. The only forces they experience are the by-products of locomotor activity. Unless it has accurate inertial senses (see pages 455–56) a flying or swimming animal must maintain some kind of sensory contact with external references. For most animals this is usually visual contact with the substrate (see pages 192–95).

As an animal moves with respect to the substrate, the images of patterns in the substrate move with it. The animal can, by comparing the direction and speed of the movement of images with its own direction and effort of locomotion, deduce the direction and speed of the current by making a calculation equivalent to course heading + wind drift = ground track (David 1986). For this to be possible, the animal's visual system must be able to resolve and distinguish patterns in the substrate. This necessity imposes strict requirements on the organism's visual system and maximum allowable distance from the substrate. However, the latter is usually not an important restriction, because the source is attached to the substrate and the current is parallel to the substrate, so that the plume will not be far above the substrate.

The need for image-forming eyes is more consequential. Organisms that swim but do not have image-forming eyes or are nocturnal probably cannot follow a chemical plume unless they can use alternative strategies. For example, bottom-dwelling catfish can probably maintain enough mechanical contact with the substrate to gain information about the current. Fish in open water have a more difficult situation, however, and it has been suggested that they are able to make use of characteristic patterns of fine structures in the vertical stratification of temperature or chemical stimuli to determine current direction (Westerberg 1984). Density gradients usually cause bodies of water to flatten out to an axial ratio on the order 1 : 1,000. As a result, a short vertical search can often sample chemical stimuli from sources representing a broad horizontal range.

Certain insects respond to airborne chemicals by flying into the wind in short hops interspersed with periods of walking, during which they can orient themselves to the wind (Kennedy 1986). This strategy should work if the hops are short enough that the insect's alignment is not lost. It has also been suggested that the dipping flight of some mosquitoes allows them to sense the wind's direction from the wind shear near the surface by rapidly changing their distance from that surface and thus affecting the strength of the wind (Gillett 1979), but this is highly speculative.

In general, animals that follow plumes have well-developed eyes or else move along the substrate. The ability to maintain an upstream orientation in response to chemical stimulation has been frequently reported for flatworms, snails (in both air and water), insects (flying and walking), crustaceans like crabs and lobsters, and fish (Bell and Tobin 1982).

A great deal of research has focused on how male moths follow plumes of sex pheromone's to locate females. Most detailed studies of this behavior have been carried out in wind tunnels, because of the vagaries of natural winds and the difficulties of recording flight paths in natural settings (Kennedy 1983; Cardé 1984; Baker 1985; Kennedy 1986). However, there exists one elaborate outdoor experiment performed with gypsy moths (David et al. 1983). In this experiment the position of the plume was followed with soap bubbles. Individual bubbles moved straight away from the source, and successive bubbles formed straight lines as a result of a constant wind, but bent lines due to shifts in wind direction. When a moth was in the plume, it flew upwind with shallow zigzags, which took it directly toward the source. But when it lost contact with the plume because of a shift in wind direction, it flew back and forth across the new wind direction, **casting** until it encountered the plume again (Figure 17-5). Because each parcel of the plume moved in a straight line from the

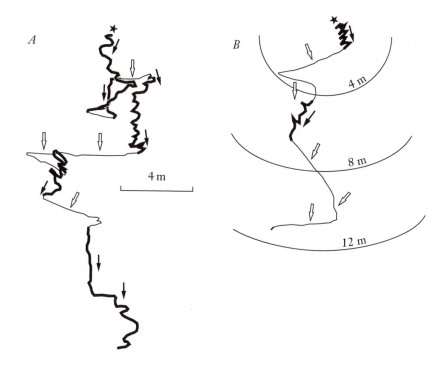

Figure 17-5   An example of a moth track in a plume. The source of the pheromone and the bubbles used to trace the air movements is indicated by the star at the top. Here gypsy moths starting downwind were tracked and their paths traced. When the path is apparently among the bubbles, it is indicated by a thick line, the associated wind direction by solid arrows. When it is not among bubbles, the path is indicated by a thin line, with the associated wind direction shown by open arrows. In (A) the moth made most of its progress toward the goal while flying upwind in the presence of the pheromone. In (B) another individual made most of its progress while flying across the wind outside the plume. (From David 1983.)

source, casting occurred at an angle inclined to that of the source, and the geometry was such that the plume was to the side in which the cast was closer to the source. Thus, even when a moth was outside the instantaneous plume it made progress on average toward the source; indeed, about a quarter of all progress was made in this way.

The responses observed in this experiment are similar to those seen in a variety of moth species in wind tunnels (Kennedy 1983; Baker 1985; Baker 1989) as well as in a detailed laboratory study of the oriental fruit moth in relation to a shift in wind direction (Baker and Haynes 1987). In the latter study, moths made rapid (0.2 s) adjustments to their flight paths, as determined by changes in the pheromone concentration experienced. In particular, they appeared to behave differently when experiencing increasing as opposed to decreasing concentrations, a response reminiscent of the indirect guiding of bacteria seen in Section 17-2.

The outdoor observations just described suggest that the moth's strategy can take advantage of the subtle aspects of plume geometry. Much more research will be necessary, however, to determine to what extent the important features of a plume change with the wind speed, topography, and atmospheric conditions and to discover whether animals can adjust their strategies for various conditions. Less-detailed evidence suggests that fish employ similar strategies (Johnsen 1984).

17-5 CHEMICAL TRAILS

A chemical trail is distinguished from a plume by its being formed by the movement of the source instead of the medium carrying the chemical stimulus. Consequently, there is no simple way for a searching animal to determine which direction along a trail a source is located. The prime questions in studying trail following therefore concern whether the polarity of the trail can be determined and, if so, how this is done.

Although trail following in ants has been much studied, there appears to be no detailed information on how they actually do it. The best hypothesis is that they sense the gradient of an airborne stimulus at the edge of a trail by bilateral comparison of the stimulation levels of their two antennae (Hölldobler and Wilson 1990, 268–69). If this were their only cue, however, the direction of the source could not be determined, and there is no evidence that they do in fact determine direction.

Dogs are well known for their ability to follow scent trails. Dogs cannot use simultaneous sampling to determine a gradient, and a dramatic demonstration that they wander back and forth across a trail in order to locate its edges is given in Gibbons (1986).

Another type of trail is the one left in the wake of a source moving through water. A good example of such behavior is the observation by Hamner and Hamner (1977) that shrimp range by swimming in horizontal linear paths and, upon contacting a chemical trail left by a sinking piece of food, follow it downward rapidly and accurately. They apparently always follow a scent trail in the downward direction which suggests that sinking food must be more common than rising food. This method eliminates the problem of determining the polarity of the trail. These researchers estimated that trails as long as 20 m could be followed by shrimp. This nevertheless leaves open the question of whether scent trails having random orientations in water can be followed efficiently.

17-6 ELECTRIC DIPOLES

Electric dipoles are of interest to many kinds of fish because they may originate from potential prey or mates or competing males. In any of these cases it may be advantageous for an organism to move to the position of the dipole, and it is worth considering strategies for guiding their locomotion appropriately.

Dipoles are peculiar stimulus goals in that they consist of both a source and a sink of the stimulus. With electric dipoles the current flows away from the source and toward the sink. In between, the current generally follows a curved path, which makes it difficult to determine the location of a dipole from measuring local intensities at a distance (Hopkins 1988a; Kalmijn 1988a).

One simple strategy for an organism to move to a dipole is to align itself with the direction of the electrical current and travel along its path, which will eventually lead it to the dipole. An ability to orient themselves with respect to uniform electric fields has been demonstrated in several species: in fact, freshwater catfish can orient themselves with an accuracy of 30° or better (Kalmijn 1988a). Recent observations of the paths taken by a male electric fish in response to an electric dipole mimicking a rival male (Schluger and Hopkins 1987) suggest that in fact the fish do follow the line of current flow (Figure 17-6).

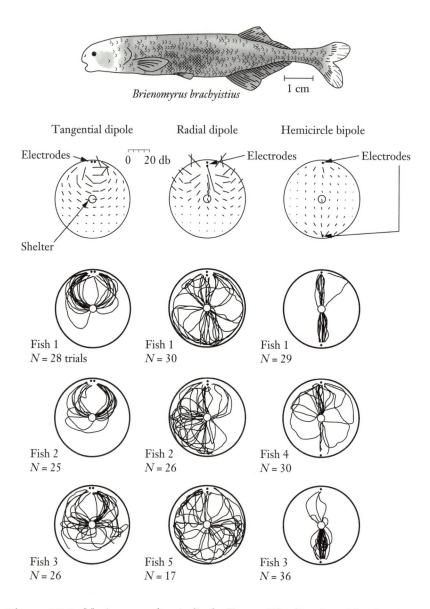

Figure 17-6 Moving to an electric dipole. *Top row*: The direction of the electric current for three electrode arrangements. *Lower three rows*: The paths of three individual fish moving to the electrodes in repeated tests. In most cases the fish followed the lines of the current's flow, even when this did not lead them directly to the goal. This is particularly clear for the tangential dipole on the left. (From Schluger and Hopkins 1987.)

CHAPTER 18

Navigating

Approach Using Stimuli
Not Connected to the Goal

■ ■

The sun is their guide by day and the stars at night.
When these are obscured, they have recourse to the points
from which winds and waves come upon the vessel.
—Captain Cook, ca. 1770, referring to the sailors of Tonga

Navigation, as the term is used here, is the process of finding a spatial goal by making use of stimuli that have no predictable relationship to the goal. Thus, animal navigation may make use of landmarks, the position of the sun, or the earth's magnetic field, just as human sailors do. The spatial goal is often some type of home such as a nest, roost, or burrow. Homing animals are favorable experimental subjects because the investigator has a good idea of what the goal is, of where the animal "wants" to go. (See Figure 18-1 for an example.) Experiments on animals' abilities to navigate usually test the ability of a species to home under certain conditions after being displaced by an experimenter. The other common behavior in which navigation plays an important role is seasonal migration, which is discussed in the next chapter.

Unraveling the question of how homing or migrating animals find their way has been surprisingly difficult. In spite of extensive experimentation with birds, much of the data remain confusing and seemingly contradictory (Baker 1984, 54). A fundamental problem is that it is rarely practical to trace the detailed path followed by an individual. A further difficulty is that many species use multiple strategies in navigating and

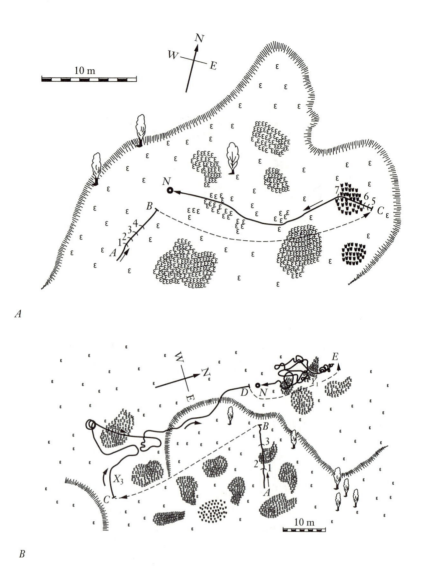

Figure 18-1 An example of homing: the path of an individual of the wasp *Ammophila pubescens* while dragging its prey toward the nest (*N*). The heavy line is the path taken by the insect; the broken line shows the displacement by the experimenter. The numbers indicate places where a metal screen was placed in the insect's path, causing it to make a detour. (From Thorpe 1963.)

switch between them according to the prevailing conditions. Given this redundancy, determining the navigation abilities of any one species is difficult, and in no case do we have a thorough understanding (Able 1980).

Table 18-1 Basic Concepts in Animal Navigation

Nature of Information Used to Determine Present Position	Required Sensory Abilities	Required Mental Abilities	Name of Strategy
Direction and distance from previous positions	Log and compass	Vector storage and addition	Dead reckoning
Present stimuli that are specific and local	Landmark recognition	Mental map with landmarks	Piloting
Present stimuli that are global	Map sense	Mental map related to map sense	Charting

Therefore, this chapter focuses on the basic strategies that have been suggested. The interrelationships of these basic concepts are presented in Table 18-1.

18-1 ACCURACY OF NAVIGATION

One of the basic questions to ask about navigation is how accurate is it. Clear evidence is difficult to obtain. One problem is that one rarely has good knowledge about the range at which goal-connected stimuli are encountered and guiding begins. Ascertaining the accuracy of celestial navigation is similarly complicated by a lack of knowledge about the range from the goal at which familiar landmarks become available. In addition, there are few data on the frequency with which a given navigational problem is successfully solved. Clearly, homing is not a completely precise and reliable behavior; whenever it is studied in detail it becomes clear that many individuals make mistakes, as seen in Figure 18-2. On the other hand, there are also examples of birds flying in a constant direction for hundreds of kilometers without deviation by local topography (Figure 18-3).

Flying insects that have been experimentally displaced often return to a nest from distances of many kilometers. The record is held by seven of twelve orchid bees that returned from 23 km away, some within two hours (Wehner 1981, 463). Success often depends on the experience of the individual, as can be seen in Figure 18-4, which shows the homing success

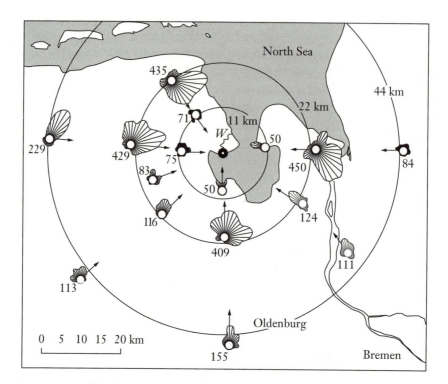

Figure 18-2 Vanishing directions of homing pigeons at different release sites. The home loft is near the center of the concentric rings. (From Wallraff 1959.)

of honeybees of different ages. Bees searching for a feeding site appear to remember landmarks with an accuracy of roughly ± 20° (Cartwright and Collett 1983). The communicating of position by dance language can apparently be sufficiently accurate to guide recruits to within ≈ 100 m (S.D.) of the goal (see page 340). An angular error of less than 8° can be achieved (Towne and Gould 1988). Whether the limitations are in the accuracy of the navigation or of the communication process is not known.

Desert ants forage as much as 100 m from their inconspicuous nests. Even in the absence of landmarks they can return to the position of their nest with an error of about 10 percent of the homing distance (Wehner and Srinivasan 1981). This corresponds to an error in direction of only 6°. Measured another way, these ants are 90 percent efficient, as the minimum distance to their home is only 10 percent shorter than the actual track

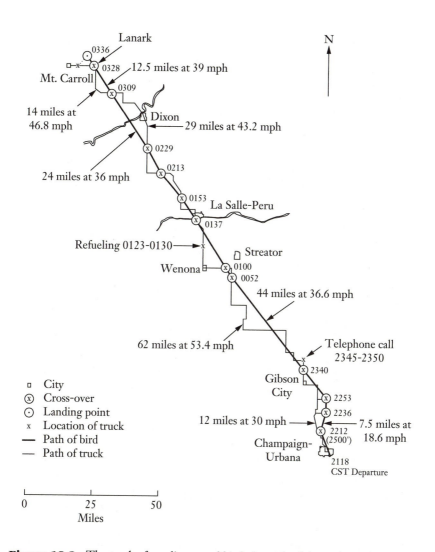

Figure 18-3 The track of a radio-tagged bird: the path of the nighttime flight of a migrating veery (heavy line) and of the truck used to track it (thin line). Once aloft, the bird keeps a constant direction of travel that is independent of topographical features. (From Cochran, Montgomery, and Graber 1967.) More detailed analysis [Cochran and Kjos 1985] of this sort of study indicates that spring-migrating thrushes maintain a nearly constant heading but their path usually wanders, because of varying crosswinds.

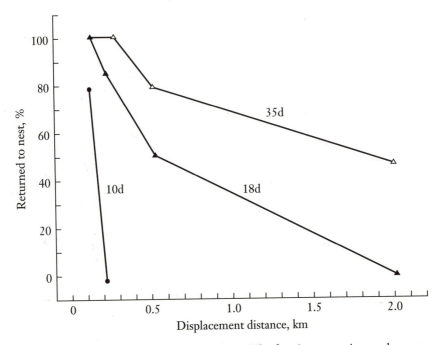

Figure 18-4 Homing success in honeybees. The fraction returning to the nest declines with the distance of the experimental displacement from the nest and is inversely related to the age of the bees. Here ages of ten, eighteen, and thirty-five days are presented. (After Wehner 1981.)

(Wehner and Flatt 1972). Extensive experiments indicate that this accuracy is accomplished primarily by the ants' recording the distance and direction of the outgoing track, then calculating a straight course back to the nest. If previous returns have been made from the same site, visual landmarks may also be used. However, if the ants are experimentally displaced only 2 m from the nest, they search over a median track length of 10 m; making it clear that they do not have a general mental map of landmarks surrounding the nest.

Salmon migrate thousands of kilometers, from the freshwater lakes and streams where they develop to oceans, where they spend several years growing before returning to their place of origin to spawn and die. Tagging studies have shown that among those that return about 95 percent return accurately to their place of origin (Jones 1981; Quinn 1984).

Whales migrate thousands of kilometers, with some returning to very specific locations. Perhaps the most impressive from the navigational point of view are the humpback whales that return to the Hawaiian Islands in the middle of the Pacific Ocean (Bauer et al. 1985).

Many studies have been performed on homing pigeons. In some of these, pigeons have been accurately tracked from aircraft. A particularly revealing case was that of a flock of pigeons released in a valley at an altitude of 1,700 m in the middle of the Alps and on the other side of a high (3,000 m) mountain range from their loft (Figure 18-5) (Wagner 1972). The pigeons followed the valley in the direction closer to the homeward direction (70° deviation) for 8 km, then climbed over the mountains in a series of switchbacks, flying close to the ground. Once they were well on their way, they were deflected relatively little from the homeward direction by topographical features. They crossed several valleys while flying 1,000 m above ground. After 93 minutes the pigeons landed near a water rill, that was less than 2 km from the direct route home. They had flown 125 km, had made 80 km of progress along the direct route (efficiency = 80/125 = 0.64), and had never been more than 7 km away from the direct route.

In tracking a large number of individual pigeons (Michener and Walcott 1967), it was found that experienced birds homed by paths that typically deviated from the direct path by 1 to 10 percent of that path's length. The efficiency of the different paths typically varied in the range of 80 to 98 percent. However, some tracks were much less well oriented. Even in unfamiliar territory, however, some birds could maintain paths within 10° of their homeward direction.

Instead of tracking pigeons, it is much more common simply to record the direction in which they disappear from view at the release site. These so-called **vanishing directions** are widely scattered but are clearly biased in the direction of the loft (see again Figure 18-2) (Matthews 1968, 78). In a series of five hundred releases, 56 percent of the vanishing directions were within ± 45° of the home direction, compared to the 25 percent that would be expected for random orientation.

More revealing are the studies in which pigeons were fitted with translucent contact lenses that prevented their recognizing objects more than a few meters away (Schmidt-Koenig and Walcott 1978). Although most subjects refused to fly very far, many did manage to fly toward their loft and a few got to within a kilometer or two of it. This result suggests that pigeons have a true navigational ability accurate to within a few kilometers, which is accurate enough to get normal pigeons within range of visible landmarks near their loft.

Several considerations complicate the interpretation of experiments with homing pigeons. For one thing, pigeons first need experience in

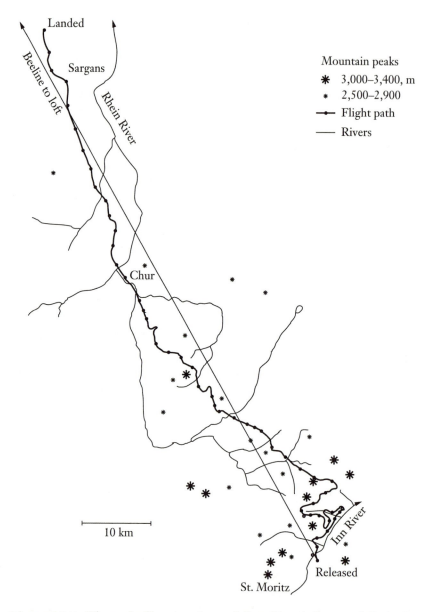

Figure 18-5 The track of homing pigeons followed by a helicopter. A flock of pigeons was released in the valley of the Inn river, in the middle of the Alps. The direct path to their loft, some 125 km distant, is indicated by the straight thin line passing over high mountains. The observed path of the pigeons is indicated by the line with dots to the point where they landed for water after traveling 125 km within 93 min. (From Wagner 1972.)

homing from moderate distances in order to be successful at homing from long distances. This experience might be used to establish familiar landmarks or to learn local gradients to develop a map sense (Lednor 1982). Vanishing directions provide the most convenient behavioral indication of their navigational abilities, but factors that influence vanishing directions often do not influence their homing success (Papi et al. 1978). In addition, pigeons released at particular sites often tend to vanish in directions that are not toward home (note Figure 18-2). Consequently, experiments that are not carefully designed can lead to misleading interpretations.

Other evidence for accurate navigation is the ability of many species of birds to make long-distance migrations but to return to a small nesting area. This feat is particularly impressive to humans when "home" is a small island in the midst of open ocean, which presumably carries few landmarks. For example, greater shearwaters range over thousands of kilometers but nest in a small group of islands only 50 km in extent in the middle of the South Atlantic (Matthews 1968, 1). Unfortunately, our ability to interpret their navigational skill is clouded by the possibility that information about the location of the islands is passed from one individual to another, if only by their observing the direction of flight of large numbers of birds. In addition, birds flying at high altitudes can see 100 km or more under favorable conditions.

Manx shearwaters are a particularly interesting case (Perrins et al. 1973). Some 70,000 birds breed on a 100 ha island off the coast of Wales. The newly hatched birds gain weight rapidly to reach a peak weight 50 percent greater than that of the adults. The adults then leave the island and about ten days later the young depart. There is little data on where they go, but one bird was found 9,600 km away on the coast of Brazil within thirteen days after being ringed. Since the east coast of South America is the normal wintering range for this species, it seems probable that most young birds fly directly to this area on the basis of innate information. At an age of two to four years (but not before), most of the birds return to nest within 50 m of the site where they fledged. These observations suggest their having remarkable navigational abilities that must be based on information acquired both individually and genetically. Furthermore, experiments in which incubating birds were displaced hundreds of kilometers to unfamiliar inland locations have demonstrated that they can often determine the direction toward the nest within a matter of minutes and return to it within only a few hours (Matthews 1953). One bird, flown to Boston and released, returned the 5,000 km to its nest in thirteen days.

Similarly, eighteen Laysan albatrosses taken from their nests on Midway Atoll were displaced in different directions around the North Pacific to distances of 2,116 to 6,630 km, then released (Kenyon and Rice 1958). Fourteen returned—ten of them in less than thirteen days. Their return times correspond to average speeds over the shortest possible paths in the range of 200 to 500 km/day for the eleven fastest returns. It is worth noting that several of the release sites were well out of the normal range of this species and thus were certainly unfamiliar to the individual birds.

In addition to the well-known abilities of birds, a homing ability seems also to be present in all terrestrial mammals (Baker 1982, 34). Homing distances have been demonstrated ranging from a few hundred meters for rodents to over 100 kilometers for some bats (see Figure 18-6). Some invertebrates also exhibit homing abilities, particularly the social insects.

Humans can obtain reliable map information about the positions of objects visible from one location and use that to navigate to and around the objects without needing further external sensory information. The errors are about 0.2 m in a ten-meter path (Thomson 1980). However, map

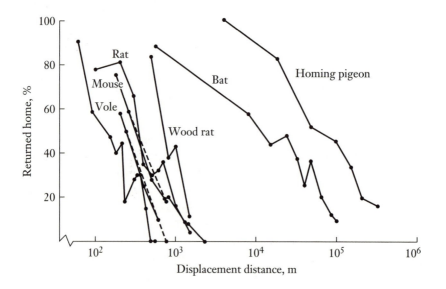

Figure 18-6 The homing distances of various animals, indicating the fraction of the animals that returned home after experimental displacement to various distances. Flying animals returned from much larger displacements than ground animals. (From Bovet 1978.)

information is available for only about 8 s after vision is obscured, making this ability useful only for navigation over short distances.

What are the abilities of humans to navigate over large distances without using instruments? The native Pacific Islanders probably had the most highly developed skill of this kind. They have frequently undertaken voyages across thousands of kilometers of open ocean, but "no navigational instruments or artifacts of any importance are recorded as ever having been used at sea in Oceania" (Lewis 1972, 2).

Several modern voyagers have attempted to test the traditional techniques. In 1965 David Lewis sailed 3,020 kilometers from Rarotonga to New Zealand without instruments, except for sighting along the mast to estimate the positions of the stars overhead. His maximum error in judging his position was 70 km (Lewis 1966). In 1976 a re-creation of an ancient double canoe was sailed some 11,000 km from Hawaii to Tahiti in five weeks, using no instruments (Finney 1979). The maximum error in its estimated position was about 100 km (Finney 1979, 230). Two subsequent voyages were completed within one day of the same duration and made landfall within 100 km of the first voyage (Stroup 1985).

Although there are cases of voyagers that have been lost at sea (Thompson and Taylor 1980; Thomas 1987; Feinberg 1988, 27) this seems to have been infrequent. Each season fifteen big sailing canoes voyaged out of Puluwat, but no one was lost at sea from that island in more than twenty years (Lewis 1970). Even when Oceanian voyagers have been storm driven for a month, they have often retained an astonishing sense of location (Lewis 1972, 145).

In order to assess the accuracy of Oceanian navigators, it is important to recognize that they used a variety of connected stimuli to detect land from over the horizon (Thompson and Taylor 1980, 27). For instance, the heating of land during the day will frequently lead to high cloud formations that do not drift with the wind. If the tops of such clouds are some thousands of meters high, they might be visible from distances of hundreds of kilometers (see Equation 8-8). Furthermore, sunlight reflected upward off the earth's surface and then downward off the bottoms of cloud layers can reveal a wooded island as a dark patch or a shallow lagoon as a bright patch, compared with the reflection off of deep water. Such cloud colors are probably useful out to about 25 km (Lewis 1972, 174).

The presence and behavior of birds at sea can also provide information about the proximity of and direction to land. The most useful are the birds that roost on land but fish at sea. White terns and noddies range out about 30 km from land, boobies and frigatebirds about twice that far, although the latter are less reliable indicators (Gladwin 1970, 196–99; Lewis 1972, 162–72).

Patterns of ocean swells can also reveal islands by the reflection and refraction of the waves by shallow water, although the ability to detect these waves requires a great deal of knowledge and careful study (Gladwin 1970, 177). The usefulness of this information is probably limited to from 40 km (Feinberg 1988, 114) to 80 km (Lewis 1972, 195) of land. Careful attention may reveal the "loom" of land as a pattern of lighter or darker light scattering on the horizon (Lewis 1972, 179). There is also an unexplained phenomenon of flashing luminescence that sometimes occurs a meter or two below the surface between 10 and 200 km from land. This glow appears to move like lightning in either direction along the bearing to land, thus providing a good guide on dark, rainy nights (Lewis 1972, 208).

Given these stimuli connected to land, Lewis (1972, 222–32) estimates that most voyages known to have occurred across open water have a navigational target encompassing an angle of 10 to 20°. On a 700-km voyage guided by an indigenous navigator without instruments they were off their target by less than 1° (Lewis 1971). Lewis seems to agree with Gladwin (1970) that Oceanian navigators could expect an accuracy of better than 5° (Lewis 1972, 231). Although this is an impressive degree of accuracy, it should be kept in mind that it is based on cultural transmission of information through extensive training (Thomas 1987) and may thus not be relevant to other animals.

18-2   MENTAL MAPS

A **map** is a mechanism for storing and retrieving information about the relationships between the locations of goals and landmarks, perhaps with additional information such as compass directions (Wiltschko and Wiltschko 1982). For example, the primary information in the mental map used by the Pacific Island navigators was the direction in which to steer in order to sail from one particular island to another. Similarly, desert ants have a mental map containing information about the familiar routes between resources (Wehner et al. 1983).

A simple map is necessary for piloting using landmarks, and many animals—both vertebrates and some invertebrates—must possess some kind of mental map (Baker 1982). A major gap in our understanding of animal behavior is in knowing how detailed and accurate these maps are and how the information in them is organized.

This concept of a mental map should not be restricted to the two-dimensional array represented by a drawing on paper. Without such tools or an equivalent mental image humans continue to navigate successfully. Indeed, the traditional Pacific Island navigators had difficulty comprehending nautical charts (Lewis 1972).

A map may be constructed by storing different kinds of information, with the types of information stored influencing the ease with which the map can be used to solve different kinds of problems. For example, the simplest strategy might be to store a sequence of the landmarks encountered in moving from home to a feeding site. If the landmarks are sufficiently close together, the animal can make direct sensory contact from one to the next, so that no calculations are required. Such a strategy has been called **steeple-chasing.** When returning home, it is only necessary to travel from one landmark to the next in reverse order, guided by sensory information from the next landmark. Flying insects are thought to sometimes employ this strategy (Wehner 1981).

The dance language of the honeybee (see pages 336–47) suggests that they store spatial information primarily as the direction and distance (that is, a *vector*) from the nest. Their mental map also includes information about the apparent height, width, and color of landmarks (Cartwright and Collett 1983; Gould 1987). Wehner and associates (Wehner et al. 1983) suggest that the desert ant has vector information only about its nest when foraging, or about the last position at which food was found when at home. Perhaps it can store information for only one vector at a time. Other animals, including ants and birds, also seem to have map information that relates landmarks to the direction and distance to home. These maps might be described as being based on a polar coordinate system with its origin centered at home. This kind of information stored in a mental map would make it easy to use to travel to and from home. Many animals in fact show an ability to determine a direct path home after traveling a tortuous outward path (Figure 18-7).

However, if an animal wanted to travel between two positions that did not include home, taking a direct path would require performing the equivalent of vector subtraction on the vectors to the two positions (Figure 18-8). Extensive experiments with the desert ant indicate that it does in fact perform vector addition (of the previous vector to its home plus the vector of its most recent movement) in order to maintain information about its current position from the nest. It seems that vector subtraction should not be any more difficult to perform, but the mechanism for it would have to be capable of accepting information for a variety of posi-

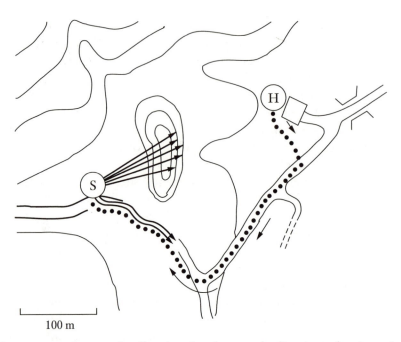

100 m

Figure 18-7 An example of homing via a shortcut: the directions taken (*arrows*) by several geese after experimental displacement (*dotted lines*) in an uncovered wagon. (From von Saint Paul 1982.)

tions. There do not seem to have been many observations suggesting that animals can plot direct courses between a variety of goals. Is this a difficult function to perform, is there little use for it, or has it simply not been studied? There is evidence that humans can move between two positions based on the experience of moving to those locations from a third position, but the distances were limited to 3 m in these experiments (Landau et al. 1981).

18-3 MOTIONS OF THE EARTH

The information available for navigation from celestial sources depends on the motions of the earth, the basic facts of which are presented here. It is convenient first to define some terms for measuring positions in the sky. The **zenith** is the position directly overhead. The position of an object at any other point in the sky can be specified by defining its elevation and

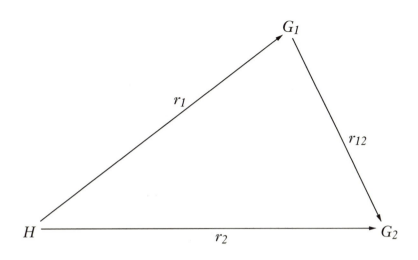

Figure 18-8 Vector subtraction to navigate between map sites. If an animal has a home-centered map that stores the positions of two goals G_1 and G_2 as distance and direction (a vector) from home H, then in order to travel from G_1 to G_2 it must calculate the appropriate vector, which is the vector subtraction of r_1 and r_2 ($r_{12} = r_2 - r_1$). It is not clear how many animals can perform this operation, or how accurately.

azimuth. The **elevation** is the height above the horizon; which varies from 0° at the horizon to 90° at the zenith. The **azimuth** describes the direction on the horizon where the object would be if its elevation were to be reduced to zero. The azimuth is not defined for an object at the zenith.

To begin, then, the earth spins like a top around its axis of rotation in a period of twenty-four hours. As occurs with a top, the axis of rotation *precesses* around an axis of symmetry (Figure 18-9). The earth's axis of rotation makes an angle of 23.5° with the axis of symmetry and rotates around it with a period of 25,800 years. This means that the earth's axis points toward different stars at different times during the precession cycle. As a result, the pole stars change from one millennium to another.

In addition, the earth orbits around the sun with a period of about 365.2 days. The plane of this orbit is perpendicular to the axis of symmetry and is thus inclined to the axis of the earth's rotation by 23.5°. Consequent-

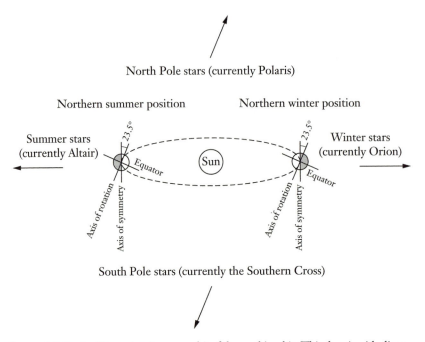

Figure 18-9 An illustration (not to scale) of the earth's orbit. This drawing idealizes the earth's orbit as a circle. The names of the stars and constellations in parentheses are approximate examples for the current millennium As viewed from the north, the earth's rotation and orbit around the sun are both counterclockwise. The earth's axis of rotation precesses around the axis of symmetry with a period of 25,800 years, and pole stars change accordingly.

ly, the elevation of the sun at any point on the earth's surface varies by ± 23.5° during an orbit, which is what produces the seasonal changes in climate.

Our solar system of the sun, earth, and other planets remains relatively stable with respect to the rest of the universe. Accordingly, the stars other than the planets have a stable orientation to the earth's axis of symmetry and are often referred to as the fixed stars, to distinguish them from the planets. There is no indication that any animal uses the planets for information; here the fixed stars are meant wherever stars are mentioned.

A given position on the earth's surface is usually described with respect to the earth's axis of rotation. The **latitude** is the angle between the local vertical (i.e., toward the zenith) and a plane perpendicular to the axis of rotation. Latitude is 0° at the equator and 90° north or south at the respective poles. Latitude is fundamentally important to organisms, because climate varies with latitude. At the equator, summer and winter have identical day lengths, but near the poles summer has continuous sunshine and winter continuous darkness.

The angle of 23.5° that the earth's axis of rotation is inclined with the axis of its orbit defines the major divisions of latitude. The tropics, defined as the highest latitudes at which the sun reaches the zenith at least one day out of the year, are located at latitudes of 23.5° North (Tropic of Cancer) and 23.5° South (Tropic of Capricorn). The polar circles, defined as the lowest latitudes at which the sun is above the horizon for a whole day at least one day out of the year, are located at latitudes of 90° − 23.5° = 66.5°: the Arctic Circle around the North Pole, and the Antarctic Circle around the South Pole. Between the tropics and the polar circles are the north and south temperate zones (Bowditch 1984, 377).

The **longitude** of a particular location is a measure of the angle the earth must rotate to bring a reference point that is at the same latitude as the location in question to face in the same direction with respect to the stars. Determining longitude thus requires simultaneously comparing the direction the two locations are facing or measurement of the time interval between their facing in the same direction. Consequently, determining longitude requires long-distance rapid communication such as a radio or the use of a very accurate clock (chronometer). It is generally thought to be infeasible for animals or humans without sophisticated instruments to determine longitude. In addition, no important characteristic of the environment varies systematically with longitude. Therefore, determining longitude would be useful only as an aid in finding a specific location.

Latitude can be determined by the orientation of the local vertical (the zenith in the sky) with respect to the sun or the stars. To use the sun, corrections must be made for the current season and time of day. To use recognized stars the time during the precession cycle must be known. However, this cycle is so long that it does not change appreciably during an individual's—or for that matter a civilization's—lifetime, and once it has been learned it can be assumed to be constant. Nonetheless, precession is sufficiently rapid that the positions of recognized stars probably cannot be converted to genetic information. If instead of animals recognizing particular stars or constellations they were to recognize the poles from the pattern of rotation, then correction for the phase of precession would be automatic and this strategy could become genetically fixed.

18-4 NAVIGATIONAL PERCEPTIONS

An ability to navigate depends on possession of abilities to obtain and use certain kinds of information. In discussing these abilities, it is useful to distinguish between the perceptual abilities required to obtain the information and the strategies for using the information. The former are discussed in this section and the latter will be discussed in the following section.

LOG

In navigation it is frequently important to determine the distance that has been traveled. Sailors traditionally call a device for providing this information a log (Bowditch 1984, 11). Some implementations of a log provide a direct reading of the distance traveled, whereas others give a direct indication of speed, information that can be combined with knowledge of the time interval to estimate distance. Here **log** refers to an animal's ability to obtain information on the distance traveled, whether directly or indirectly via an estimate of speed. There seems to have been relatively little research on the mechanisms by which animals estimate their distance traveled, which is probably considered a relatively trivial question by most researchers.

A straightforward mechanism for estimating the distance traveled is from the total effort exerted in locomotion. Even modern sailors use the number of engine revolutions for estimating the distance traveled (Bowditch 1984, 154). Ants and bees clearly have access to log information, although the mechanisms they use are not clear (Seeley 1977). Desert ants apparently calculate the time spent walking, as do humans. A variety of evidence suggests that honeybees use their effort or energy consumption to measure travel distance (Lindauer 1976).

The other common mechanism for estimating distance traveled is from speed indicated by referring to external stimuli. Animals with good vision often use patterns of visual flow to estimate their speed of travel (see pages 192–95).

In homing experiments an important consideration is whether the subjects can obtain log and compass information during their experimental displacement. Whether they can depends on the stimuli used by the subject and the conditions of the displacement. A log mechanism based on effort would not be effective during the normal conditions of displacement, but one based on external stimuli might be.

COMPASS

In ordinary speech a **compass** is a device for determining direction by using information from some global stimulus such as the sun, stars, or the earth's magnetic field, or some inertial reference. Here the term is used to refer to an animal's ability to do the same thing. It is common practice in studying animal behavior to limit this category to references to celestial sources or the earth's magnetic field. What matters to an animal's strategy for navigation, however, is whether the source is sufficiently far away that its direction will not appear to change as the animal moves. Thus we may include here discussion of some terrestrial sources. The subsections that follow will discuss these constant-direction stimuli and give some evidence that they are used by particular species for navigation.

INERTIAL REFERENCES   In principle, an animal's displacement can be determined without referring to external stimuli by employing an inertial reference that permits angular and linear accelerations to be sensed and converted to information about the direction and distance traveled (Barlow 1964). Such an inertial mechanism can provide both log and compass information and would have the advantage of automatically including any unintended displacements caused by winds or currents. However, the distance moved can easily be determined by other means, and an inertial reference is probably most relevant to compass information, which is why it is discussed in this section.

One test for an animal's using an inertial compass is based on the fact that errors should increase with the time passed since the previous directional fix using external stimuli. In the simplest model, these errors can be thought of as a random walk in one dimension (angle), and the net error should increase with the square root of time (see pages 93, 391).

This kind of information, which is available from the vestibular apparatus in vertebrates, is very important for small movements such as those for maintaining balance when walking on two legs or for navigation over distances of a few steps. A detailed study (Mittelstaedt and Mittelstaedt 1982) has indicated that gerbils can use inertial stimuli to return to a nest from distances of about ten body lengths. This strategy is also employed in modern navigation technology (Bowditch 1984, 1129–57). However, technological instruments are a thousand times more sensitive to acceleration than are vertebrates.

Numerous experiments with birds have failed to demonstrate a role for inertial navigation over substantial distances (Matthews 1968, 96–100). On the other hand, the tracking of fish swimming in midwater has demonstrated their surprising ability to swim in a constant direction, although they presumably lack visual cues from either the bottom or the surface of the sea. Jones (1984) has suggested that they might use an inertial compass to maintain their direction after choosing a direction while on the bottom. However, information from an electrical field produced by their movement through the earth's magnetic field (see Section 10-3) seems a more likely explanation.

TERRESTRIAL REFERENCES   One of the most important human discoveries was the magnetic compass, which allows one to determine quickly the approximate direction of the earth's axis. Its utility in human navigation naturally led to the hypothesis that migrating animals might somehow sense the earth's magnetic field and orient themselves using it. However, it was not until the 1960s that F. W. Merkel and his colleagues acquired convincing evidence that an animal, the European robin, oriented itself weakly—but significantly—via the magnetic field in its vicinity (Able 1980; Keeton 1981; Baker 1984, 95; Wiltschko and Wiltschko 1988). Since then, evidence has been obtained that a number of other species of birds that migrate at night have similar abilities. There are also reports (Keeton 1981; Wiltschko and Wiltschko 1988) that homing pigeons have become disoriented on cloudy days by altered magnetic fields around their heads (Figure 18-10).

The birds seem to be able to use magnetic fields for orientation only within a narrow range of intensities, but they can adapt to a wide range of intensities within a few days (Baker 1984, 104; Wiltschko and Wiltschko 1988). Some of the evidence suggests that magnetic field information may take a relatively long time to acquire, and it has been proposed that birds may use this information to choose star patterns to follow (Able 1980). There is evidence that the orientations both of migrating birds and of homing pigeons are altered by not only magnetic anomalies in the earth's crust but also magnetic storms (Richardson 1978; Baker 1984, 154–64; Alerstam 1987).

Strategies used to determine direction have been suggested by experiments that manipulated the magnetic field in a variety of ways (Wiltschko and Wiltschko 1972 and 1988; Baker 1984, 99). These experiments, surprisingly, indicate that the birds *cannot* distinguish between north and

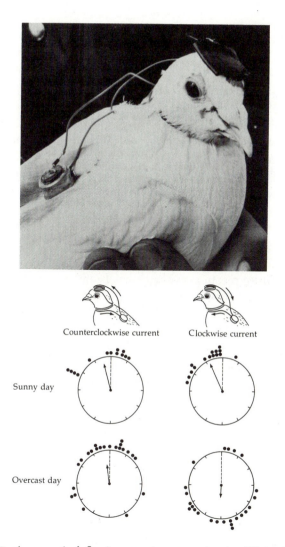

Figure 18-10 A magnetic deflection experiment. A pair of Helmoltz coils fitted on a pigeon's head and neck can generate a magnetic field through the bird's head when the coils are connected to a battery. The vanishing directions (*solid circles*), at about 16 km, of birds carrying a radio beacon, as compared to the direction to their loft (*vertical dotted line*). The arrows show the mean direction for all birds of a given type. Under sunny skies, the coils had no apparent effect on the directions in which the birds flew. However, on overcast days a clockwise current, which reverses the vertical component of the earth's magnetic field, caused most of the birds to fly away from the homeward direction. [From *The Mystery of Pigeon Homing*, by William Keeton. Copyright © 1974 by Scientific American, Inc. All rights reserved. Original data from Walcott and Green 1974.]

south magnetic directions but instead use the vertical inclination of the magnetic field to distinguish north from south. Such a strategy would not work near the equator, but it has the advantage that the direction it points to will not be altered by reversals in the polarity of the earth's magnetic field (Skiles 1985), which has in fact occurred several times since the evolution of birds (Wiltschko and Wiltschko 1972).

A variety of other evidence indicates that free-flying birds sometimes use the magnetic field for orientation purposes (Able 1980). Radar studies show that birds flying under cloud cover, which eliminates the opportunity to obtain celestial information, can maintain an appropriate orientation, but their tracks are more scattered and less linear than those not hampered by the clouds (Wiltschko and Wiltschko 1988). Experiments have shown that homing pigeons and ring-billed gulls become disoriented by magnets attached to their heads when they are either inexperienced or are under cloudy skies. More recently, evidence has been acquired that a variety of other animals also obtain some limited amount of information about direction from magnetic fields (Gould 1980; Mather and Baker 1981; Mather 1985; Baker 1987). On the other hand, many migrants seem to become completely disoriented in fog (Matthews 1968, 8), which suggests that the magnetic field is not enough by itself to provide for orientation. A possibility is that birds need a visual horizon against which to measure the inclination of the magnetic field.

Robin Baker (1984) has carried out numerous experiments on people's abilities to orient themselves without visual information. He originally found that the subjects were surprisingly accurate and their orientation was disrupted by changing the magnetic field around the subjects' heads (Baker 1982, 132). However, in his later experiments the accuracy of the orientation was much lower (Baker 1985a). His statistical analyses have since been criticized (Dayton 1985), and other experimenters have failed to find evidence enough to convince them of a magnetic compass in humans. This controversy is well presented in a single volume (Kirschvink et al. 1985).

The basic problem is that the degree of orientation found in subsequent experiments is so small that it might have been produced by using subtle stimuli of ordinary sensory modalities (Adler and Pelkie 1985). And the impact of experiments with magnetic fields is blunted by the fact that the apparent effect has changed from disorientation to improved orientation as more data have been collected. It should also be pointed out that some authorities remain skeptical about whether any animal can obtain information from magnetic fields (Gould 1980; Griffin 1982).

Some animals make use of visual features on the horizon. If these so-called **horizon features** are actually located at a distance well beyond the range of movement of an animal, such as when a mountain is viewed by an ant, it can be used as a directional reference, since its apparent direction does not change with the subject's position. In this situation such a feature plays the role of a compass, although the same feature might also be a landmark for a migrating bird. Insects seem commonly to focus on the skyline for landmarks (Wehner 1981, 465–69); which is sensible, because it has very high contrast in ultraviolet light. Even humans in many cultures have used the horizon's features to define direction, particularly for astronomical purposes (Aveni 1981).

A situation intermediate between a horizon feature and a landmark occurs when an animal uses an unfamiliar feature for a **temporary goal.** For example, some compass information may not be as precise as is desirable, which may make it efficient to use a compass to choose a visual feature on the horizon to move toward. Doing so may permit traveling in a straighter path, and it will also correct any sideways drift as the animal curves toward the goal from downwind, a behavior observed in a butterfly (Nielsen 1961, 45). Temporary goals are also commonly employed by humans using a magnetic compass to navigate cross-country. It has also been observed that animals often follow features called **leading lines** that are aligned in their own general direction of travel; some examples are coastlines (Able 1980) and roads (Nielsen 1961; Baker 1982, 170). This technique may also help avoid sideways drift. For honeybees leading lines seem to be the most influential type of landmark (Marler and Hamilton 1966, 559; Lindauer 1976).

The movements of air or water are often stable enough to make them useful compass references. Wind direction, for instance, can provide a useful reference, at least over short times. It is easily determinable from the force it produces on an animal in contact with a substrate. Flying animals may detect wind direction by the visual flow of the ground's features or perhaps by some aspect of air turbulence. The desert ant uses wind to resolve the 180° ambiguity (see page 461) inherent in polarized light (Wehner and Duelli 1971). And night-migrating songbirds deprived of celestial information by cloud cover still seem able to fly downwind (Able et al. 1982a). Furthermore, birds wearing frosted contact lenses that were released in midair from balloons flew downwind (Able et al. 1982b).

These birds could not see visual features on the ground, which suggests that they had some other way of determining wind direction. Pacific island navigators also make use of wind direction when no other cues are available, but this cue is the least reliable of the ones they use (Lewis 1972, 93).

Some ocean currents are relatively constant or predictable, and many fish can probably detect their direction of flow by the electric fields they induce in the earth's magnetic field (Able 1980) (see also Section 10-3), or else by the mechanical stimulation created when the fish are in contact with the bottom.

Ocean swells often have a nearly constant direction that could provide compass information. Waves over shallow water produce alternating currents along the bottom, which spiny lobsters may use for orientation during seasonal movements (Able 1980). It has also been suggested that fish may use swells as a compass reference, because the direction of waves can in principle be determined by the pattern of accelerations experienced by a fish. Assuming for fish sensitivities to acceleration that are close to those now known for humans, the direction of a typical ocean swell could be determined to a depth of about 30 m (Cook 1984). Pacific island navigators make extensive use of the direction of swells to maintain direction when celestial cues are not available (Feinberg 1988). Ground swells are the most reliable, because they are produced by prevailing winds over large areas and thus are relatively stable. Expert navigators are able with careful study to distinguish between swells from several different sources, by using visual and inertial cues (Lewis 1972, 84; Stroup 1985).

CELESTIAL REFERENCES   The manx shearwaters that as we have seen (page 445) homed so well do so only when released under sunny skies (Matthews 1953). Heavily clouded skies eliminated all tendency to fly immediately in the direction of the nest and increased their time to return to it. Similar effects are found with homing pigeons (Matthews 1968, 112–116), turtles (DeRosa and Taylor 1980), and monarch butterflies (Kanz 1977). Such observations led to the hypothesis that the position of the sun is an important aid to navigation for many animals.

The sun is generally in the south in temperate zones of the Northern Hemisphere so that it can be used as a crude compass simply by orienting oneself with respect to it and ignoring the fact that its direction changes during the day. Some butterflies are thought to orient themselves in this way, flying predominantly away from the sun in the spring (taking them north to cooler climates) and toward the sun in the fall, taking them south to warmer climates (Baker 1969). In other species, individual butterflies fly at various fixed angles to the sun but again do not compensate for its movement. The latter behavior is probably using the sun as a collimating stimulus (see pages 397) rather than for navigation. The use of the sun as a compass can be tested by changing the apparent direction of the sun with mirrors (Baker 1982, 121) or looking for disorientation of the subjects on cloudy days, assuming no other mechanism of orientation is available.

The eyes of birds contain a peculiar structure called the *pecten oculi.* Pettigrew (1978, 78) has suggested that it functions to aid in determining the position of the sun, by causing a shadow on the retina from light scattered from the sun's image elsewhere on the retina. The position and shape of the pecten are such that a sharp-edged shadow can be produced in the vicinity of the image of the horizon. This characteristic should permit accurate comparison of the position of the sun with respect to that of the horizon or other visual landmarks.

When the sun is not directly visible, its position can in principle be inferred indirectly from the patterns of intensity, color, or polarization of light from the sky. Among these possibilities only that of skylight polarization is known to be used (see pages 186–87). The sun is located on the great circle through the point in sky that is observed and perpendicular to the direction of polarization. Thus, a single small patch of sky provides only ambiguous information about the sun's location, but this uncertainty can be resolved by reference to similar information from a patch of sky in another direction or other kinds of information. At sunset there is a 180° ambiguity in the position of the sun because of the symmetry of the polarization pattern.

Information from the polarization of skylight appears to be commonly employed by arthropods. For instance, mosquitoes and some other flying insects often use the polarization of the sky above for a compass and cease to fly when this information is obstructed by clouds or by the sun's being within 30° of the zenith (Wellington 1974). Experiments indicate that honeybees determine the direction of polarization by sensing the modulation of intensity as they rotate around a vertical axis (Rossell and Wehner 1986).

There is evidence that even birds that migrate at night make use of the direction of the setting sun (Moore 1985). A solidly overcast sky for several days sometimes results in drastic increases in the spread of flight directions taken by migrants in the field (Able 1980). And the nocturnally migrating white-throated sparrow seems to use the orientation of light polarization from the sky at dusk to determine its orientation during the night (Able 1982b). At sunset the polarization direction of light from the sky directly overhead is north and south, when the sun sets due west. This condition is approximately what prevails during spring and fall migrations. The birds could simply orient themselves in the direction of polarization, although they would need additional information such as the sun's position to resolve the 180° ambiguity of polarization. In such cases it appears that the birds use the orientation of the setting sun to calibrate some other reference used to guide flight during the night.

The crude sun compass can be greatly improved upon by using an internal clock to compensate for the movement of the sun across the sky. Many animals are indeed capable of such increased sophistication. The mechanism of a **time-compensated sun compass** was discovered simultaneously in the honeybee and a bird (Able 1980). As was predictable from this widespread phylogenetic occurrence, numerous other species have since been shown to possess similar capabilities. Time compensation has been clearly demonstrated by artificially shifting the phase of the circadian rhythms of the subject, then observing the occurrence of predictable changes in the subject's direction of orientation. This type of experiment is particularly potent because it leads to a predicted shift of the orientation direction rather than simply to disorientation.

The species in which this effect has been demonstrated include many arthropods and vertebrates. In both birds and bees, the position of the sun at various times of the day is learned (Marler and Hamilton 1966, 557; Matthews 1968, 29–30; Lindauer 1976; Keeton 1981; Baker 1982, 122; Dyer and Gould 1983), but fish may have an innate mechanism (Wallraff 1984). There is evidence that young homing pigeons initially use a magnetic compass, but at around three months of age, depending on their experience, they learn to use a sun compass (Wiltschko and Wiltschko 1988). Birds can use the sun even when it is as close to the zenith as 3° (Marler and Hamilton 1966, 550). So far the time-compensated sun compass is the most sophisticated navigational perception that has been well established.

Under cloudy skies, honeybees orient on the basis of landmarks and do not have exceptional abilities to detect the sun through clouds (Dyer and Gould 1983). Furthermore, homing pigeons are often reluctant to fly when the sun is not visible if they are far from the loft (Michener and Walcott 1967).

Although the sun moves along its path through the sky at a steady 15°/hr, the change in azimuth is not steady. Azimuth changes slowly when the sun is low in the sky, rapidly when it is high, the exact relationship varying with the latitude and the season. These complications raise questions about how accurate time compensation actually is. Carefully studied ants do not in fact make accurate compensation (Wehner 1984). Honeybees compensate either from extrapolating the rates earlier in the day, similar to taking a forty-minute running average, or by remembering the rate at the same time on previous days (Dyer and Gould 1983).

During the day, Pacific island navigators use the sun for orientation, but accurately compensating for its movement takes practice (Lewis 1972). The sun is clearly used as a secondary reference subordinate to the stars.

At night an obvious strategy would be to use the moon as the sun is used. However, the moon is less useful, because it is only above the horizon about half of nighttime and its direction with respect to the time of day varies on a monthly cycle. Nonetheless, there is evidence that at least a few species do make use of the moon.

Some moths, for instance, respond to the moon when it is available (Sotthibandhu and Baker 1979). They apparently use it as a collimating stimulus, not compensating for its motion. This is similar to the way some butterflies use the sun. Beach amphipods seem to use a lunar compass, but the question of their time compensation is not yet settled (Wehner 1984). The mallard duck also seems to have a time-compensated moon compass (Baker 1982, 122).

It has recently been recognized by students of animal behavior (Danthanarayana 1986) that light reflected from the moon is polarized and this polarization may provide a useful compass when the moon is visible. For one thing, it has the advantage that time compensation is not necessary. The polarization is north and south ($\pm 28°$, depending on the season and the phase of the eighteen-year precession of the moon's orbit), except for periods within two days of the full moon, when it changes to a weak polarization in the perpendicular direction.

Mosquitoes migrating at night are often most active when the moon provides a high degree of polarization (Danthanarayana 1986). A few bird species seem to migrate only on nights when the moon is visible (Richardson 1978), but this choice could be for visibility rather than to use the moon for navigation. In other bird species migration may be inhibited by moonlight.

For animals that are active or migrate at night it is natural to ask whether they can make use of the stars to provide directional information. In the case of arthropods, it is not clear that they can in fact see any stars (Waterman 1989, 110); the most sensitive (superposition) eyes can probably just detect the brightest stars (Wehner 1984).

White-throated sparrows that were carried aloft by balloon under clear skies during their spring migration, released, and then tracked by radar were able to start flying in the appropriate direction (north) within less than a minute (Figure 18-11) (Emlen and Demong 1978). They did not seem to respond to features on the surface or to wind direction.

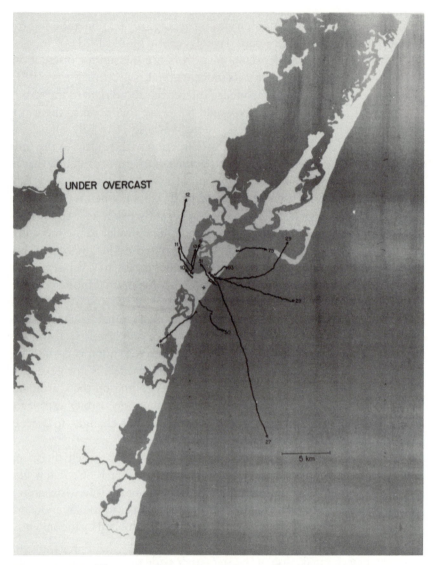

Figure 18-11 The tracks of birds released from balloons. Individual white-throated sparrows trapped during spring migration were carried aloft by a balloon and released at an appropriate altitude. They were then tracked by radar and the tracks superimposed on a map of the area with water in gray. Under clear skies in 1971 and 1972, most birds rapidly chose the appropriate northeastly direction and maintained a rather straight course without any apparent tendency to follow the coastline. Under overcast skies, however, the tracks were much less well oriented and showed a zigzag pattern due to flying in circles and drifting downwind. (From Emlen and Demong 1978.)

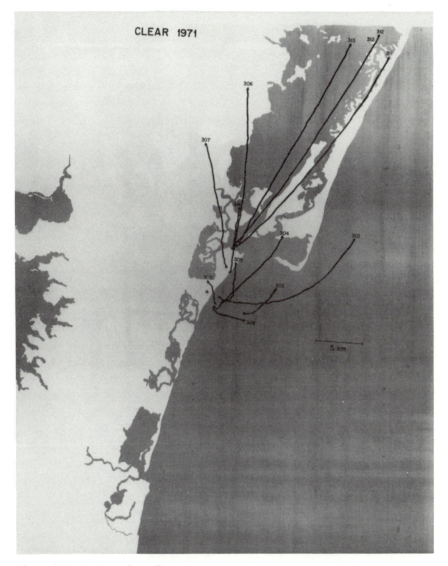

Figure 18-11 (*Continued*)

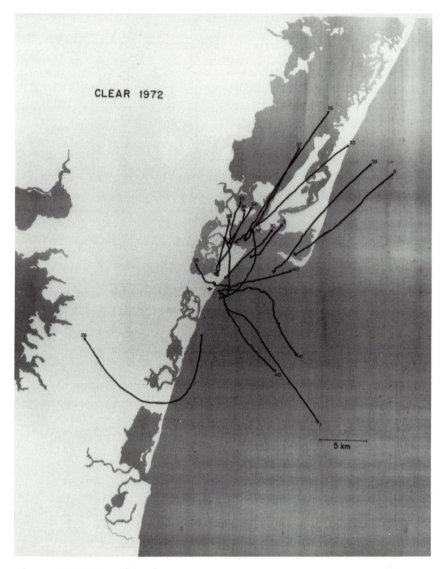

CLEAR 1972

5 km

Figure 18-11 (*Continued*)

Under cloudy skies they flew slower, less straight, and on average did not initially choose the appropriate direction, suggesting their use of the stars for direction finding. When disoriented, these birds were observed to move with sinusoidal deviations from a straight path that are probably due to the birds flying in circles and drifting downwind. This observation suggests that flying in circles may aid these sparrows in compass orientation.

Some birds have been demonstrated to use stars by showing their orientation varied with the orientation of the star pattern projected on the ceiling of a planetarium (Emlen 1967; Emlen 1970; Able 1980). Indigo buntings have been studied the most thoroughly. During a sensitive period proceeding their first migratory season they learn to associate patterns of stars (bird constellations) with the axis of rotation. These learned patterns are then used as references for orientation (Emlen 1970; Keeton 1981) (Figure 18-12). This strategy has the advantage of accommodating the precession of the axis of rotation of the earth without introducing the errors of up to 47° that would occur using a particular star or constellation for a reference (Mead 1983, 130).

Birds that migrate across the equator, unlike the indigo bunting, could not rely on this strategy and might employ instead a time-compensated compass based on constellations near the celestial equator (Emlen 1975). Different constellations would have to be employed during the spring and autumn migrations.

The various stimuli that provide compass information differ in their reliability and in the accuracy and ease with which they can be read. Therefore, it may be expedient to use a more reliable compass to calibrate one that is more convenient to use. This is done, for instance, when a magnetic compass is used to choose a visual feature on the horizon as a temporary goal. In this regard it has been suggested that moths use a magnetic compass to calibrate a moon compass (Baker 1987), which is presumably more accurate over short time periods, and that birds use the magnetic field (Wiltschko and Wiltschko 1988) or polarized sky light at sunset (Able 1982) to calibrate sun or star compasses. In contrast, it has also been suggested that birds use the sun to calibrate a magnetic compass (Cochran 1987). Similarly, desert ants use wind direction or the sun's azimuth to resolve the ambiguity of skylight polarization near sunset (Duelli and Wehner 1973).

COMPASS INFORMATION USED BY HUMANS WITHOUT INSTRUMENTS
Pacific ocean navigators determine direction primarily by where defined stars rise and set (Thompson and Taylor 1980, 26). Travel from one island to another is accomplished by following a **star path,** a succession of stars

Outdoor   Artificial   Reversed   Normal   Reversed   Control

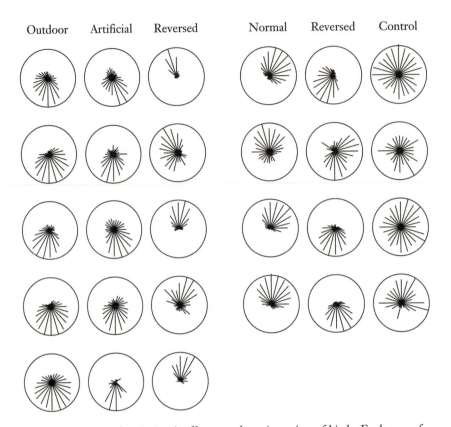

Figure 18-12 A planetarium's effects on the orientation of birds. Each row of three circles shows the behavior of one individual indigo bunting recorded by an Emlen funnel under three different conditions. *Left*: The results of experiments in which the three conditions tested were exposure to the natural night sky (*outdoor*), to an appropriate planetarium sky (*artificial*), and to a planetarium sky in which the north–south axis was *reversed*. *Right*: The results in which the three conditions were a *normal* planetarium sky, a planetarium sky with a *reversed* axis, and a *control* in which the planetarium stars were extinguished. These experiments indicate that the birds tested oriented themselves by using some information from the pattern of stars in the night sky. (From Emlen 1967.)

that rise or set in the appropriate direction as the earth rotates (Feinberg 1988, 111–113, 171–172). Some islanders make use of the more abstract concept of a **star compass**. With this technique directions are defined by the rising and setting of thirteen stars plus stationary Polaris and the Southern Cross at five positions: its rising and setting, when it is at ± 45°

or at the rising and setting of α-Centauri, and when the Cross is upright—for a total of thirty-two directions (Lewis 1970). Although these directions are defined by the azimuth of the rising and setting of certain stars, and boats are optimally steered by one of these stars when it is about 15° above the horizon ahead, it is important that accomplished navigators be sufficiently familiar with the overall relationships of the stars to determine their direction from seeing only a small patch of sky through the clouds (Lewis 1972, 52). It seems that the Polynesians had individual names for about two hundred stars and recognized many more as associates of those (Dodd 1972, 52).

Stars have the advantage that the azimuth of their rising or setting does not change with the seasons, in contrast to the sun, whose rising and setting varies by ± 23.5°, even at the equator (Lewis 1972, 79). However, the azimuth of rising or setting of a given star does change with the latitude, as determined by

$$\cos \alpha = \frac{\sin d}{\cos L} \qquad (18\text{-}1)$$

where α is the azimuth of the star's rising, d is the astronomical declination of the star, and L is the latitude of the observer (Lewis 1972, 311). The star sets in the opposite direction, azimuth 360° − α. For an observer on the equator ($L = 0°$), the azimuth is 90° − d. As one moves away from the equator, the azimuth changes slowly at first, then rapidly as the latitude approaches that at which the star never crosses the horizon. Consequently, star directions are much more useful close to the equator, where most of the Pacific islanders live. Stars on the celestial equator, with a declination of 0° like the northernmost star in the belt of Orion, rise due east and set due west as observed from any latitude.

MAP SENSE

The ability to sense one's position directly without reference to previous positions may be called a **map sense.** Sailors have this ability; modern sailors use radio signals and sailors in the nineteenth century used the sextant and chronometer. The possession of such an ability permits correcting for drift created by cross-currents or for any errors made in determining previous positions. This ability is an important advantage over other methods.

Although many insects have some sort of mental map, none are thought to have a map sense (Wehner 1981, 493). In contrast, birds have been postulated to possess a map sense based on a bicoordinate or grid map capable of locating objects and one's current position in a two-dimensional coordinate system covering the whole surface of the earth, analogous to the latitude/longitude system of human navigation. Several strategies for using the sun's motion and an accurate sense of time have been proposed (Matthews 1968, 122–31). One's latitude can more easily be determined from the stars by the height of the pole stars above the horizon or the identities of stars passing directly overhead. Alternatively, animals might be able to obtain latitude information from the inclination of the earth's magnetic field to the vertical (Bowditch 1984, 208; Wiltschko and Wiltschko 1988).

Longitude is more difficult to determine. Despite a great deal of effort, human sailors could not determine their longitude until they could carry highly accurate clocks (Bowditch 1984, 34). The possible methods by which animals may obtain partial longitude information include the discrepancy between the orientations of the axis of rotation of the earth, as determined from the already established pole stars, and the axis of the earth's magnetic field, as determined from a magnetic sense (Keeton 1981). Some have argued that magnetic-field information can provide the basis of a map sense (Gould 1982b; Walcott 1982; Gould 1985b) but others maintain that magnetic anomalies preclude its usefulness (Lednor 1982). Specific methods of developing a grid map have been discussed by Baker (1984, 45–52). The Wiltschkos (1988) argue that the magnetic field provides some kind of map information, but not simply one based on gradients. On the other hand, some authorities doubt that any animal has a map sense (Emlen 1975; Able 1980; Baker 1982, 141). The fundamental argument against a map sense is that animals have a much wider range of familiarity than is usually appreciated (Baker 1982, 141).

Although seasonal migration could be based on innate orientation mechanisms (see Chapter 19), homing cannot, because the displacement is arbitrary. Displacements are normally the result of responses to situations in the environment, such as the need to search for food, or the experimental manipulations of scientists. If familiar landmarks are available, homing can be worked out by their use, but otherwise there are two basic strategies for solving the problem. One is to record the directions and distances moved during the displacement, a strategy that is often called **route-based navigation**. The other is to use a map sense to relate one's current position to that of home. Although numerous experiments have been carried out with homing pigeons, there is yet no clear consensus for or against either strategy (Baker 1984, 171–200), and it is likely that both types of information contribute to the sucess of homing (Lednor 1982; Wiltschko and Wiltschko 1988).

Most tests of the hypothesis that necessary information is acquired during displacement (Matthews 1968, 98; Keeton 1974; Gould 1982b) have proven negative. However, in a few experiments some information seemed to be acquired during displacement (Papi et al. 1978). In the experiment depicted in Figure 18-13, the birds acted as if they had information about direction during the initial stages of the displacement and preferentially flew in the opposite direction. Other evidence indicates that during displacement they sometimes make use of chemical cues carried on the wind from distant landmarks and sometimes use the earth's magnetic field.

Evidence favoring the existence of a map sense in animals comes from tracking pigeons during their homing flights over unfamiliar territory (Michener and Walcott 1967; Wagner 1972). These studies show that incorrect directions initially chosen by the birds are usually corrected, contrary to the hypothesis that all the important information about the direction home is obtained during displacement. Even more revealing are experiments with pigeons wearing frosted contact lenses that preclude them from seeing objects more than a few meters away (Schmidt-Koenig and Walcott 1978). Although most such birds refused to fly, many did fly in the direction of the loft, and a few circled or landed within a few kilometers of it (Figure 18-14). These results demonstrate that at least these individuals had information about the position of the loft in two dimensions, not simply information about its direction. Furthermore, the experiments suggest that the information was accurate to within a few kilometers. The use of some kind of map sense seems to be the best explanation for these observations.

Many seemingly contradictory observations have been reconciled by a hypothesis that young pigeons rely on information obtained during displacement, before they learn how the stimuli that contribute to a map sense varies with direction from their loft. It has also been suggested that in some regions like northern Italy animals put less reliance on a map sense, presumably because the stimuli available favor other mechanisms (Wiltschko and Wiltschko 1982). There is also evidence that the map sense in animals has a long-time constant, which further complicates experiments. In one set of tracking experiments (Michener and Walcott 1967), individual birds took a median time of two hours to orient themselves toward the loft. The best guess thus far for the sensory basis of a map sense is that at least one of its components is magnetic (Gould 1982b; Wiltschko and Wiltschko 1988).

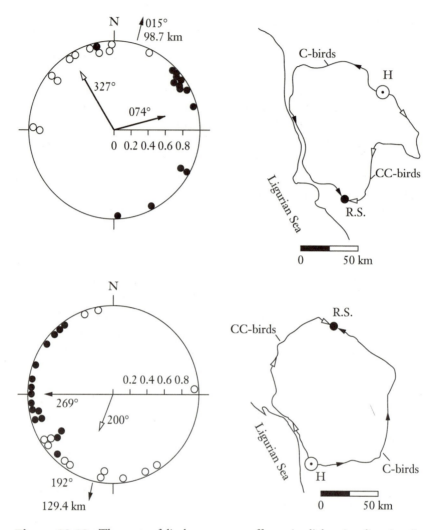

Figure 18-13 The route of displacement can affect animal's homing direction. In at least some experiments the route over which an animal was displaced made a difference in the initial direction it started out in when homing. Here two groups of pigeons from the same loft (H) were transported to a release site (R.S.) by two different routes (*arrows*). The dots in the circle indicate the vanishing directions of individual birds, the arrows the mean directions for each group. The birds that had been transported via a clockwise path (*open symbols*) tended to fly to the right of the true homeward direction, while those transported by the counterclockwise path (*solid symbols*) tended to start out to the left of true homeward. (From Papi et al. 1978.) Not all the experiments of this type showed such clear-cut effects, however, and other experimenters could find no such effects (Keeton 1974).

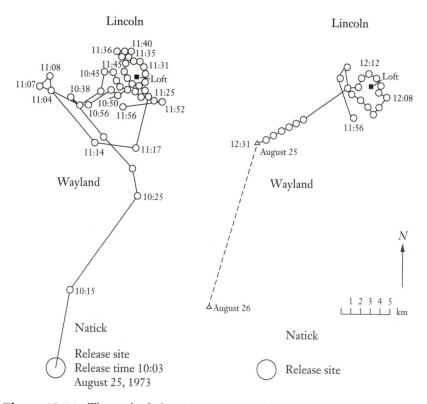

Figure 18-14 The track of a homing pigeon with obscured vision. A pigeon wearing frosted contact lenses and a radio beacon was released 16 km south of its loft (*solid square*) and tracked from an airplane. As can be seen, the pigeon made a fairly direct flight to the vicinity of the loft, then circled around it several times, passing within 0.5 km of it before setting down about 7 km from the loft. The next day it was found sitting at a site 15 km from the loft that was within 5 km of the release site. (From Schmidt-Koenig and Walcott 1978.)

In experiments with amphibians (Baker 1982, 128), displacement in open containers in view of the sun led to movement in the direction opposite to the displacement (that is, homeward). However, displacement in a covered container did not lead to homeward migration.

Obviously, humans with the aid of various tools have the equivalent of a map sense, but what of humans limited to their biological capabilities? In principle, latitude can be estimated by the identities of the stars passing directly overhead or by the height of the pole stars above the horizon. Lewis (1972, 246) estimated that, by sighting along the mast of a ship, one could use zenith stars to estimate latitude to within 0.5°, which would correspond to an error in distance of only 50 km. The Pacific island navigators he reported on know the identities of the zenith stars for particular islands, associating for example the star Sirius with the island of Ra'iatea. As a result, the latitude of these islands is defined for them. However, the evidence that they used this information for navigation is poor (Lewis 1972, 235). In contrast to latitude, there is no known way to determine longitude without an accurate clock. The Pacific island navigators traveled from one known island to another by employing their cultural knowledge of the direction and distance in which to proceed, managing without a map sense.

18-5 NAVIGATIONAL STRATEGIES

The basic strategies for animal navigation may now be summarized. If the displacement is within a familiar area, landmarks may be used in the strategy called **piloting**. Some authorities restrict the use of the term *navigation* to strategies for finding goals across unfamiliar territory and thus do not consider piloting to be a form of navigation (Grzimek 1977, 695; Baker 1984, 9). However, the problems that are solved and the information used in both piloting and navigation are so similar that it seems best to consider the two together. Also, in sailing the activity of piloting is often considered a form of navigation (Bowditch 1984). In its narrowest sense, navigation, excluding piloting, may be called **true navigation** (Able 1980).

In a homing experiment that requires returning from an unfamiliar area, a direct return requires that the subject solve two fundamental problems. The first is to determine which direction is home from the present location. The second involves how to identify that direction. And a third problem is that it may also be useful to have an estimate of the distance home. The second problem can be solved with a compass. The first problem and the third can in principle be solved either by recognizing landmarks of known position (*piloting*), by recording the distance and direction traveled from home (*dead reckoning*), or by comparing information from a map sense that identifies the current position with information from a mental map having information on the position of home (*charting*). Another possibility is to pursue some strategy of *indirect navigation* that leads home by an indirect path. These strategies, summarized in Figure 18-15, are discussed in the following sections.

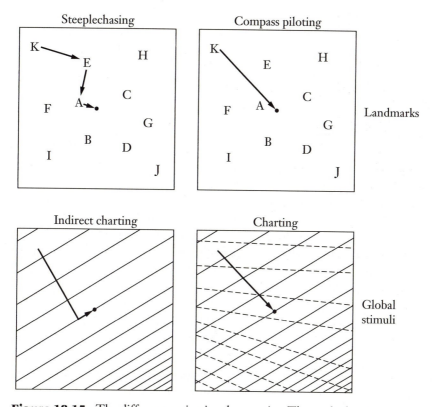

Figure 18-15 The different navigational strategies. The goal of each one is represented by the central dot. The letters represent landmarks. The lines are contours of equal value for some stimulus. The arrows show the paths of animals employing the strategies indicated.

PILOTING

The most straightforward way to move toward a goal is to make use of familiar landmarks along the way. In sailing this strategy is called piloting (Bowditch 1984, 282). A **landmark** is a feature of the environment whose position can be sensed and that generally is within the range of locomotion or at least close enough so that its direction varies significantly with the position of the navigator. Some scholars of animal navigation (Able 1980) have included the use of stimuli connected to the goal in this category, but that case is more a strategy of guiding (see Chapter 17). A landmark is assumed to have no predictable relationship to the goal and so must be learned. Strategies based on stimuli that have important systematic relationships are covered later.

Piloting is the easiest form of navigation to accept, because it is the form most often used by humans. However, it is a demanding technique in that it requires the ability somehow to memorize the relative locations of landmarks and goals. The difficulty of doing so is easy to overlook, because this process is so automatic for humans.

Two important strategies within this category of piloting may be distinguished. The simplest, **steeplechasing** (Matthews 1968, 86), requires having distinguishable landmarks within the sensory field of one another and having the animal know their relative distances from the goal. The animal then simply moves toward the landmark it perceives that is closest to the goal, using any strategy of guiding (Chapter 17) it chooses. A related strategy for piloting employs **image matching** to guide the travel. In this method the animal stores successive images of its surroundings from previous trips and moves so as to make the current image match the stored ones, in the appropriate order.

Another strategy is one I call **compass piloting.** This procedure requires that the animal have access to compass information, along with a mental map that includes information on compass directions from landmarks to the goal or to other landmarks closer to the goal. The animal navigates by identifying a landmark within its range of detection, looking up its map information to find the direction from the landmark to the goal, and using its compass to guide movement in the appropriate direction.

The evidence that insects use visual landmarks has been reviewed in detail by Wehner (1981). It is clear that ants (Hölldobler 1976) and flying insects (Tinbergen 1972, 146–96) often use visual landmarks to find the location of a nest from distances of a few meters, probably by image matching. On leaving a nest for the first time or after nearby landmarks have changed, a bee or wasp often flies in specific patterns that seem to serve to obtain visual information about landmarks surrounding the nest. These flight patterns often involve scanning and backward flight, followed by a phase of circular and spiral flight paths centered on the nest, to increasing radii (up to 30 m, with honeybees). The first orientation flight of a honeybee typically lasts for about two minutes. These flights are necessary to make successful returns to the nest.

In most insects that have been studied the most important landmarks are those that affect their visual horizon (Wehner 1981). Moreover, arthropods that live in flat environments have eyes specialized for high resolution in the horizontal plane (Wehner 1987). However, the examples cited are probably a biased sample, as it is easier to study animal behavior in an open environment. In a forest, for instance, there is no visual horizon, and important landmarks may appear only as patches of sky visible through openings in the canopy (Hölldobler 1980; Wehner 1983). In both cases, the boundary between sky and land or sky and vegetation would stand out in high contrast for the ultraviolet receptors of insects.

Experiments (Gould 1987) have indicated that honeybees use the direction, shape, and color of landmarks to return to a source of nectar. Computer modeling shows that relatively simple strategies allow stored images to be used to guide locomotion to the appropriate site (Cartwright and Collett 1983). If landmarks are available that form a leading line to the goal, honeybees will follow them (piloting), in preference to using compass information (dead reckoning) (Dyer and Gould 1983).

Some behaviors suggest methods are used to minimize the burden of information processing in using visual landmarks. For example, in their final approach to a nest many flying insects always face in the same direction (Wehner 1981, 482), which simplifies the problem of matching a pattern of landmarks to a memorized pattern (Wehner 1981, 468), this in fact seems to be the mechanism flying insects use (Wehner 1981, 469–81). In addition, insects searching for food (Waddington 1983) or a mate often repeatedly follow an individually defined path or *trapline* that may extend for many kilometers and seem to be followed by memorized landmarks. Adopting this type of foraging in which the resource sites are always visited in the same order minimizes the demands on the animal's mental map.

Insects act as if their mental map contains two-dimensional images of landmarks and employ a template-matching process to determine the direction to a goal (Wehner 1983). Honeybees could not be trained to distinguish between landmarks on the basis of their physical size (Cartwright and Collett 1983), suggesting that their visual information is not reduced to a three-dimensional interpretation of the physical world as it is for humans (see pages 190–92). In insects the matching process apparently requires the same orientation of the eyes in both the current image and the memorized one. Long-distance movements guided by piloting are presumed to be based on a sequence of stored images. With this mechanism it is not simple to reverse the process for the return trip, and indeed insects often do return by different pathways. Even humans planning to return by the same path find it helpful to look back to memorize the view as seen from the return direction.

The abilities of vertebrates to memorize landmark information are not yet well documented. However, the ability of certain birds and mammals to store and later retrieve food from large numbers of caches (Sherry 1985) suggests that they have an impressive ability to remember many locations.

In the case of animals, memorized positions could be stored using various types of information such as the latitude and longitude or instructions for moving from one landmark to another. However, we still have little knowledge of how this is done. The Pacific island navigators are known to make use of landmarks such as other islands, reefs, and submerged reefs detectable from the surface.

The evidence that birds use landmarks in navigation is complex (Baker 1984, 55–76). For instance, homing pigeons sometimes respond to visible landmarks but often ignore even familiar ones (Michener and Walcott 1967; Keeton 1981). The visual features that influence the homing pigeon's initial flight direction include an aversion to flying over water, above continuous clouds, or over high mountains, and an attraction to features characteristic of the home area (Wagner 1972; 1978). In mid-course, pigeons often do not respond to familiar landmarks (Michener and Walcott 1967). And near their loft visible landmarks appear to be necessary (Matthews 1968, 73, 85–86; Schmidt-Koenig and Walcott 1978).

Landmarks may also be based on other than visual stimuli. There are, for instance, some good experiments indicating that homing pigeons have used olfactory landmarks (Baldaccini et al. 1982; Papi 1982; Baker 1984, 63–69, 193–194). An especially intriguing suggestion is that homing pigeons may be able to detect distant sources of ultrasound and perhaps determine their direction of propagation by perceiving Doppler shifts as they fly in circles (Quine and Kreithen 1981). However, there is no evidence that such sources have a sufficiently stable frequency to make them useful or that the birds can distinguish such sounds from local noise, especially noise created in flying.

How landmarks may be used by birds is less clear. There is evidence that both steeplechasing and compass piloting are used by pigeons (Matthews 1968, 86). The use of compass piloting by pigeons is revealed by the observation that a shift in a pigeon's physiological clock causes it to misread the time-compensated sun compass and fly in an inappropriate, but predictable, direction.

DEAD RECKONING

In sailing, an approach based simply on making estimates of the direction steered (using a compass) and the distance traveled (using a log) is called

dead reckoning (Bowditch 1984, 260). In animal behavior, the term **vector navigation** is sometimes used to mean the same thing. This strategy is distinguished by the fact that in using it one's position is determined only relative to some previous position. No information is obtained to correct errors in previous positions, and errors accumulate. In particular, this strategy suffers from an inability to correct for unperceived movements such as are caused by wind drift or ocean currents. In fact, this characteristic is one test for the use of dead reckoning. Because of this limitation, the strategy may work best for walking animals, which are not as easily displaced as those in air or water. In fact, many walking arthropods do make extensive use of dead reckoning (Wehner 1984).

The basic test for the use of dead reckoning is to displace an animal experimentally. In many cases the subject will continue in the same direction for the distance that would have taken it to its goal, then initiate local search behavior (Figure 18-16). This behavior has been demonstrated in ants (Wehner and Srinivasan 1981), fish (Ogden and Quinn 1984), and some migrating birds (Emlen 1975). Quite large banding experiments with starlings have indicated that young birds on their first autumn migration do not correct for experimental displacement and thus are presumed to be employing dead reckoning (Emlen 1975).

Another indication of navigation by dead reckoning is sideways displacement by cross-currents. Honeybees flying over water lacking in visible features will be drifted by crosswinds—but not if visible features are provided (Schöne 1984, 247). This finding is evidence that they employ dead reckoning when visible features are absent. Similarly, homing pigeons (Michener and Walcott 1967) and migrating birds correct for wind drift when flying over land, presumably using vision to do so, but the latter do not do so over water, which provides no visual objects to refer to (Alerstam 1981).

This characteristic is perhaps the basic disadvantage of dead reckoning. There would in fact be a great advantage to an animal for it to have some sort of mechanism for detecting its own unintended displacement. An additional reason for considering such mechanisms is that many experiments on navigation make use of experimental displacements.

The major question with regard to displacement is whether animals use any of the less obvious cues, as for example movement through the earth's magnetic field. There is in fact evidence suggesting that homing pigeons are subject to magnetic influences during artificial displacement, but the information they obtain during displacement is apparently not necessary for their accurate homing (Presti 1985). Marine fish are proba-

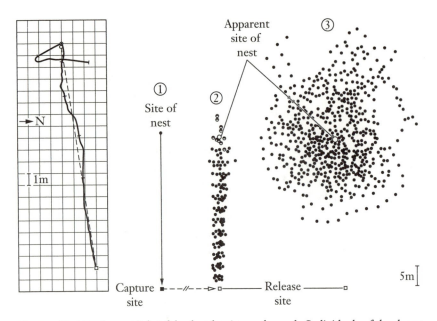

Figure 18-16 An example of dead reckoning and search. Individuals of the desert ant, *Cataglyphis albicans*-A, were captured while they were foraging at a distance from their nest and were then displaced to an unfamiliar area. Their path was recorded as they attempted to return to their nest. *Left*: The complete track (*solid line*) of one individual; the dashed line would be the direct path back to the apparent position (*open circle*) of the nest if it had not been displaced. The ant travels a nearly straight path to what should be the vicinity of the nest, then appears to initiate a local search pattern. *Right*: The behavior of ten individual ants. The dots under 2 indicate the positions at which the local search appeared to start. The dots under 3 show the positions of the ants during the first 15 min of searching. (From Wehner and Srinivasan 1981.)

bly able to detect water currents carrying them through the geomagnetic field, by the electric fields they generate (see Section 10-3). It now seems that during displacement experiments nearly as much attention should be focused on what the subject is doing during its displacement as during its homing performance.

The indigenous navigators of Oceania sail almost entirely by dead reckoning and so must be much concerned about unintended drift. Sideways movement caused by wind is estimated from the angle between the centerline of a boat and its wake (Lewis 1972, 116). Drifting caused by water currents is much harder to detect, however, because in the open ocean all the visible points of reference are also drifting with the current.

Knowledge of the common currents in the home area is part of the lore handed down between generations of navigators. In addition, when departing an island they commonly use landmarks to detect cross-currents (Lewis 1972, 106; Thomas 1987, 138). Some navigators claim to be able to detect a current from the shape of the waves, although this is probably possible only in certain circumstances (Lewis 1972, 49, 110–13).

CHARTING

The strategy of navigation most frequently employed by modern sailors is to use a map sense to gain information about their current position in conjunction with a nautical chart, the equivalent of a mental map, to obtain information about the position of the goal and then plot an efficient course to that goal. This course is then followed by using a compass. Let us call this strategy **charting.**

The use of this technique is inevitably linked to the use of a map sense, and the discussion of the latter includes about all that can be said about charting. Since the nature of any map sense in animals is unknown, any charting strategy also remains uncertain. Nonetheless, it is important to define this strategy because it is potentially quite important to animals. Furthermore, it is important to distinguish between the required perceptual ability (the map sense) and the strategy for using it (charting).

INDIRECT NAVIGATION

The indirect strategies include techniques for navigating to a goal from unfamiliar territory without knowledge about one's current position or sufficient information to determine the direction to the goal. Such strategies have not often been proposed by experimenters, but they should be considered as possible mechanisms in homing experiments where the path followed by the subject is not observed. They are particularly pertinent where the homing process is slow. Discussions of homing have tended to concentrate on either a random search and guiding by stimulus fields at the low efficiency extreme or else on a map sense and charting at the high efficiency extreme. It would clearly be well to also consider strategies having intermediate efficiencies. For one thing, it only requires information sufficient to provide a small bias in the correct direction to dramatically improve homing speed over that of a random search (see Section 17-2) (Adler and Pelkie 1985).

An example of such a strategy would be *parallel sailing*, which was practiced by sailors before chronometers were available to determine longitude (Bowditch 1984, 33). In this method ships sailed north or south to the latitude of their destination, then east or west as appropriate, according to their knowledge of the destination.

Another example of indirect navigation is the suggestion that birds may compare the height of the sun with its height at the same instant (not the time of day) as at home, then fly toward it if the sun were too low but away from it otherwise. Such a strategy would eventually lead the bird home via a series of arcs (Matthews 1968, 128). However, it would require a clock with accuracy and stability not yet found in any organism.

A more general example of such a strategy is that if one knows the value of some parameter such as altitude at home and can determine it locally, for example, by carrying an altimeter, then a potentially useful strategy would be to travel up or down the altitude gradient until the altitude of home is reached and then follow that altitude contour. However, this strategy does not work in all situations. Choosing the wrong direction along the contour may lead far from home. In addition, if the altitude varies roughly over the area of the displacement, there is a good chance of getting caught on a contour that circles around a mountain or lake without ever connecting to the contour passing through home. Thus, stimuli that vary smoothly over the area of interest would be the most useful.

The earth's magnetic field is one such obvious possibility. Both the intensity and the inclination to the vertical vary rather smoothly over most of the earth's surface (Skiles 1985). Knowing the magnitude of one parameter at home and possessing an ability to determine its value at another location would allow a subject to use this method to locate home from almost any point in its hemisphere. An ability also to distinguish north from south would permit finding home from either hemisphere. It should be noted that this strategy, if used alone, might lead to one's traveling nearly around the world before getting home. Nonetheless, it would work, given enough accuracy and assuming that neither home nor the starting location are in areas of local abnormalities. In addition, other information would usually be available to aid in choosing the correct direction along the contour.

Some experiments with homing pigeons suggest that they might be following gradients in the magnetic field intensity toward the value at the loft (Wiltschko and Wiltschko 1988). Also, analysis of whale strandings suggests that they occur most often at locations of minima of the local magnetic field strength (Kirschvink et al. 1986). Whales might follow minima in the magnetic field, because minima generally run north–south and are less variable than the maxima.

Other examples of indirect strategies for navigation include the use of hitching (see Chapter 15). Given reliable currents, navigation could be controlled entirely by the correct timing of starting and stopping. A return to one's previous positions could occur in ocean gyres, in tidal currents (by switching between ebb and flood tides), or in appropriately selected winds.

Another example is Kiepenheuer's model for bird migration (Box 18-1). Kiepenheuer's model accounts for the general observation that many bird species nesting in the Northern Hemisphere migrate in the fall to the southeast or southwest at high latitudes but shift toward more southerly directions at low latitudes. Table 18-2 shows some details for a common migration route through Western Europe.

If this model were to be followed indefinitely, the animal would end up orbiting around one of the magnetic poles. Thus, its use must be controlled by other information. Kiepenheuer's assertion that shearwaters could migrate in a circular path around the Pacific Ocean by following a single value for the projected angle is misleading. They would need at least to switch between the alternative solutions to the ambiguity at the extreme latitudes of their migration.

Nonetheless, Kiepenheuer shows that this model predicts paths that in many cases approximate great-circle routes (the shortest possible paths between points on the spherical earth), and many actual migration routes

Table 18-2 Representative Parameters for Kiepenheuer's Model of Bird Migration

	Position		Earth's Field		Control Variables		Heading	
Location	Longitude	Latitude	Declination	Dip Angle (γ)	Projected Angle (γ')	Side*	Magnetic (α_m)	True
Germany	9° E	50° N	– 3°	66°	–69.3°	Right	238°	235°
France	0°	45° N	– 8°	57°	–69.3°	Right	216°	207°
Spain	6° W	37° N	–10°	48°	–69.3°	Right	205°	195°
Morocco	10° W	28° N	–12°	40°	–69.3°	Right	193°	180°

* Side refers to the direction in which a change of the heading increases the projected angle.

Box 18-1 Kiepenheuer's Model for Bird Migration.

Kiepenheuer (1984) has proposed a simple model for indirect navigation in bird migration that explains why many migration routes are curved. His hypothesis is that the birds measure the projection of the earth's magnetic field on some vertical plane of their body and select a horizontal heading that causes the angle of the projection from the horizontal (γ') to match some preset value (γ_c') as closely as possible. If the plane is perpendicular to the direction of travel, the preferred heading with respect to magnetic north (α_m) is given by

$$\sin \alpha_m = \frac{\tan \gamma}{\tan \gamma_c'}$$

where γ is the inclination or dip angle of the earth's magnetic field at the location. (This formula is equivalent to Kiepenheuer's Equation 2, if a misprint there is corrected, except that the formula given here preserves the sign of the projection.)

This model does not completely specify a heading. If we assume that the polarity or sign of the projection can be determined as in the equation above, then there is ambiguity between the courses α_m and $180° - \alpha_m$.

Further information is needed to resolve this ambiguity. A simple way to distinguish between the two possible courses is to determine whether the projection increases or decreases with a turn toward the right

---

do fit well with its predictions. Although in one case a thrush that was tracked 1,500 km by radiotelemetry did not change course as predicted by this model (Cochran 1987), and blackcaps in cages under constant magnetic fields shifted their orientation (Helbig et al. 1989), such a simple model that seems to fit so many cases deserves further study.

Finally, many of the characteristics of the various different navigational strategies are summarized in Table 18-3.

18-6　HOMING

The fundamental challenge in navigation is determining the location of the goal with respect to present position. The likely solutions to this problem are

or the left. However, experiments with birds in cages (see pages 455–56) suggest that the polarity of the magnetic field cannot be determined. In this case the signs in the equation above should be eliminated by taking the absolute value. This leads to a fourfold ambiguity ($\pm \alpha_m$ and $180° \pm \alpha_m$), which must be resolved by additional information. Cues indicating the direction of north versus south combined with determining the effect of rotation of the heading would suffice.

In this model, at a given location and γ, a heading of magnetic east ($\alpha_m = 90°$) or west ($\alpha_m = -90°$) produces an extreme projected angle of $\gamma' = \pm 90° \pm \frac{\gamma}{2}$. Rotation of the heading toward north or south causes a reduction in the horizontal component of the projected magnetic field and a change in the projected angle toward $\pm 90°$.

Near the poles, the dip angle may be too extreme to obtain the preferred projected angle, making the best available course magnetic east or west, which would result in continuously circling around the pole.

Approaching the equator the dip angle decreases, so that the preferred heading becomes more northerly ($\alpha_m = 0°$) or southerly ($\alpha_m = 180°$). Right on the magnetic equator, the earth's magnetic field has no vertical component and this model does not define a preferred direction. It may be supposed that other information indicating north or south is employed in this area.

different for the two common behaviors of navigational interest—homing and seasonal migration. Some brief comments on homing will be made here; migration is treated in the next and final chapter.

Homing is the most common behavior in which navigation plays a part. The basic problem in homing is how to return to the location of a goal previously visited after having displaced from it. The displacement might be either voluntary or imposed (for instance by an experimenter). Because the animal has previously visited the goal, one strategy for homing is to obtain the necessary information during the displacement. The other strategy is to obtain the necessary information at the location where it is needed. The first strategy would probably be based on log and compass information. The second strategy might be based on a map sense or more likely recognition of familiar landmarks.

Table 18-3 Navigational Strategies

	Piloting	Dead Reckoning	Charting	Indirect Navigation
Required abilities	Mental map	Log and Compass	Map sense and mental map	Varies with strategy
Environmental requirements	Landmarks	None for inertial stimuli; otherwise depends on type	Two stimuli that vary independently and in predictable ways with position	Appropriate stimuli lack local minima or maxima in the area of interest
Advantages	Correct for previous errors*	Simple requirements	Correct for previous errors	Correct for previous errors and do not need to know landmarks or direction to goal
Disadvantages	Requires familiarity with the area	No correction for previous errors	Appropriate stimuli may not exist	Longer path to goal and appropriate stimuli may not exist
Experimental tests	Not successful in unfamiliar areas	Displacement† does not alter direction or length of subsequent path	Path is direct to goal before and after displacement	Path is altered after displacement but is usually not direct to goal either before or after displacement
Organisms that have been shown to use it	Insects, birds, mammals	Bees, birds	None	None

* Correction for previous errors includes correction for unintended displacements such as crosswinds
† Displacement refers to an experimental displacement during which it can be assumed no positional information is obtained

If log and compass information are available during displacement, the direction and distance to home can be calculated. Desert ants seem to employ this strategy (page 442). It is not necessary to record the information of each step of the outward path; one can simply keep track of the direction and distance (vector) to home and update it by vector addition of the previous home vector and the vector to the previous position. The home vector strategy would allow an animal to return home at any time but not be able to return to previously visited sites.

CHAPTER 19

The Mystery of Migration

■ ■

The wild goose is more cosmopolite than we;
he breaks his fast in Canada,
takes luncheon in the Susquehanna,
and plumes himself for the night in a Louisiana bayou.
—Henry David Thoreau, 1840

Some of the great spectacles of nature are the seasonal migrations of birds and fish. How do they know where to go? Even more amazing is the migration of monarch butterflies. Each October on clear, sunny days, looking north out of my third-story office window in Atlanta I see a few lonely individuals flying toward and over the building. They do not fly in what seems a very precise direction, and it is difficult to imagine how they can fly thousands of kilometers to find a winter refuge in central Mexico. This concluding chapter makes some general comments about migration and presents some specific examples of migration in terms of the information that may be used by migrating species to guide their travel.

19-1 GENERAL COMMENTS ON MIGRATION

POPULATION MOVEMENTS

Repeated migrations between such fixed sites as geographically separate breeding and feeding grounds require that a population—or more specifically each deme—include migrations in its life history that form a complete loop, so that the appropriate population at each site is renewed. For some species, including many birds, individual adults traverse the loop each year on seasonal migrations. In other species, particularly fish, different stages of the life-cycle traverse different legs of the migratory loop. And in still other species, involving some insects, different individuals traverse separate legs of the migratory loop. In all cases there must be a mechanism for genetic information to traverse the loop, to make it evolutionarily stable (Walker 1979).

The longest seasonal migrations are those of certain birds that migrate between the Arctic and the Antarctic (>10,000 km) and thus experience two summers per year. The arctic tern breeds from latitude 50° N to as far north as bare ground extends (80° N), and winters at the edge of the Antarctic pack ice ($\approx$ 60° S) (Salomonsen 1967). It spends a greater proportion of its life in daylight than any other animal that lives more than a few months. Most whales likewise migrate long distances and must be well oriented during their migrations, because they cover great distances at average speeds ($\approx$ 10 km/hr) that are not much less than their speed over short distances (Lockyer and Brown 1981). The gray whale makes a one-way migration of about 10,000 km, the longest of any mammal (Meier and Fivizzani 1980). A squid has been known to migrate 1,000 km, and lobsters and blue crabs may migrate more than 100 km (Gauthreaux 1980; Meier and Fivizzani 1980). Some grasshoppers and butterflies migrate thousands of kilometers (Meier and Fivizzani 1980).

Even plant populations migrate, albeit over periods of many years, in response to climatic variations (Gauthreaux 1980). Climate varies incessantly, and the resulting migrations continue even in our own time. During glacial episodes the climatic zones can move faster than many plant species can migrate, making it possible to estimate the maximum rate of migration. For trees the rate averages about 300 m/yr (Gauthreaux 1980).

Animals as well as plants must contend with climatic change. In particular, the seasonal migrations of animals that we know so well must be modified as the earth's climate changes. This change in pattern has implications for the mechanisms that determine where and when populations migrate, because climate can change radically within a thousand years, but that is not much time for evolution to adjust animals' innate mechanisms for specific locations. Thus we may predict that strategies are used that are based on common physiological mechanisms, and suggest that genetic information will not specify a particular latitude and longitude.

TIMING

The arrival of summer bird species in the spring is an occasion frequently noted by humans living in temperate areas and much studied by scientists. In some cases the arrival date is extremely regular. For instance, the arrival of swallows at the mission of San Juan Capistrano in California is so regular that a yearly celebration is planned for a specific date. Several studies have indicated that, in general, species that arrive early in the spring (noninsectivores and omnivores) are more variable in their date of arrival than are insectivorous species such as swallows which arrive late in the spring (Gauthreaux 1980).

The clue for determining the season that is best established by researchers is day length or **photoperiod** (see Section 2-2). This method is clearly used in timing the migrations of birds in the temperate zone, where day length varies widely. The use of photoperiod to determine the season to an accuracy of one day requires determining day length to an accuracy of about three minutes, which seems fantastic.

In the less-studied tropics, photoperiod and temperature have little variation and other information such as cues from monsoons may be more important (Gauthreaux 1980). Although which aspects of the monsoons may be detected is unclear, low daytime light intensity created by cloud cover is a good hypothesis, since light, in the form of the photoperiod, is widely used for timing in other circumstances.

There is increasing evidence that some animals have an annual clock or *circannual rhythm* that helps time their seasonal migrations. The existence of such a mechanism has been clearly demonstrated in some birds (Meier and Fivizzani 1980). This device probably plays an important role in timing migrations and is likely to be particularly important in the tropics.

Tidal rhythms are also thought to exist (Meier and Fivizzani 1980). These clocks are set or entrained by stimuli such as temperature, mechanical stimulation, and hydrostatic pressure.

PATHFINDING

The navigational problem that is of particular interest with regard to seasonal migrations is how the information is acquired that specifies the spatial goal during an individual's first migration. Three modes for the acquisition of this information may be envisioned. The most direct one is based on self exploration and identification of a location having appropriate resources. This strategy, discussed shortly, has been emphasized by

Robin Baker. The next mode is via communication from contemporaries of the same species. Such *cultural transmission* is a likely mechanism for use by species such as caribou and geese in which the young migrate in herds or flocks accompanied by experienced adults. The third mode is from information acquired by ancestors and involves *genetic transmission*. The differential survival of animals that have moved to appropriate places at appropriate times of the year is sufficient to develop the required information in the genome. This mode plays at least a partial role in the seasonal migration of many birds. Some even show a sort of "migratory restlessness" at appropriate seasons with appropriate directions when in the laboratory (Emlen 1967; Wiltschko and Wiltschko 1988).

The other basic question is what kind of information directs the migration along the way. There is little clear evidence, for instance, to indicate that there is a series of defined positions en route. Massive experiments involving the displacement of migrating starlings and recapturing of marked birds indicate that young, inexperienced birds simply fly a certain distance in a particular direction without correcting for the experimental displacement. This suggests that it is genetically acquired information that specifies direction and distance to them. In contrast, experienced birds will correct for the experimental displacement. They presumably have acquired through their own experience information about specific locations having appropriate resources.

Some authorities have emphasized simple responses for directing migration. For example, Kenneth Able (1980) has said "in the absence of all other compass information, a bird might well be able to select a night for migration on the basis of ambient weather, orient solely with respect to wind direction, and have a high probability of flying in the correct seasonal direction." This general view has been supported by others (Emlen 1975) and is consistent with radar tracking studies of autumnal bird migrations over the western North Atlantic, although in that case the birds seem to fly on a particular, southeasternly, compass heading (Williams et al. 1977).

At another extreme, R. Robin Baker (1982) has emphasized the importance of **exploration** by animals, for them to develop a detailed mental map of large areas. In the terms used throughout this book, his idea of exploration requires the use of *search*, *stimulus fields*, and *navigation*. For exploration to be useful, the animal must be able to retain significant information by forming a *mental map*. This hypothesis explains observations that many birds travel extensively in random directions after fledging but before the fall migration. Baker argues that migration is often best understood as being exploration with a preferred directional component

(Baker 1982, 84). Further evidence to support the concept of exploration as a contribution to migration is that, even in bird species that make long-distance migrations, there is commonly movement in directions other than the presumed goal, and young birds sometimes make "practice" migrations (Baker 1982, 88). If exploration is in fact a major element in migration, it is easy to understand how migrations can adapt rapidly to changing climatic conditions.

Others have dismissed the hypothesis of exploration as an explanation of migration because it assumes that animals have abilities to memorize spatial relationships, which have not been clearly demonstrated. However, Baker argues that we know that humans have these abilities and it is therefore parsimonious to assume that other mammals—perhaps even other vertebrates—also have them.

Much of Baker's argument (1982) that all migrating birds find their way primarily by exploration and the use of familiar landmarks is based on his suggestion that there are useful landmarks (seamarks?) at sea, such as waters of varying colors. However, it has not been established that these are perceptible. Baker's statement (1984, 75) that Polynesian navigators have access to landmarks at sea is supported neither by his citation (Lewis 1972) nor by other sources (Gladwin 1970, 162; Feinberg 1988). In addition, radar studies indicate that migrating birds compensate for wind drift while flying over land—but not over water (Alerstam and Pettersson 1976; Alerstam 1981), which indicates that the birds have no landmarks at sea. Thus, we must conclude that exploration and piloting are not the only strategies used in bird migration.

It is remarkable that even after decades of extensive experimentation two such different hypotheses remain viable explanations for the migration of most animals. This situation is a testament to the complexity of animal behavior in the natural environment. Dramatic progress probably awaits making many more observations in which the detailed movements of individuals are recorded and analyzed.

19-2 SOME EXAMPLES OF MIGRATION

THERMOTAXIS IN SOIL

The soil environment varies in many important ways over small vertical distances from the surface, which suggests that mobile microorganisms living in the topsoil would benefit from having an ability to move to a particular depth. For example, an organism preparing to release spores would best do so at the surface, where air currents could disperse the spores, even when conditions at the surface might not be favorable for growth. Similarly, mobile plant parasites, while searching for the roots of a host plant might find it efficient to search at a particular depth.

The main question to determine is what information could be used to guide an organism's travel to the appropriate depth. One clear possibility, at least in temperate latitudes, is the spatiotemporal temperature pattern in the soil that is caused by daily and seasonal temperature changes at the surface (see Section 6-3).

Some soil-inhabiting organisms have been demonstrated to be extraordinarily sensitive to temperatures that allow them to guide their travel in temperature gradients of 0.05° C/cm or less (see Section 6-2). The direction, whether up or down the gradient, in which they move is determined by their previous experience of temperature. When this behavior is modeled by computer in a typical soil thermal environment, some apparently adaptive movements are predicted.

In the slime mold *Disctyostelium discoideum*, the pseudoplasmodia that migrate before forming spores move away from a temperature near their acclimation temperature. Considering their rates of locomotion and of thermal acclimation, this behavior is predicted to cause them to migrate toward the surface (Dusenbery 1988a). Their response to temperature gradients thus makes sense as a method of guiding their travel to the surface.

Infective juveniles of the root-knot nematode *Meloidogyne incognita* migrate toward a temperature that is several degrees Centigrade above the temperature to which they are acclimated (Diez and Dusenbery 1989). According to the best available estimates of their rate of locomotion, rate of thermal acclimation, and temperature range of locomotion, this behavior is predicted (Dusenbery 1989e) to cause any individual starting between 0 and 15 cm deep in soil to migrate to a depth of about 5 cm (Figure 19-1). This depth might well be an optimal one in which to search for chemical stimuli such as carbon dioxide leading to host roots (Pline and Dusenbery 1987).

The use of a spatiotemporal temperature pattern to guide migration to a particular depth involves a stimulus that does not have a simple relationship to the goal. Furthermore, the stimulus is general and is not specific to the goal. Consequently, such behavior might be classified as true navigation (see page 366). What makes this behavior noteworthy is that such simple organisms are not usually considered capable of any type of navigation. Although some may find this conclusion hard to accept, this example

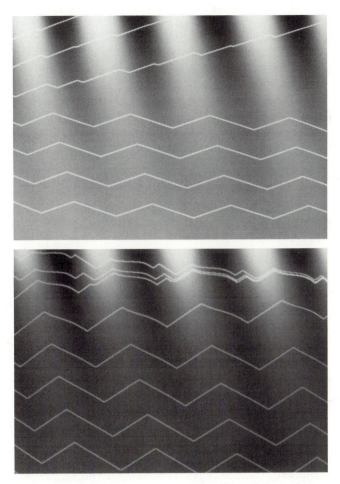

Figure 19-1 Nematode movement in the spatiotemporal temperature pattern of soil. In each picture the soil surface is represented at the top and the bottom corresponds to a depth of 40 cm. Time increases rightward by 4.2 days. The temperature is represented by gray levels, with white corresponding to high and black to low temperatures. The white lines represent the successive positions of model organisms starting at different depths. The upper picture represents Organisms similar to the slime mold that move away from the mean temperature, and the lower picture organisms like the a plant-parasitic nematode that moves toward a temperature that is offset a few degrees from its acclimation temperature. In both cases any organisms starting below a certain depth gradually move deeper. Those starting closer to the surface move gradually toward the surface in the upper case example and toward a specific depth in the lower one. (From Dusenbery 1988a; Dusenbery 1989e.)

may stimulate thinking about what are really the fundamental types of strategies for traveling to spatial goals.

LOCUSTS

One of the most famous examples of migration is of the swarms of desert locusts that plague Africa and the Mideast (Figure 19-2), which have been extensively studied because of their impact on agriculture. Their life history strategy involves exploiting the rich vegetation that temporarily follows a rainfall, after which they migrate to new sites with fresh rainfall. Their migration strategy is to hitch to winds, which they do in great swarms. Individual locusts start flying during daylight, when the temperature is appropriate (17–21° C) and the wind not too strong (Johnson 1974; Rainey 1979). The presence of flying insects seems to stimulate others to fly.

Once airborne, locusts are often carried to heights of a kilometer by thermal convection currents. They seem not to choose any particular direction to fly, but if they start to go away from the swarm they turn back into it (Rainey 1976). This pattern is presumably guided by visual contact with the swarm. Usually, locusts stop flying at sunset (Rainey 1979).

Since no particular direction of flying predominates, the locust swarm drifts downwind (Rainey 1976). Under appropriate circumstances swarms can be carried long distances, but they also sometimes move in loops (Rainey 1976). However, their strategy works better than would movement in random directions. They must travel to an area in which it has rained recently in order to lay eggs that can develop successfully. To do so they make use of the fact that ascending air, which is necessary for the production of rainfall, produces a low-level wind convergence (Rainey 1976). For example, the intertropical convergence zone (see page 368) is associated with monsoon rains. Swarms tend to be carried to areas of wind convergence, which are also likely to be places of rainfall. Thus, the success of locusts appears to depend on subtle features of the weather patterns and very little on making use of information from the environment.

MONARCH BUTTERFLIES

Monumental mark-and-recapture studies have clearly demonstrated that monarch butterflies often migrate thousands of kilometers from the Great Lakes region (Urquhart 1960, 298) to specific wintering sites in central

Figure 19-2 A photo of a locust swarm, which gives some indication of the density of locusts in swarms. (Courtesy U.S. Dept. of Agriculture.)

Mexico (Figure 19-3) (Urquhart and Urquhart 1978) and return from there (Urquhart and Urquhart 1979). Experiments indicate that monarchs can fly continuously for over 100 hours without feeding and that they frequently feed on nectar during their month-long migration (Johnson 1974).

Throughout most of their fall migration the monarchs fly south or southwest (Figure 19-4) (Schmidt-Koenig 1985). Those that come upon the coast of the Gulf of Mexico change direction toward the west and follow along the coast toward Mexico, although some may winter along the Gulf Coast.

Monarchs tend to avoid flying over wide bodies of water and move along a shoreline toward the southwest (Figure 19-5) (Urquhart 1960, 277). This pattern leads to concentrations of them on peninsulas project-ing south or west. There they await favorable weather conditions, at which

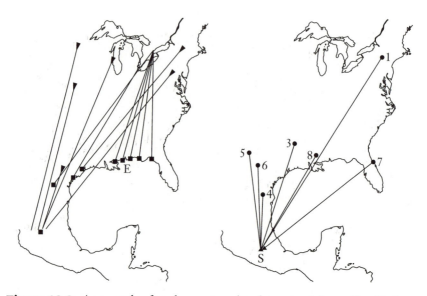

Figure 19-3   An example of mark-recapture data for monarch butterflies. Each
straight line connects the site where a marked individual was released (*triangle*) and
the site where it was recaptured (*circle* or *square*). Remember that the individual
probably did not travel along the straight line in moving from one site to the other.
Left: Data for a sample of individuals marked and released near the Great Lakes
during summer and fall, then recaptured along the Gulf Coast or in Mexico. The
data clearly indicate a pattern of migration to the south and southwest, with those
reaching the Gulf Coast moving west along the coast and many eventually reaching
the small wintering areas in central Mexico. *Right*: Similar data for individuals
marked at the overwintering site in Mexico. During spring the individuals clearly
move north and east. The one recapture in the northeastern U.S. suggests that at
least a few individuals make it all the way back to the northern limit of their range.
(From Urquhart and Urquhart 1978, 1979.)

time they fly out across the water (Schmidt-Koenig 1985) and are often
seen, for instance, crossing Lake Ontario (Urquhart 1960, 278). These
behaviors sometimes produce spectacularly large concentrations of migrating monarchs. On October 24, 1954, for example, an estimated 1
million monarchs flew over Dallas (Urquhart 1960, 280).

Monarchs seem mostly to migrate on days when they have tailwinds
(Urquhart 1960, 258–59; Gibo and Pallett 1979; Gibo 1986) and may have
a degree of weather-prediction ability (Urquhart 1960, 271). When

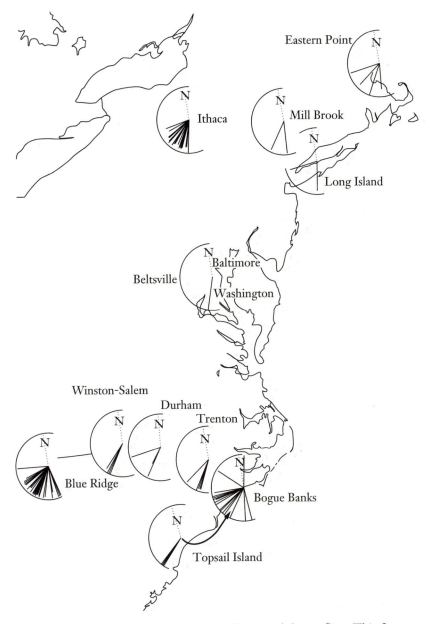

Figure 19-4 The migration directions of monarch butterflies. This figure summarizes data on the observed direction of flight of undisturbed monarchs during four fall migration periods. Each line is the mean direction of at least twenty individuals. Clearly, the direction of movement is to the southeast in this area. (From Schmidt-Koenig 1985.)

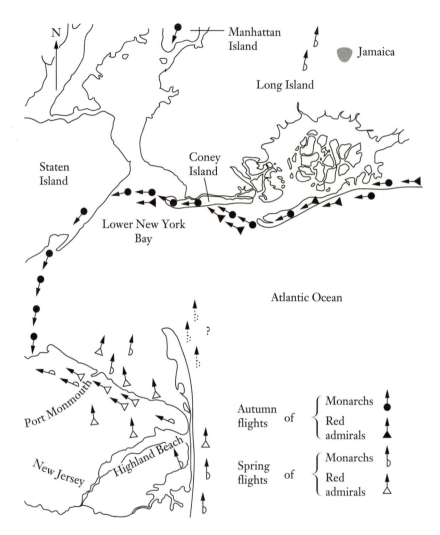

Figure 19-5 The flight paths of migrating butterflies, showing the predominant flight directions of monarch and red-admiral butterflies in the vicinity of a coastline having a complex of bays and peninsulas. (From Urquhart 1960.)

thermals are available, they make use of them to gain altitude and drift downwind (Gibo and Pallett 1979; Schmidt-Koenig 1985).

The strategy monarchs adopt for selecting a direction might simply be to fly in the azimuth of the sun. They have, for instance, been observed to fly southeast in the morning and southwest in the afternoon (Urquhart

1960, 274), and observations of caged monarchs indicate that they maintain at least a crude orientation toward the sun (Kanz 1977). Their apparent tendency to migrate toward the southwest can then be explained by
their flying late in the afternoon but not early in the morning, when they
are often too cold to fly (Urquhart 1960, 270).

However, other butterflies can maintain a constant flight direction
throughout the day (Walker 1979), and more recent observations
(Schmidt-Koenig 1985) indicate that monarchs can maintain their flight
direction accurately under cloudy conditions. This suggests that they have
some other compass sense, such as the use of the earth's magnetic field. In
addition, they seem to compensate for drift caused by crosswinds
(Schmidt-Koenig 1985), which suggests the availability of more sophisticated navigational strategies.

SALMON

Salmon are famous for the runs they make as the mature adults return
from the sea where they grew to adulthood to the freshwater streams
where they developed and will spawn and die. Indeed, salmon runs were
once so extensive that the whole culture and economy of the Indians of the
Pacific Northwest centered around them. These migrations raise some
major questions, one of the most puzzling being how they find their way
back to the particular stream in which they originated. It has been established that the return migrations are specific in their time of year as well as
location (Brannon 1984).

Blinded fish home nearly as well as do the controls (Hasler and Scholz
1978), and a variety of experiments support the hypothesis that juvenile
salmon imprint to odors in their natal stream and use these cues to return
to this stream years later (Northcote 1984). The chemicals have not been
identified but are persistent and include organic components. Artificial
chemicals can be used to imprint juveniles and later attract the adults to a
different stream than the one in which they developed.

The terminal phase of salmon's return can thus be guided by using a
stimulus field (see Chapter 17). However, it remains unclear how, after
traveling thousands of kilometers in the open ocean, the fish find their way
back close enough to home to detect the chemical stimuli. Their return to
the natal stream must be well oriented, however, because mark-recapture
studies indicate that they progress 50 km/day toward the coast, a speed
that requires a nearly straight path, given their swimming speed. Such

observations have led to suggestions that homing salmon possess a calendar, a map sense, and a compass (Quinn 1984).

Experimental evidence suggests that juvenile salmon have a time-compensated sun compass supplemented by a magnetic compass when the sun is not visible (Quinn 1984), and presumably adults have the same abilities. A map sense in salmon has been hypothesized that is based on the inclination of the earth's magnetic field and on the angle (i.e., declination) between true north as defined by the axis of rotation and magnetic north (Quinn 1984).

Simpler mechanisms are possible. Cook (1984) proposes that migrating salmon may orient to waves and perhaps simply swim in the direction of wave propagation. Leggett (1984) emphasizes the salmon's use of the strategy of hitching by making use of ocean currents. In summary, much remains to be revealed about the information that salmon use for migration.

BIRDS

The general patterns of bird migration have been summarized by Wallraff (1984):

1. Offspring of a particular population migrate, in their first autumn of life, in similar directions. Birds belonging to different populations of the same species may select different directions. The same is true for birds of different species originating from the same area.
2. Flight routes between breeding and wintering grounds may follow, by and large, a straight beeline course, yet in other cases they may include detours and thus some directional changes. Usually detours appear in some way adapted to large-scale geomorphological features.
3. Autumn and spring migrations do not always follow the same routes.

As previously mentioned, evidence suggests that an individual bird's first seasonal migration is based on an innate tendency to fly in a particular direction for a particular length of time (Wiltschko and Wiltschko 1988). However, after gaining the experience of being in a particular wintering site, a bird is able to return to that site from an unfamiliar location (Wallraff 1984). Similarly, birds return to nest in the area where they fledged only if they were able to fly about in that area before starting their autumn migration (Wiltschko and Wiltschko 1988).

In a particularly impressive migration, many species of small birds fly nonstop for several thousand kilometers over the North Atlantic in the fall. As shown in Figure 19-6, this migration is well documented by radar observations. Most impressively, the birds fly at altitudes as high as five kilometers during the second half of the flight. Also worthy of note is that they change direction in the middle of the flight while still over the open ocean. Since there are presumably no landmarks available there for reference, they probably use the elapsed time of flight to determine when to change direction.

To summarize the evidence for orientation mechanisms in birds, orientation in cages has been experimentally altered by changing the position of sun, stars, or magnetic field direction. Although the effect of magnetic fields is so weak it can be detected only with certain cage designs and specific statistical analysis techniques, it has been replicated in different laboratories (Emlen 1975). Experiments with free-flying birds demonstrate a predictable change of direction (via clock shifting) for a sun compass and for a reversal of the vertical component of the magnetic field (Baker 1984). The time-compensated sun compass is very well established, and the star compass seems highly probable. The magnetic compass also seems well established, although it does have severe limitations (Wiltschko and Wiltschko 1988). Many night-migrating birds seem to use the direction of the setting sun (Moore 1985) or the polarization of overhead light (Able 1982b) at sunset to determine their heading for the night. This heading is probably used to select star patterns for guidance during the night.

The question of how the homeward direction is established during homing experiments still remains unclear. Information obtained during displacement is probably used in some cases, but not in others. Homing from an unfamiliar area without information obtained during displacement requires having a map sense. The evidence for a map sense in homing birds suggests (Wallraff 1984) the hypothesis that (1) birds are sensitive to certain stimuli that vary smoothly over wide areas, probably some features of the earth's magnetic field, and sometimes odors; (2) they learn how these stimuli vary in the vicinity of home in relation to a compass; and (3) when they are displaced outside their familiar area they determine the value of the stimuli at the present site, then extrapolate the trends in stimuli at home to estimate their present direction from home, and then use their compass to fly in the homeward direction.

It seems that migrating birds need some visual cues (Wallraff 1984). Disorientation during seasonal migrations sometimes occurs under conditions of total overcast and fog or rain, and mass mortality may occur at lighthouses, tall buildings, or television towers, which seem to attract the

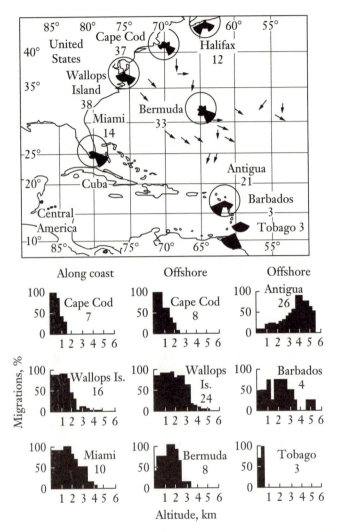

Figure 19-6 Radar tracking of fall bird migrations over the western North Atlantic. *Top:* The circular histograms show the number of moderate and heavy migrations in different directions observed by land-based radars; the arrows indicate the daily average directions of moderate and heavy migrations observed by ship radars. Note that along the northeastern United States the migrations are alongshore or southeast, out to sea. This southeasterly orientation persists past Bermuda but changes to the southward over the Lesser Antilles. *Bottom:* These histograms show the fraction of moderate and heavy migrations in which birds were detected at different altitudes. Over Antigua most were flying very high, probably to avoid the trade winds, but they have descended by the time they reach Tobago and approach the mainland. (From Williams et al. 1977.)

birds (Nisbet 1970; Emlen 1975). Radar studies of birds flying between cloud layers indicate that the birds are then less well oriented than under other conditions (Able 1982a). Songbirds in the northeastern United States migrating on nights when cloud cover has deprived them of celestial cues before sunset simply fly downwind, even if this is not their usual direction of migration (Able 1978). On the other hand, a thrush tracked by radiotelemetry for 1,500 km during the spring migration (Cochran 1987) maintained a consistent heading in spite of complete cloud cover. Limitations on visibility may cause loss of the visual horizon, which could be necessary for the use of other stimuli such as the inclination of the magnetic field.

Many birds cross the equator during their migrations, which complicates the use of most strategies. Seasonal migrations occur near the equinoxes when the sun passes directly overhead at the equator. Thus, a bird migrating across the equator must shift from flying toward the sun to away from it. Similarly, a bird referring to the inclination of the earth's magnetic field would have to shift from flying toward the upward direction to the downward direction, or vice versa. And a bird flying away from star constellations near one pole might lose sight of them as they sank below the horizon. Species making such long migrations might better learn to use constellations nearer the equator, but this would then probably require time compensation for movement of the stars. Using stars near the celestial equator is similar to using the sun, so that one's direction would have to be switched when crossing the equator.

The Wiltschkos (1988) provide a view in which the genetic information for the autumnal migration of many birds is based on a magnetic compass. In some night-migrating species—especially those that nest far to the north, where the magnetic field varies more rapidly—young birds learn a star compass, from observing the rotation of the sky. They initiate migration by flying at an angle to the learned constellations near the pole. During migration, as they move south the pole descends toward the horizon, and the magnetic field becomes more reliable. They therefore come to rely more on a magnetic compass and perhaps use it to modify their star compass. Adopting a new star compass based on constellations near the celestial equator could solve the problem of the magnetic field's becoming horizontal near the equator. This type of plasticity in compasses could easily lead to migration paths that are not straight, such as dog-leg paths around the Mediterranean.

A recent study (Helbig et al. 1989) has demonstrated that even the activity of birds in cages can shift its orientation in appropriate directions

at relevant times. For instance, a shift of about 60° from southeast to
south-southwest over a period of ten days seems appropriate for taking the
blackcaps of eastern Austria around the eastern end of the Mediterranian
and on to East Africa. Since these birds have remained in one location, the
shift cannot be a response to the local environment. It could instead be
endogenous timing or a response to a seasonal change in the photoperiod.

In summary, it appears that birds employ a variety of strategies for
guiding migration, which vary between species, phases of a migration
loop, and current conditions. This variability greatly complicates the study
of navigation mechanisms, and we do not yet have a really reliable understanding of the mechanisms used by any species.

Bibliography

■ ■

Able, K. P. 1977. "The flight behaviour of individual passerine nocturnal migrants: A tracking radar study." *Anim. Behav.* 25:924–35.

———. 1978. "Field studies of the orientation cue hierarchy of nocturnal songbird migrants." In *Animal Migration, Navigation, and Homing.* K. Schmidt-Koenig and W. T. Keeton, eds. Berlin: Springer-Verlag, pp. 228–38.

———. 1980. "Mechanisms of orientation, navigation, and homing." In *Animal Migration, Orientation, and Navigation.* S. A. Gauthreaux, Jr., ed. New York: Academic Press, pp. 283–73.

———. 1982a. "The effects of overcast skies on the orientation of free-flying nocturnal migrants." In *Avian Navigation;* F. Papi and H. G. Wallraff, eds. Berlin: Springer-Verlag, pp. 38–49.

———. 1982b. "Skylight polarization patterns at dusk influence migratory orientation in birds." *Nature* 299:550–51.

———. 1982c. "Field studies of avian nocturnal migratory orientation. I. Interaction of sun, wind and stars as directional cues." *Anim. Behav.* 30:761–67.

———, V. P. Bingman, P. Kerlinger, and W. Gergits. 1982b. "Field studies of avian nocturnal migratory orientation. II. Experimental manipulation of orientation in white-throated sparrows (*Zonotrichia albicollis*) released aloft." *Anim. Behav.* 30:768–73.

———, and W. F. Gergits. 1985. "Human navigation, attempts to replicate Baker's displacement experiment." In *Magnetite Biomineralization and Magnetoreception in Organisms.* J. L. Kirschvink, D. S. Jones, and B. J. MacFadden, eds. New York: Plenum, pp. 569–72.

Ache, B. W. 1988. "Integration of chemosensory information in aquatic invertebrates." In *Sensory Biology of Aquatic Animals.* J. Atema, R. R. Fay, A. N. Popper, and W. N. Tavolga, eds. New York: Springer-Verlag, pp. 387–401.

Ackerman, E., L. B. M. Ellis, and L. E. Williams. 1979. *Biophysical Science*. Englewood Cliffs, N. J.: Prentice-Hall.

Adam, G., and M. Delbürck. 1968. "Reduction of dimensionality in biological diffusion processes." *In Structural Chemistry and Molecular Biology*. A. Rich and N. Davidson, eds. San Francisco: W. H. Freeman and Co. , pp. 198–215.

Adler, J. 1969. "Chemoreceptors in bacteria." *Science* 166 : 1588–97.

Adler, K., and C. R. Pelkie. 1985. "Human homing orientation." In *Magnetite Biomineralization and Magnetoreception in Organisms*. J. L. Kirschvink, D. S. Jones, and B. J. MacFadden, eds. New York: Plenum, pp. 573–93.

Ahl, A. S. 1986. "The role of vibrissae in behavior: a status review." *Veterinary Research Communications* 10 : 245–68.

Aidley, D. J. 1981. "Questions about migration." In *Animal Migration*. D. J. Aidley, ed. Cambridge: Cambridge University Press, pp. 1–8.

Albone, E. S., and S. G. Shirley. 1984. *Mammalian Semiochemistry*. New York: Wiley.

Alcock, J. 1975. *Animal Behavior*. Sunderland, Mass: Sinauer Associates.

Aldrich, J. R. 1985. "Pheromone of a true bug (*Hemiptera–Heteroptera*): attractant for the predator, *Podisus maculiventris*, and kairomonal effects." In *Semiochemistry: Flavors and Pheromones*. A. C. Society, ed. New York: Walter de Gruyter, pp. 95–119.

Alerstam, T. 1981. "The course and timing of bird migration." In *Animal Migration*. D. J. Aidley, ed. Cambridge: Cambridge University Press, pp. 9–54.

———. 1987. "Bird migration across a strong magnetic anomaly." *J. Exp. Biol.* 130 : 63–86.

———, and S. -G. Pettersson. 1976. "Do birds use waves for orientation when migrating across the sea?" *Nature* 259 : 205–7.

Ali, M. A. 1978. General introduction. In *Sensory Ecology*. M. A. Ali, ed. New York: Plenum, pp. 3–8.

Alldredge, A. L., and Y. Cohen. 1987. "Can microscale chemical patches persist in the sea? Microelectrode study of marine snow, fecal pellets." *Science* 235 : 689–91.

Allison, A. C., and R. T. T. Warwick. 1949. "Quantitative observations on the olfactory system of the rabbit." *Brain* 72 : 186–97.

Altman, P. L.; J. F. Gibson, Jr.; C. C. Wang; D. S. Dittmer, and R. M. Grebe. 1958. *Handbook of Respiration*. Philadelphia: Saunders.

Amadon, D. 1953. "Migratory birds of relict distribution: Some inferences." *Auk 70 :* 461–69.

Amoore, J. E. 1967. "Specific anosmia: A clue to the olfactory code." *Nature* 214 : 1095–98.

———. 1971. "Olfactory genetics and anosmia." In *Chemical Senses*. Part 1: "Olfaction." L. M. Beidler, ed. New York: Springer-Verlag, pp. 245–56.

———. 1975. "Four primary odor modalities of man: Experimental evidence and possible significance." In *Olfaction and Taste*. D. A. Denton and J. P. Coghlan, eds. New York: Academic Press, pp. 283–89.

———. and L. J. Forrester. 1976. "Specific anosmia to trimethylamine: the fishy primary odor." *J. Chem. Ecol.* 2 : 49–56.

———, L. J. Forrester, and P. Pelosi. 1976. "Specific anosmia to isobutyraldehyde: the malty primary odor." *Chem. Senses Flavor* 2 : 17–25.

———, P. Pelosi, and L. J. Forrester. 1977. "Specific anosmias to 5a–androst–16–en–

3–one and ω–pentadecalactone: the urinous and musky primary odors." *Chem. Senses Flavor* 2 : 401–25.

Andrews, J. C. 1983. "Deformation of the active space in the low Reynolds number feeding current of calanoid copepods." *Can. J. Aquat. Sci.* 40 : 1293–1302.

Armitage, J. P. 1988. "Tactic responses in photosynthetic bacteria." *Can. J. Microbiology* 34 : 475–81.

———, and J. M. Lackie. 1990. *Biology of the Chemotactic Response.* Cambridge: Cambridge University Press.

Arnold, G. P. 1981. "Movements of fish in relation to water currents." In *Animal Migration.* D. J. Aidley, ed. Cambridge: Cambridge University Press, pp. 55–79.

———, and P. H. Cook. 1984. "Fish migration by selective tidal stream transport: first results with a computer simulation model for the European continental shelf." In *Mechanisms of Migration in Fishes.* J. D. McCleave, G. P. Arnold, J. J. Dodson, and W. H. Neill, eds. New York: Plenum, pp. 227–61.

Atema, J. 1985. "Chemoreception in the sea: adaptations of chemoreceptors and behaviour to aquatic stimulus conditions." In *Physiological Adaptations of Marine Animals.* Society of Experimental Biology, ed. Cambridge: Cambridge University Press, pp. 387–423.

———. 1988. "Distribution of chemical stimuli." In *Sensory Biology of Aquatic Animals.* J. Atema, R. R. Fay, A. N. Popper, and W. N. Tavolga, eds. New York: Springer-Verlag, pp. 29–56.

Au, W. W. L. 1988. "Sonar target detection and recognition by odontocetes." In *Animal Sonar.* P. E. Nachtigall and P. W. B. Moore, eds. New York: Plenum, pp. 451–65.

———, and D. W. Martin. 1989. "Insights into dolphin sonar discrimination capabilities from human listening experiments." *J. Acoustical Soc. of Am.* 86(5) : 1662–70.

Aveni, A. F. 1981. "Tropical archeoastronomy." *Science* 213 : 161–71.

Aylor, D. E. 1976. "Estimating peak concentrations of pheromones in the forest." In *Perspectives on Forest Entomology.* J. F. Anderson and H. K. Kaya, eds. pp. 177–88.

———, J. –Y. Parlange, and J. Granett. 1976. "Turbulent dispersion of disparlure in the forest and male gypsy moth response." *Environ. Entomol.* 5 : 1026–32.

Bailey, W. J. 1991. *Acoustic Behaviour of Insects.* London: Chapman and Hall.

Baker, R. R. 1968. "A possible method of evolution of the migratory habit in butterflies." *Phil. Trans. R. Soc.* B 253 : 309–41.

———. 1969. "The evolution of the migratory habit in butterflies." *J. Anim. Ecol.* 38 : 703–46.

———. 1978. *The Evolutionary Ecology of Animal Migration.* New York: Holmes and Meier.

———. 1981. "Man and other vertebrates: a common perspective to migration and navigation." In *Animal Migration.* D. J. Aidley, ed. Cambridge: Cambridge University Press, pp. 241–60.

———. 1982. *Migration.* London: Hodder and Stoughton.

———. 1984. *Bird Navigation.* New York: Holmes and Meier.

———. 1985a. "Magnetoreception by man and other primates." In *Magnetite Bio-*

mineralization and Magnetoreception in Organisms. J. L. Kirschvink, D. S. Jones, and B. J. MacFadden, eds. New York: Plenum.

———. 1985b. "Moths: population estimates, light-traps and migration." In *Case Studies in Population Biology.* L. M. Cook, ed. Manchester: Manchester University Press, pp. 188–211.

———. 1987. "Integrated use of moon and magnetic compasses by the heart-and-dart moth, *Agrotis exclamationis.*" *Anim. Behav.* 35 : 94–101.

Baker, T. C. 1985. "Chemical control of behavior." In *Comprehensive Insect Physiology, Biochemistry, and Pharmacology.* G. A. Kerkut and L. I. Gilbert, eds. Oxford: Pergamon, pp. 621–72.

———. 1986. "Pheromone-modulated movements of flying moths." In *Mechanisms in Insect Olfaction.* T. L. Payne, M. C. Birch, and C. E. J. Kennedy, eds. Oxford: Clarendon, pp. 39–48.

———. 1989. "Pheromones and flight behavior." In *Insect Flight.* G. J. Goldsworthy and C. H. Wheeler, eds. Boca Raton, Fla.: CRC Press, pp. 231–55.

———, and K. F. Haynes. 1987. "Manoeuvres used by flying male oriental fruit moths to relocate a sex pheromone plume in an experimentally shifted wind-field." *Physiological Entomology* 12 : 263–79.

Balan, J. 1985. "Measuring minimal concentrations of attractants detected by the nematode *Panagrellus redivivus.*" *J. Chem. Ecol.* 11 : 105–11.

Baldaccini, N. E., S. Benvenuti, V. Flaschi, P. Ioalé, and F. Papi. 1982. "Pigeon orientation: experiments on the role of olfactory stimuli perceived during the outward journey." In *Avian Navigation.* F. Papi and H. G. Wallraff, eds. Berlin: Springer-Verlag, pp. 160–69.

Baldwin, I. T., and J. C. Schultz. 1983. "Rapid changes in tree leaf chemistry induced by damage: evidence for communication between plants." *Science* 221 : 277–79.

Ballaré, C. L., A. L. Scopel, and R. A. Sánchez. 1990. "Far-red radiation reflected from adjacent leaves: an early signal of competition in plant canopies." *Science* 247 : 329–32.

Banner, A. 1972. "Use of sound in predation by young lemon sharks, *Negaprion Brevirostris* (Poey)." *Bulletin of Marine Science* 22 : 251–283.

Barfield, R. J., and D. A. Thomas. 1986. "The role of ultrasonic vocalizations in the regulation of reproduction in rats." *Ann. N. Y. Acad. Sciences* 474 : 33–43.

Barlow, J. S. 1964. "Inertial navigation as a basis for animal navigation." *J. Theoret. Biol.* 6 : 76–117.

Barrett, C. R., Jr. 1987. "Adaptive thresholding and automatic detection." In *Principles of Modern Radar.* J. L. Eaves and E. K. Reedy, eds. New York: Van Nostrand Reinhold, pp. 368–93.

Barrows, W. M. 1915. "The reactions of an orb-weaving spider, *Epeira sclopetaria Clerck,* to rhythmic vibrations of its web." *Biological Bull.* 29 : 316–33.

Barth, F. G. 1982. "Spiders and vibratory signals: sensory reception and behavioral significance." In *Spider Communication.* P. N. Witt and J. S. Rovner, eds. Princeton, N. J.: Princeton University Press, pp. 67–122.

Bastian, J. 1981. "Electrolocation. I. How the electroreceptors of *Apteronotus albifrons* code for moving objects and other electrical stimuli." *J. Comp. Physiol.* A144 : 465–79.

———. 1986. "Electrolocation." In Electroreception. T. H. Bullock and W. Heilignenberg, eds. New York: Wiley, pp. 577–612.

Batschelet, E. 1981. *Circular Statistics in Biology*. London: Academic Press.

Bauer, G. B., M. Fuller, A. Perry, J. R. Dunn, and J. Zoeger. 1985. "Magnetoreception and biomineralization of magnetite in cetaceans." In *Magnetite Biomineralization and Magnetoreception in Organisms*. J. L. Kirschvink, D. S. Jones, and B. J. MacFadden, eds. New York: Plenum, pp. 489–507.

Baylis, J. R. 1982. "Avian vocal mimicry: its function and evolution." In *Acoustic Communication in Birds*. D. E. Kroodsma, E. H. Miller, and H. Ouellet, eds. New York: Academic Press, pp. 51–83.

Becker, P. H. 1982. "The coding of species-specific characteristics in bird sounds." In *Acoustic Communication in Birds*. D. E. Kroodsma, E. H. Miller, and H. Ouellet, eds. New York: Academic Press, pp. 213–52.

Beer, C. G. 1982. "Conceptual issues in the study of communication." In *Acoustic Communication in Birds*. D. E. Kroodsma, E. H. Miller, and H. Ouellet, eds. New York: Academic Press, pp. 279–310.

Bell, W. J. 1985. "Sources of information controlling motor patterns in arthropod local search orientation." *J. Insect. Physiol.* 31 : 837–47.

———. 1991. *Searching Behaviour*. London: Chapman and Hall.

———, and T. R. Tobin. 1982. "Chemo-orientation." *Biol. Rev.* 57 : 219–60.

———, C. Tortorici, R. J. Roggero, L. R. Kipp, and T. R. Tobin. 1985. "Sucrose-stimulated searching behaviour of *Drosophila melanogaster* in a uniform habitat: modulation by period of deprivation." *Anim. Behav.* 33 : 436–48.

Benhamou, S., and P. Bovet. 1989. "How animals use their environment: a new look at kinesis." *Anim. Behav.* 38 : 375–83.

Bennet–Clark, H. C. 1971. "Acoustics of insect song." *Nature* 234 : 255–59.

Ben-Yosef, N., and A. Rose. 1978. "Spectral response of the human eye." *J. Opt. Soc. Am.* 68 : 935–37.

Berg, H. C. 1983. *Random Walks in Biology*. Princeton, N. J.: Princeton University Press.

———. 1985. "Physics of bacterial chemotaxis." In *Sensory Perception and Transduction in Aneural Organisms*. G. Colombetti, F. Lenci. and P. -S. Song, eds. New York: Plenum, pp. 19–30.

———, and D. A. Brown. 1972. "Chemotaxis in *Escherichia coli* analyzed by three-dimensional tracking." *Nature* 239 : 500–504.

———, and E. M. Purcell. 1977. "Physics of chemoreception." *Biophys. J.* 20 : 193–219.

Bergman, A. S. 1990. *Auditory Scene Analysis*. Cambridge, Mass.: MIT.

Bernard, G. D. 1979. "Red-absorbing visual pigment of butterflies." *Science* 203 : 1,125–27.

Bilsen, F. A., and R. J. Ritsma. 1969. "Repetition pitch and its implication for hearing theory." *Acustica* 22 : 63–73.

Bjostad, L., C. Linn, W. Roelofs, and J. -W. Du. 1985. "Identification of new sex pheromone components in *Tricholplusia ni* and *Argyrotaenia velutinana*, predicted from biosynthetic precursors." In *Semiochemistry: Flavors and Pheromones*. A. C. Society, ed. New York: Walter de Gruyter, pp. 223–50.

Black, C. A. 1968. *Soil–Plant Relationships*. New York: Wiley.

Blakemore, C. 1973. "The baffled brain." In *Illusion in Nature and Art.* R. L. Gregory and E. H. Gombrich, eds. London: Gerald Duckworth, pp. 9–47.

Blakemore, R. P. 1975. "Magnetotactic bacteria." *Science* 190 : 377–79.

———. 1982. "Magnetotactic bacteria." *Ann. Rev. Microbiol.* 36 : 217–38.

Blaxter, J. H. S. 1975. "Fish vision and applied research." In *Vision in Fishes.* M. A. Ali, ed. New York: Plenum, pp. 757–73.

———. 1988. "Sensory performance, behavior, and ecology of fish." In *Sensory Biology of Aquatic Animals.* J. Atema, R. R. Fay, A. N. Popper, and W. N. Tavolga, eds. New York: Springer-Verlag, pp. 203–32.

———, E. J. Denton, and J. A. B. Gray. 1981. "Acousticolateralis system in clupeid fishes." In *Hearing and Sound Communication in Fishes.* W. N. Tavolga, A. N. Popper, and R. R. Fay, eds. New York: Springer-Verlag, pp. 39–59.

Bleckmann, H. 1988. "Prey identification and prey localization in surface-feeding fish and fishing spiders." In *Sensory Biology of Aquatic Animals.* J. Atema, R. R. Fay, A. N. Popper, and W. N. Tavolga, eds. New York: Springer-Verlag, pp. 620–41.

Blest, A. D. 1957. "The function of eyespot patterns in the Lepidoptera." *Behaviour* 11 : 209–56.

Block, S. M., J. E. Segall, and H. C. Berg. 1982. "Impulse responses in bacterial chemotaxis." *Cell* 31 : 215–26.

———. 1983. "Adaptation kinetics in bacterial chemotaxis." *J. Bacteriol.* 154 : 312–23.

Blum, M. S. 1985. "Alarm pheromones." In *Comprehensive Insect Physiology, Biochemistry, and Pharmacology.* G. A. Kerkut and L. I. Gilbert, eds. Oxford: Pergamon, pp. 193–224.

Bonner, J. T., W. W. Clarke, Jr., C. L. Neely, Jr., and M. K. Slifkin. 1950. "The orientation to light and the extremely sensitive orientation to temperature gradients in the slime mold Dictyostelium discoideum." *J. Cell. Comp. Physiol.* 36 : 149–58.

———. and M. R. Dodd. 1962. "Evidence for gas-induced orientation in the cellular slime molds." *Devel. Biol.* 5 : 344–61.

———, H. B. Suthers, and G. M. Odell. 1986. "Ammonia orients cell masses and speeds up aggregating cells of slime molds." *Nature* 323 : 630–32.

Bossert, W. H. 1968. "Temporal patterning in olfactory communication." *J. Theor. Biol.* 18 : 157–70.

———, and E. O. Wilson. 1963. "The analysis of olfactory communication among animals." *J. Theor. Biol.* 5 : 443–69.

Bovet, J. 1978. "Homing in wild myomorph rodents: current problems." In *Animal Migration, Navigation, and Homing.* K. Schmidt-Koenig and W. T. Keeton, eds. Berlin: Springer-Verlag, pp. 405–12.

Bowditch, N. 1984. *American Practical Navigator.* Washington, D. C.: Defense Mapping Agency Hydrographic/Topographic Center.

Bowmaker, J. K. 1977. "The visual pigments, oil droplets and spectral sensitivity of the pigeon." *Vision Res.* 17 : 1129–38.

Brackenbury, J. H. 1982. "The structural basis of voice production and its relationship to sound characteristics." In *Acoustic Communication in Birds.* D. E. Kroodsma, E. H. Miller, and H. Ouellet, eds. New York: Academic Press, pp. 53–73.

Bradburne, J. A., M. J. Kasperbauer, and J. N. Mathis. 1989. "Reflected far-red light effects on chlorophyll and light-harvesting chlorophyll protein LHC–II contents under field conditions." *Plant Physiol.* 91 : 800–803.

Bradshaw, J. W. S., and P. E. Howse. 1984. "Sociochemicals of ants." In *Chemical Ecology of Insects.* W. J. Bell and R. T. Cardé, eds. London: Chapman and Hall, pp. 128–57.

Brannon, E. L. 1984. "Influence of stock origin on salmon migratory behavior." In *Mechanisms of Migration in Fishes.* J. D. McCleave, G. P. Arnold, J. J. Dodson, and W. H. Neill, eds. New York: Plenum, pp. 103–11.

Brenowitz, E. A. 1982. "The active space of red-winged blackbird song." *J. Comp. Physiol.* 147 : 511–22.

Brines, M. L., and J. L. Gould. 1982. "Skylight polarization patterns and animal orientation." *J. Exp. Biol.* 96 : 69–91.

Brodie, E. D., III. 1989. "Genetic correlations between morphology and antipredator behaviour in natural populations of the garter snake *Thamnophis ordinoides.*" *Nature* 342 : 542–43.

Brower, L. P., W. H. Calvert, L. E. Hedrick, and J. Christian. 1977. "Biological observations on an overwintering colony of monarch butterflies (*Danaus plexippus, Danaidae*) in Mexico." *J. Lep. Soc.* 31 : 232–42.

Brown, C. H. 1982. "Ventroloquial and locatable vocalizations in birds." *Z. Tierpsychol.* 59 : 338–50.

———. 1989a. "The acoustic ecology of East African primates and the perception of vocal signals by grey-cheeked mangabeys and blue monkeys." In *The Comparative Psychology of Audition: Perceiving Complex Sounds.* R. J. Dooling and S. H. Hulse, eds. Hillsdale, N. J.: Lawrence Erlbaum, pp. 201–39.

———. 1989b. "The active space of blue monkey and grey-cheeked mangabey vocalizations." *Anim. Behav.* 37 : 1023–34.

———. 1989c. "The measurement of vocal amplitude and vocal radiation pattern in blue monkeys and grey-cheeked mangabeys." *Bioacoustics* 1 : 253–71.

———, and P. L. Schwagmeyer. 1984. "The vocal range of alarm calls in thirteen-lined ground squirrels." *Z. Tierpsychol.* 65 : 273–88.

———. 1986. "Cliff swallow colonies as information centers." *Science* 234 : 83–85.

Brownell, P., and R. D. Farley. 1979. "Orientation to vibrations in sand by the nocturnal scorpion *Paruroctonus mesaensis:* mechanism of target localization." *J. Comp. Physiol.* 131 : 31–38.

———. 1977. "Compressional and surface waves in sand: used by desert scorpions to locate prey." *Science* 197 : 479–82.

Bruderer, B. 1978. "Effects of alpine topography and winds on migrating birds." In *Animal Migration, Navigation, and Homing.* K. Schmidt-Koenig and W. T. Keeton, eds. Berlin: Springer-Verlag, pp. 252–65.

Buck, J. B. 1978. "Functions and evolutions of bioluminescence." In *Bioluminescence in Action.* P. J. Herring, ed. New York: Academic Press, pp. 419–60.

Buck, L., and R. Axel. 1991. "A novel multigene family may encode odorant receptors: a molecular basis for odor recognition." *Cell* 65 : 175–87.

Budelmann, B. -U. 1988. "Morphological diversity of equilibrium receptor systems in aquatic invertebrates." In *Sensory Biology of Aquatic Animals.* J. Atema, R. R. Fay, A. N. Popper, and W. N. Tavolga, eds. New York: Springer-Verlag, pp. 758–82.

Bullock, T. H., and F. P. J. Diecke. 1956. "Properties of an infra-red receptor." *J. Physiol.* 134 : 47–87.

———, and W. Heiligenberg. 1986. *Electroreception.* New York: Wiley.

———, R. Orkand, and A. Grinnell. 1977. *Introduction to Nervous Systems.* San Francisco: W. H. Freeman.

Buning, T. D. C. 1983. "Thermal sensitivity as a specialization for prey capture and feeding in snakes." *Amer. Zool.* 23 : 363–75.

Burda, H., V. Bruns, and M. Müller. 1990. "Sensory adaptations in subterranean mammals." In *Evolution of Subterranean Mammals at the Organismal and Molecular Levels.* New York: Alan R. Liss, pp. 269–93.

Burr, A. H. 1984a. "Evolution of eyes and photoreceptor organelles in the lower phyla." In *Photoreception and Vision in Invertebrates.* M. A. Ali, ed. New York: Plenum, pp. 131–78.

———. 1984b. "Photomovement behavior in simple invertebrates." In *Photoreception and Vision in Invertebrates.* M. A. Ali, ed. New York: Plenum, pp. 179–215.

———. 1985. "The photomovement of *C. elegans,* a nematode which lacks ocelli. Proof that the response is to light, not radiant heating." *Photochemistry and Photobiology* 41 : 577–82.

Bursell, E., and D. W. Ewer. 1950. "On the reactions to humidity of *Peripatopsis moseleyi* (wood–mason)." *J. Exp. Biol.* 26 : 335–53.

Butler, R. A., and K. Belendiuk. 1977. "Spectral cues utilized in the localization of sound in the median sagittal plane." *J. Acoust. Soc. Am.* 61 : 1264–69.

Buwalda, R. J. A., A. Schuijf, and A. D. Hawkins. 1983. "Discrimination by the cod of sounds from opposing directions." *J. Comp. Physiol.* 150 : 175–84.

Cade, W. 1975. "Acoustically orienting parasitoids: fly phonotaxis to cricket song." *Science* 190 : 1312–13.

Cahn, P. H. 1972. "Sensory factors in the side-to-side spacing and positional orientation of the tuna, *Euthynnus affinis,* during schooling." *Fishery Bull.* 70 : 197–204.

Camhi, J. M. 1980. "The escape system of the cockroach." *Scientific Am.* 243 : 158–72.

———, and W. Tom. 1978. "The escape behavior of the cockroach *Periplaneta americana.* I. Turning response to wind puffs." *J. Comp. Physiol.* 128 : 193–201.

Campbell, G. S. 1977. *An Introduction to Environmental Biophysics.* New York: Springer-Verlag.

Cardé, R. T. 1984. "Chemo-orientation in flying insects." In *Chemical Ecology of Insects.* W. J. Bell and R. T. Cardé, eds. Sunderland, Mass.: Sinauer Associates, pp. 111–24.

Cardé, R. T., A. M. Cardé, A. S. Hill, and W. L. Roelofs. 1977. "Sex pheromone specificity as a reproductive isolating mechanism among the sibling species *Archips argyrospilus* and *A. mortuanus* and other sympatric tortricine moths (Lepidoptera: Tortricidae)." *J. Chem. Ecol.* 3 : 71–84.

Carlile, M. J. 1980. "Positioning mechanisms—the role of motility, taxis and tropism in the life of microorganisms." In *Contemporary Microbial Ecology.* D. C. Ellwood, M. J. Latham, J. N. Hedger, J. M. Lynch, and J. H. Slater, eds. London: Academic Press, pp. 55–74.

Carlow, P. 1976. *Biological Machines: A Cybernetic Approach to Life.* London: Edward Arnold.

Carr, W. E. S. 1988. "The molecular nature of chemical stimuli in the aquatic environment." In *Sensory Biology of Aquatic Animals.* J. Atema, R. R. Fay, A. N. Popper, and W. N. Tavolga, eds. New York: Springer-Verlag, pp. 3–27.

Carslaw, H. S., and J. C. Jaeger. 1959. *Conduction of Heat in Solids.* Oxford: Clarendon .

Cartwright, B. A., and T. S. Collett. 1983. "Landmark learning in bees." *J. Comp. Physiol.* 151 : 521–43.

Carvell, G. E., and D. J. Simmons. 1990. "Biometric analyses of vibrissal tactile discrimination in the rat." *J. Neuroscience* 10 : 2638–48.

Cerdá-Olmedo, E., and E. D. Lipson. 1987. *Phycomyces.* Cold Spring Harbor, N. Y.: Cold Spring Harbor Laboratory.

Chalfie, M., and J. White. 1988. "The nervous system." In *The Nematode Caenorhabditis elegans.* W. B. Wood, ed. Cold Spring Harbor, N. Y.: Cold Spring Harbor Laboratory, pp. 337–91.

Chamberlin, T. C. 1965. "The method of multiple working hypotheses." *Science* 148 : 754–59.

Chapman, C. J., and O. Sand. 1974. "Field studies of hearing in two species of flatfish *Pleuronectes platessa* (L.) and *Limanda limanda* (L.) (Family Pleuronectidae)." *Comp. Biochem. Physiol.* 47A : 371–85.

Churchill, C. 1774. *The Works of C. Churchill.* Book 3: *Gotham.* London: John Churchill.

Clark, C. W., and D. A. Levy. 1988. "Diel vertical migrations by juvenile sockeye salmon and the antipredation window." *Am. Nat.* 131 : 271–90.

Clausen, C. P. 1976. "Phoresy among entomophagous insects." *Ann. Rev. Entomol.* 21 : 343–68.

Clayton, R. K. 1965. "Phototaxis in microorganisms." In *Photophysiology.* A. C. Giese, ed. New York: Academic, pp. 51–77.

Clewell, D. B., and K. E. Weaver. 1989. "Sex pheromones and plasmid transfer in *Enterococcus faecalis.*" *Plasmid* 21 : 175–84.

Cochran, W. W. 1987. "Orientation and other migratory behaviours of a Swainson's thrush followed for 1500 km." *Anim. Behav.* 35 : 927–29.

———, and C. G. Kjos. 1985. Wind drift and migration of thrushes: a telemetry study." Illinois Natural History Survey Bulletin 33 : 297–330.

———, G. G. Montgomery, and R. R. Graber. 1967. "Migratory flights of Hylocichla thrushes in spring: a radiotelemetry study." *Living Bird* 6 : 213–25.

Cohen, L. M., H. Neimark, and L. K. Eveland. 1980. "*Schistosoma mansoni*: response of cercariae to a thermal gradient." *J. Parasitol.* 66 : 362–64.

Cohen, R. J., Y. N. Jan, J. Matricon, and M. Delbrück. 1975. "Avoidance response, house response, and wind responses of the sporangiophore of Phycomyces." *J. Gen. Physiol.* 66 : 67–95.

Coleman, P. D. 1963. "An analysis of cues to auditory depth perception in free space." *Psychological Bull.* 60 : 302–15.

Colin, P. L., D. W. Arneson, and W. F. Smith-Vaniz. 1979. "Rediscovery and redescription of the Caribbean anomalopid fish *Kryptophanaron alfredi* Silvester and Fowler (Pisces: Anomalopidae)." *Bull. Marine Science* 29 : 312–19.

Collett, C. S. 1983. "Sensory guidance of motor behaviour." In *Animal Behavior.* Vol. 1: *Causes and Effects.* T. R. Halliday and P. J. B. Slater, eds. New York: W. H. Freeman, pp. 40–74.

Conner, J., S. Camazine, D. Aneshansley, and T. Eisner. 1985. "Mammalian breath:

trigger of defensive chemical response in a tenebrionid beetle (*Bolitotherus cornutus*)." *Behav. Ecol. Sociobiol.* 16 : 115–18.

Conner, W. E., T. Eisner, R. K. Vander Meer, A. Guerrero, D. Ghiringelli, and J. Meinwald. 1980. "Sex attractant of an arctiid moth (*Utethesia ornatrix*): a pulsed chemical signal." *Behav. Ecol. Sociobiol.* 7 : 55–63.

Cook, P. H. 1984. "Directional information from surface swell: some possibilities." In *Mechanisms of Migration in Fishes.* J. D. McCleave, G. P. Arnold, J. J. Dodson, and W. H. Neill, eds. New York: Plenum, pp. 79–101.

Coombs, S., J. Janssen, and J. F. Webb. 1988. "Diversity of lateral line systems: evolutionary and functional considerations." In *Sensory Biology of Aquatic Animals.* J. Atema, R. R. Fay, A. N. Popper, and W. N. Tavolga, eds. New York: Springer-Verlag, pp. 553–93.

Cott, H. B. 1941. *Adaptive Coloration in Animals.* New York: Oxford.

Cox, A. W. 1974. *Sonar and Underwater Sound.* Lexington, Mass.: D. C. Heath.

Cox, C. S. 1974. "Refraction and reflection of light at the sea surface." In *Optical Aspects of Oceanography.* N. G. Jerlov and E. S. Nielsen, eds. New York: Academic, pp. 51–75.

Crank, J. 1975. *The Mathematics of Diffusion.* Oxford: Clarendon.

Croll, N. A. 1970. *The Behaviour of Nematodes.* New York: St. Martin's.

Cronin, T. W., and R. B. Forward, Jr. 1979. "Tidal vertical migration: an endogenous rhythm in estuarine crab larvae." *Science* 205 : 1020–22.

Culotti, J. G., and R. L. Russell. 1978. "Osmotic avoidance defective mutants of the nematode *Caenorhabditis elegans.*" *Genetics* 90 : 243–56.

Cummings, W. C., and D. V. Holliday. 1987. "Sounds and source levels from bowhead whales off Port Barrow, Alaska." *J. Acoust. Soc. Am.* 82 : 814–21.

Curcio, J. A., and C. C. Petty. 1951. "The near infrared absorption spectrum of liquid water." *J. Opt. Soc. Am.* 41 : 302–4.

Dahlquist, F. W., R. A. Elwell, and P. S. Lovely. 1976. "Studies of bacterial chemotaxis in defined concentration gradients." *J. Supramolecular Structure* 4 : 329–42.

Danthanarayana, W. 1986. "Lunar periodicity of insect flight and migration." In *Insect Flight.* W. Danthanarayana, ed. Berlin: Springer-Verlag, pp. 88–119.

David, C. T. 1986. "Mechanisms of directional flight in wind." In *Mechanisms in Insect Olfaction.* T. L. Payne, M. C. Birch, and C. E. J. Kennedy, eds. Oxford: Clarendon, pp.49–67.

David, C. T., and J. S. Kennedy. 1987. "The steering and zigzagging flight by male gypsy moths." *Naturwissenschaften* 74 : 194–6.

David, C. T., J. S. Kennedy, and A. R. Ludlow. 1983. "Finding of a sex pheromone source by gypsy moths released in the field." *Nature* 303 : 804–6.

David, C. T., J. S. Kennedy, A. R. Ludlow, J. N. Perry, and C. Wall. 1983. "A reappraisal of insect flight towards a distant point source of wind-borne odor." *J. Chem. Ecol.* 8 : 1207–15.

Davidson, N. 1962. *Statistical Mechanics.* New York: McGraw-Hill.

Dawkins, R., and J. R. Krebs. 1978. "Animal signals: information or manipulation?" In *Behavioural Ecology.* J. R. Krebs and N. B. Davies, eds. Oxford: Blackwell, pp. 282–309.

Dayton, T. 1985. "Statistical and methodological critique of Baker's chapter." In *Mag-*

netite Biomineralization and Magnetoreception in Organisms. J. L. Kirschvink, D. S. Jones, and B. J. MacFadden, eds. New York: Plenum, pp. 563–68.

DeAngelis, D. L., and G. T. Yeh. 1984. "An introduction to modeling migratory behavior of fishes." In *Mechanisms of Migration in Fishes.* J. D. McCleave, G. P. Arnold, J. J. Dodson, and W. H. Neill, eds. New York: Plenum, pp. 445–69.

de Boer, K., and A. T. van Urk. 1941. "Some particulars of direction hearing." *Philips Technical Rev.* 6 : 359–64.

DeLisi, C. 1982. "A theory of measurement error and its implications for spatial and temporal gradient sensing during chemotaxis." *Cell Biophysics* 4 : 211–29.

DeLisi, C., F. Marchetti, and G. Del Grosso. 1982. "A theory of measurement error and its implications for spatial and temporal gradient sensing during chemotaxis." Cell Biophysics 4 : 211–29.

Delius, J. D., and J. Emmerton. 1978. "Sensory mechanisms related to homing in pigeons." In *Animal Migration, Navigation, and Homing.* K. Schmidt-Koenig and W. T. Keeton, eds. Berlin: Springer-Verlag, pp. 36–41.

Denton, E. J., and J. A. B. Gray. 1988. "Mechanical factors in the excitation of the lateral lines of fishes." In *Sensory Biology of Aquatic Animals.* J. Atema, R. R. Fay, A. N. Popper, and W. N. Tavolga, eds. New York: Springer-Verlag, pp. 595–617.

Derby, C. D., and J. Atema. 1988. "Chemoreceptor cells in aquatic invertebrates: peripheral mechanisms of chemical signal processing in decapod crustaceans." In *Sensory Biology of Aquatic Animals.* J. Atema, R. R. Fay, A. N. Popper, and W. N. Tavolga, eds. New York: Springer-Verlag, pp. 365–85.

DeRosa, C. T., and D. H. Taylor. 1980. "Homeward orientation mechanisms in three species of turtles (*Trionyx spinifer, Chrysemys picta,* and *Terrapene carolina*)." *Behav. Ecol. Sociobiol.* 7 : 15–23.

Devreotes, P. N., and T. L. Steck. 1979. "Cyclic 3',5' AMP relay in *Dictyostelium discoideum.* II. Requirements for the initiation and termination of the response." *J. Cell Biol.* 80 : 300–9.

Diez, J. A., and D. B. Dusenbery. 1989. "Preferred temperature of *Meloidogyne incognita.*" *J. Nematology* 21 : 99–104.

Dijkgraaf, S. 1962. "The functioning and significance of the lateral-line organs." *Biol. Rev.* 38 : 51–105.

Dingle, H. 1980. "Ecology and evolution of migration." In *Animal Migration, Orientation, and Navigation.* S. A. Gauthreaux, Jr., ed. New York: Academic, pp. 1–101.

———. 1985. "Migration." In *Comprehensive Insect Physiology, Biochemistry, and Pharmacology.* G. A. Kerkut and L. I. Gilbert, eds. Oxford: Pergamon, pp. 375–415.

Dixon, A. F. G., and D. R. Mercer. 1983. "Flight behaviour in the sycamore aphid: factors affecting take-off." *Ent. Exp. and Appl.* 33 : 43–49.

Doane, J. F., Y. W. Lee, J. Klingler, and N. D. Westcott. 1975. "The orientation response of *Ctenicera destructor* and other wireworms (Coleoptera: Elateridae) to germinating grain and to carbon dioxide." *Can. Entomologist* 107 : 1233–51.

Dodd, E. 1972. *Polynesian Seafaring.* New York: Dodd, Mead.

Dooling, R. J. 1982. "Auditory perception in birds." In *Acoustic Communication in Birds.* D. E. Kroodsma, E. H. Miller, and H. Ouellet, eds. New York: Academic, pp. 95–130.

———. 1989. "Perception of complex, species-specific vocalizations by birds and humans." In *The Comparative Psychology of Audition: Perceiving Complex Sounds.*

R. J. Dooling and S. H. Hulse, eds. Hillsdale, N. J.: Lawrence Erlbaum Associates, pp. 423–44.

Doucet, P. G., and A. N. Wilschut. 1987. "Theoretical studies on animal orientation. III. A model for kinesis." *J. Theor. Biol.* 127 : 111–25.

Døving, K. B., H. Westerberg, and P. B. Johnsen. 1985. "Role of olfaction in the behavioral and neuronal responses of Atlantic salmon, *Salmo salar*, to hydrographic stratification." *Can. J. Fish. Aquat. Sci.* 42 : 1658–67.

Drickamer, L. C., and S. H. Vessey. 1982. *Animal Behavior.* Boston: Willard Grant.

Duelli, P. 1975. "A fovea for e–vector orientation in the eye of *Cataglyphis bicolor* (Formicidae, Hymenoptera)." *J. Comp. Physiol.* 102 : 43–56.

———, and R. Wehner. 1973. "The spectral sensitivity of polarized light orientation in *Cataglyphis bicolor* (Formicidae, Hymenoptera)." *J. Comp. Physiol.* 86 : 37–53.

Duffey, E. 1956. "Aerial dispersal in a known spider population." *J. Anim. Ecol.* 25 : 85–111.

Dulka, J. G., N. E. Stacey, P. W. Sorensen, and G. J. Van Der Kraak. 1987. "A steroid sex pheromone synchronizes male–female spawning readiness in goldfish." *Nature* 325 : 251–3.

Dunn, G. A. 1990. "Conceptual problems with kinesis and taxis." In *Biology of the Chemotactic Response.* J. P. Armitage and J. M. Lackie, eds. Cambridge: Cambridge University Press, pp. 1–13.

Dusenbery, D. B. 1976. "Attraction of the nematode *Caenorhabditis elegans* to pyridine." *Comp. Biochem. Physiol.* 53C : 1–2.

———. 1980a. "Behavior of free-living nematodes." In *Nematodes as Biological Models.* B. M. Zuckerman, ed. New York: Academic, pp. 127–58.

———. 1980b. "Responses of the nematode *Caenorhabditis elegans* to controlled chemical stimulation." *J.Comp. Physiol.* 136 : 327–31.

———. 1983. "Chemotactic behavior of nematodes." *J. Nematol.* 15 : 168–73.

———. 1987. "Theoretical range over which bacteria and nematodes locate plant roots using carbon dioxide." *J. Chem. Ecol.* 13 : 1617–24.

———. 1988a. "Avoided temperature leads to the surface: computer modeling of slime mold and nematode thermotaxis." *Behav. Ecol. Sociobiol.* 22 : 219–23.

———. 1988b. "Behavioral responses of *Meloidogyne incognita* to small temperature changes." *J. Nematol.* 20 : 351–55.

———. 1988c. "Limits of thermal sensation." *J. Theor. Biol.* 131 : 263–71.

———. 1989a. "Calculated effect of pulsed pheromone release on range of attraction." *J. Chem. Ecol.* 15 : 971–77.

———. 1989b. "Efficiency and the role of adaptation in klinokinesis." *J. Theor. Biol.* 136 : 281–93.

———. 1989c. "Optimal search direction for an animal flying or swimming in a wind or current." *J. Chem. Ecol.* 15 : 2511–19.

———. 1989d. "Ranging strategies." *J. Theor. Biol.* 136 : 309–16.

———. 1989e. "A simple animal can use a complex stimulus pattern to find a location: nematode thermotaxis in soil." *Biol. Cybernetics* 60 : 431–38.

———. 1989f. "Upwind searching for an odor plume is sometimes optimal." *J. Chem. Ecol.* 16 : 1971–76.

———. 1989g. "The value of asymmetric signal processing in klinokinesis." *Biol. Cybernetics* 61 : 401–4.

Dworkin, M., and D. Kaiser. 1985. "Cell interactions in myxobacterial growth and development." *Science* 230 : 18–24.

Dyer, F. C., and J. L. Gould. 1983. "Honey bee navigation." *Am. Scientist* 71 : 587–97.

Eaves, J. L. 1987. "Introduction to radar." In *Principles of Modern Radar.* J. L. Eaves and E. K. Reedy, eds. New York: Van Nostrand Reinhold, pp. 1–27.

Eberhard, W. G. 1977. "Aggressive chemical mimicry by a bolas spider." *Science* 196 : 1173–75.

Echard, J. D. 1987. "Detection in noise." In *Principles of Modern Radar.* J. L. Eaves and E. K. Reedy, eds. New York: Van Nostrand Reinhold, pp. 253–80.

Eckert, R., D. Randall, and G. Augustine. 1988. *Animal Physiology.* New York: Freeman.

Edwards, J. S. 1986. "Derelicts of dispersal: arthropod fallout on Pacific Northwest volcanoes." In *Insect Flight.* W. Danthanarayana, ed. Berlin: Springer-Verlag, pp. 172–184.

Ehret, G. 1989. "Hearing in the mouse." In *The Comparative Psychology of Audition: Perceiving Complex Sounds.* R. J. Dooling and S. H. Hulse, eds. Hillsdale, N. J.: Lawrence Erlbaum Associates, pp. 3–32.

Eisner, T. 1970. "Chemical defense against predation in arthropods." In *Chemical Ecology.* E. Sondheimer and J. B. Simeone, eds. New York: Academic, pp. 157–217.

Eisner, T., R. E. Silberglied, D. Aneshansley, J. E. Carrel, and H. C. Howland. 1969. "Ultraviolet video-viewing: the television camera as an insect eye." *Science* 166 : 1172–74.

Elkinton, J. S., and R. T. Cardé. 1984. "Odor dispersion." In *Chemical Ecology of Insects.* W. J. Bell and R. T. Cardé, eds. Sunderland, Mass.: Sinauer Associates, pp. 73–91.

Ellin, R. I., R. L. Farrand, F. W. Oberst, C. L. Crouse, N. B. Billups, W. S. Koon, N. P. Musselman, and F. R. Sidell. 1974. "An apparatus for the detection and quantitation of volatile human effluents." *J. Chromatography* 100 : 137–52.

Elliott, E. J. 1986. "Chemosensory stimuli in feeding behavior of the leech *Hirudo medicinalis.*" *J. Comp. Physiology* A159 : 391–401.

El-Sherif, M., and W. F. Mai. 1969. "Thermotactic response of some plant parasitic nematodes." *J. Nematology* 1 : 43–48.

Elsner, B. 1978. "Accurate measurements of the initial track (radius 1,500 m) of homing pigeons." In *Animal Migration, Navigation, and Homing.* K. Schmidt-Koenig and W. T. Keeton, eds. Berlin: Springer-Verlag, pp. 194–98.

Emlen, S. T. 1967. "Migratory orientation in the indigo bunting, *Passerina cyanea.* Part I: evidence for use of celestial cues." *Auk* 84 : 309–42.

———. 1969a. "Bird migration: influence of physiological state upon celestial orientation." *Science* 165 : 716–18.

———. 1969b. "The development of migratory orientation in young indigo buntings." *Living Bird* 8 : 113–26.

———. 1970. "Celestial rotation: its importance in the development of migratory orientation." *Science* 170 : 1198–1201.

———. 1975. "Migration: orientation and navigation." In *Avian Biology.* D. S. Farner, J. R. King, and K. C. Parkes, eds. New York: Academic, pp. 129–219.

———. and N. J. Demong. 1978. "Orientation strategies used by free-flying bird migrants: a radar tracking study." In *Animal Migration, Navigation, and Homing.* K. Schmidt-Koenig and W. T. Keeton, eds. Berlin: Springer-Verlag, pp. 283–93.

Esch, H., and J. A. Bastian. 1970. "How do newly recruited honey bees approach a food site?" *Zeitschrift fur vergleichende Physiologie* 68 : 175–81.

Evans, R. C., D. T. Tingey, M. L. Gumpertz, and W. F. Burns. 1982. "Estimates of isoprene and monoterpene emission rates in plants." *Bot. Gaz.* 143 : 304–10.

Evans, W. E. 1988. "Natural history aspects of marine mammal echolocation: feeding strategies and habitat." In *Animal Sonar.* P. E. Nachtigall and P. W. B. Moore, eds. New York: Plenum, pp. 521–34.

Fantino, E. and C. A. Logan. 1979. *The Experimental Analysis of Behavior.* San Francisco: Freeman.

Fay, R. R. 1974. "Auditory frequency discrimination in vertebrates." *J. Acoust. Soc. Am.* 56 : 206–9.

———. 1988. "Peripheral adaptations for spatial hearing in fish." In *Sensory Biology of Aquatic Animals.* J. Atema, R. R. Fay, A. N. Popper, and W. N. Tavolga, eds. New York: Springer-Verlag, pp. 711–31.

Feinberg, R. 1988. *Polynesian Seafaring and Navigation.* Kent, Ohio: Kent State.

Fernald, R. D. 1988. "Aquatic adaptations in fish eyes." In *Sensory Biology of Aquatic Animals.* J. Atema, R. R. Fay, A. N. Popper, and W. N. Tavolga, eds. New York: Springer-Verlag, pp. 435–66.

Finch, S. 1980. "Chemical attraction of plant-feeding insects to plants." In *Applied Biology.* T. H. Coaker, ed. New York: Academic, pp. 67–143.

Finney, B. R. 1979. *Hokule'a: The Way to Tahiti.* New York: Dodd, Mead.

Fisher, E. S., and D. A. Lauffenburger. 1987. "Mathematical analysis of cell-target encounter rates in two dimensions." *Biophys. J.* 51 : 705–16.

———, and R. P. Daniele. 1988. "The effect of alveolar macrophage chemotaxis on bacterial clearance from the lung surface." *Am. Rev. Respir. Dis.* 137:1129–34.

Fisher, R. A., and D. O. Greenbank. 1979. "A case study of research into insect movement: spruce budworm in New Brunswick." In *Movement of Highly Mobile Insects: Concepts and Methodology in Research.* R. L. Rabb and G. G. Kennedy, eds. Raleigh, N. C.: N. C. State University Dept. of Entomology, pp. 220–29.

Foelix, R. F. 1982. *Biology of Spiders.* Cambridge, Mass.: Harvard University Press.

Fontolliet, P. –G. 1986. *Telecommunication Systems.* Dedham, Mass.: Artech House.

Ford, N. B. 1986. "The role of pheromone trails in the sociobiology of snakes." In *Chemical Signals in Vertebrates.* D. Duvall, D. Müller-Schwarze, and R. M. Silverstein, eds. New York: Plenum, pp. 261–78.

———, and J. R. Low. 1984. "Sex pheromone source location by garter snakes: a mechanism for detection of direction in nonvolatile trails." *J. Chem. Ecol.* 10 : 1193–99.

Ford, R. D. 1970. *Introduction to Acoustics.* Amsterdam: Elsevier.

Foster, K. W., J. Saranak, N. Patel, G. Zarilli, M. Okabe, T. Kline, and K. Nakanishi. 1984. "A rhodopsin is the functional photoreceptor for phototaxis in the unicellular eukaryote Chlamydomonas." *Nature* 311 : 756–59.

———, and R. D. Smyth. 1980. "Light antennas in phototactic algae." *Microbiol. Revs.* 44 : 572–630.

Fraenkel, G. S., and D. L. Gunn. 1961. *The Orientation of Animals.* New York: Dover .

Frankel, R. B., R. P. Blakemore, F. F. T. de Araujo, D. M. S. Esquivel, and J. Danon. 1981. "Magnetotactic bacteria at the geomagnetic equator." *Science* 212 : 1269–70.

Fromm, J. E., and W. J. Bell. 1987. "Search orientation of *Musca domestica* in patches of sucrose drops." *Physiol. Entomol.* 12 : 297–307.

Fullard, J. H., and D. W. Thomas. 1981. "Detection of certain African, insectivorous bats by sympatric, tympanate moths." *J. Comp. Physiol.* 143 : 363–68.

Galland, P., and E. D. Lipson. 1987. "Light physiology of Phycomyces sporangiophores." In *Phycomyces.* E. Cerdá-Olmedo and E. D. Lipson, eds. Cold Spring Harbor, N. Y.: Cold Spring Harbor Laboratory, pp. 49–92.

Galler, S. R., K. Schmidt-Koenig, G. J. Jacobs, and R. E. Belleville. 1972. *Animal Orientation and Navigation.* Washington, D.C.: National Aeronautics and Space Administration.

Gardner, M. B., and R. S. Gardner. 1973. "Problem of localization in the median plane: effect of pinnae cavity occlusion." *J. Acustic. Soc. Am.* 53 : 400–408.

Garland, G. F. 1971. *Introduction to Geophysics.* Philadelphia: Saunders.

Gates, D. M., H. J. Keegan, J. C. Schleter, and V. R. Weidner. 1965. "Spectral properties of plants." *Applied Optics* 4 : 11–20.

Gatlin, L. L. 1972. *Information Theory and the Living System.* New York: Columbia University Press.

Gauthreaux, S. A., Jr. 1978. "Importance of the daytime flights of nocturnal migrants: redetermined migration following displacement." In *Animal Migration, Navigation, and Homing.* K. Schmidt-Koenig and W. T. Keeton, eds. Berlin: Springer-Verlag, pp. 220–27.

———, 1980. "The influences of long-term and short-term climatic changes on the dispersal and migration of organisms." In *Animal Migration, Orientation, and Navigation.* S. A. Gauthreaux, Jr., ed. New York: Academic, pp. 103–74.

———, and K. P. Able. 1970. "Wind and the direction of nocturnal songbird migration." *Nature* 228 : 476–77.

Geier, R. 1965. *The Climate Near the Ground.* Cambridge, Mass.: Harvard University Press.

Gerhardt, H. C., B. Dickamp, and M. Ptack. 1989. "Inter-male spacing in choruses of the spring peeper *Pseudacris crucifer.*" *Animal Behaviour* 38 : 1012–1024.

Gerisch, G. 1982. "Chemotaxis in Dictyostelium." *Ann. Rev. Physiol.* 44 : 535–52.

Gibbons, A. 1991. "Déja vu all over again: chimp-language wars." *Science* 251 : 1561–62.

Gibbons, B. 1986. "The intimate sense of smell." *National Geographic* 170 : 324–61.

Gibo, D. L. 1986. "Flight strategies of migrating monarch butterflies (*Danaus plexippus* L.) in southern Ontario." In *Insect Flight.* W. Danthanarayana, ed. Berlin: Springer-Verlag, pp. 172–84.

———, and M. J. Pallett. 1979. "Soaring flight of monarch butterflies, *Danaus plexippus* (Lepidoptera: Danaidae), during the late summer migration in southern Ontario." *Can. J. Zool.* 57 : 1393–1401.

Gillett, J. D. 1979. "Out for blood: flight orientation up-wind in the absence of visual clues." *Mosquito News* 39 : 221–29.

Gillies, M. T. 1980. "The role of carbon dioxide in host-finding by mosquitoes (Dipteria: Culicidae): a review." *Bull. Ent. Res.* 70 : 525–32.

———, and M. T. Wilkes. 1974. "Evidence for downwind flights by host-seeking mosquitoes." *Nature* 252 : 388–89.

Gladwin, T. 1970. *East Is a Big Bird.* Cambridge, Mass.: Harvard University Press.

Gogala, M. 1985a. "Vibrational communication in insects (biophysical and behavioural aspects)." In *Acoustic and Vibrational Communication in Insects.* K. Calmring and N. Elsner, eds. Berlin: Verlag Paul Parey, pp. 117–26.

————, 1985b. "Vibrational songs of land bugs and their production." In *Acoustic and Vibrational Communication in Insects.* K. Calmring and N. Elsner, eds. Berlin: Verlag Paul Parey, pp. 143–150.

Golden, J. W., and D. L. Riddle. 1984. "The *Caenorhabditis elegans* dauer larva: developmental effects of pheromone, food, and temperature." *Devel. Biol.* 102 : 368–78.

Goode, M., and D. B. Dusenbery. 1985. "Behavior of tethered *Meloidogyne incognita.*" *J. Nematology* 17 : 460–64.

Gould, J. L. 1975. "Honey bee recruitment: the dance-language controversy." *Science* 189 : 685–93.

————. 1980. "The case for magnetic sensitivity in birds and bees (such as it is)." *Am. Scientist* 68 : 256–67.

————. 1982a. *Ethology.* New York: Norton.

————. 1982b. "The map sense of pigeons." *Nature* 296 : 205–11.

————. 1985a. "Absence of human homing ability as measured by displacement experiments." In *Magnetite Biomineralization and Magnetoreception in Organisms.* J. L. Kirschvink, D. S. Jones and B. J. MacFadden, eds. New York: Plenum, pp. 595–99.

————. 1985b. "Are animal maps magnetic?" In *Magnetite Biomineralization and Magnetoreception in Organisms.* J. L. Kirschvink, D. S. Jones, and B. J. MacFadden, eds. New York: Plenum, pp. 257–68.

————. 1987. "Landmark learning in honey bees." *Anim. Behav.* 35 : 26–34.

————, and C. G. Gould. 1988. *The Honey Bee.* New York: Scientific American.

Gould, J. L., M. Henerey, and M. C. MacLeod. 1970. "Communication of direction by the honey bee." *Science* 169 : 544–54.

Gower, D. B. 1972. "16-unsaturated C19 steroids, a review of their chemistry, biochemistry and possible physiological role." *J. Steroid Biochem.* 3 : 45–103.

Green, C. D. 1966. "Orientation of male *Heterodera rostochiensis* Woll. and *H. schachtii* Schm. to their females." *Ann. Appl. Biol.* 58 : 327–39.

Greenbank, D. O., G. W. Schaefer, and R. C. Rainey 1980. "Spruce budworm Lepidoptera, Tortricidae. moth flight and dispersal: new understanding from canopy observations, radar, and aircraft." *Memoirs Entom. Soc. Can.* No. 110:1–49.

Greenewalt, C. H. 1968. *Bird Song: Acoustics and Physiology.* Washington, D.C.: Smithsonian Institution.

Gregory, R. L. 1973. "The confounded eye." In *Illusion in Nature and Art.* R. L. Gregory and E. H. Gombrich, eds. London: Duckworth, pp. 49–95.

Griffin, D. R. 1953. "Acoustic orientation in the oil bird, Steatornis." *Proc. Natl. Acad. Sciences* 39 : 884–93.

————. 1971. "The importance of atmospheric attenuation for the echolocation of bats (Chiroptera)." *Anim. Behav.* 19 : 55–61.

————. 1982. "Ecology of migration: is magnetic orientation a reality." *Q. Rev. Biol.* 57 : 293–94.

Grossweiner, L. 1989. "Photophysics." In *The Science of Photobiology.* K. C. Smith, ed. New York: Plenum, pp. 1–45.

Grüsser, O. -J., and U. Grüsser-Cornehls. 1978. "Physiology of vision." In *Fundamen-*

tals of Sensory Physiology. R. F. Schmidt, ed. New York: Springer-Verlag, pp. 126–79.

Grzimek, B. 1977. *Grzimek's Encyclopedia of Ethology.* New York: Van Nostrand Reinhold.

Hagstrum, D. W., and L. R. Davis, Jr. 1982. "Mate-seeking behavior and reduced mating by *Ephestia cautella* (Walker) in a sex pheromone-permeated atmosphere." *J. Chem. Ecol.* 8 : 507–15.

Halliday, D., and R. Resnick. 1988. *Fundamentals of Physics.* New York: Wiley.

Halliday, T. R., and P. J. B. Slater. 1983. *Causes and Effects.* New York: Freeman.

Häder, D. -P. 1987. "Photosensory behavior in procaryotes." *Microbiological Reviews* 51: 1–21.

Häder, D. -P. 1988. "Ecological consequences of photomovement in microorganisms." *J. Photochem. Photobiol. B* : 1 : 385–414.

Hamner, P., and W. M. Hamner. 1977. "Chemosensory tracking of scent trails by the planktonic shrimp *Acetes sibogae australis.*" *Science* 195 : 886–88.

Hancock, J. C. 1961. *An Introduction to the Principles of Communication Theory.* New York: McGraw-Hill.

Harden Jones, F. R., M. Greer Walker, and G. P. Arnold. 1978. "Tactics of fish movement in relation to migration strategy and water circulation." In *Advances in Oceanography.* H. Charnock and S. G. Deacon, eds. New York: Plenum, pp. 185–207.

Hartline, P. H. 1974. "Thermoreception in snakes." In *Electroreceptors and Other Specialized Receptors.* A. Fessard, ed. Heidelberg: Springer-Verlag, pp. 297–312.

Hasler, A. D., and A. T. Scholz. 1978. "Olfactory imprinting in Coho salmon (*Oncorhynchus kisutch*)." In *Animal Migration, Navigation, and Homing.* K. Schmidt-Koenig and W. T. Keeton, eds. Berlin: Springer-Verlag, pp. 356–69.

Hawkins, A. D. 1981. "The hearing abilities of fish." In *Hearing and Sound Communication in Fishes.* W. N. Tavolga, A. N. Popper, and R. R. Fay, eds. New York: Springer-Verlag, pp. 109–33.

———, and A. D. F. Johnstone. 1978. "The hearing of the Atlantic salmon, *Salmo salar.*" *J. Fish. Biol.* 13 : 655–73.

———, and A. A. Myrberg, Jr. 1983. "Hearing and sound communication under water." In *Bioacoustics.* B. Lewis, ed. New York: Academic Press, pp. 347–405.

Haynes, K. F., and M. C. Birch. 1985. "The role of other pheromones, allomones and kairomones in the behavioral responses of insects." In *Comprehensive Insect Physiology, Biochemistry, and Pharmacology.* G. A. Kerkut and L. I. Gilbert, eds. Oxford: Pergamon, pp. 225–55.

Heath, M. E., and J. A. Downey. 1990. "The cold face test (diving reflex) in clinical autonomic assessment: methodological considerations and repeatability of responses." *Clinical Science* 78 : 139–47.

Hecht, S., S. Shlaer, and M. H. Pirenne. 1942. "Energy, quanta, and vision." *J. Gen. Physiol.* 25 : 819–40.

Hedgecock, E. M., and R. L. Russell. 1975. "Normal and mutant thermotaxis in the nematode *Caenorhabditis elegans.*" *Proc. Natl. Acad. Sciences* 72 : 4061–65.

Helbig, A. J., P. Berthold, and W. Wiltschko. 1989. "Migratory orientation of blackcaps

(*Sylvia atricapilla*): population-specific shifts of direction during the autumn." *Ethology* 82 : 307–15.

Helversen, D. V., and O. V. Helversen. 1983. "Species recognition and acoustic localization in acridid grasshoppers: a behavioral approach." In *Neuroethology and Behavioral Physiology*. F. Huber and H. Markl, eds. Berlin: Springer-Verlag, pp. 95–107.

Hensel, H. 1981. *Thermoreception and Temperature Regulation*. London: Academic Press.

———. 1982. *Thermal Sensations and Thermoreceptors*. Springfield, Ill.: Charles C. Thomas.

Hergenröder, R., and F. G. Barth. 1983. "Vibratory signals and spider behavior: how do the sensory inputs from the eight legs interact in orientation?" *J. Comp. Physiol.* 152 : 361–71.

Herring, P. J. 1987. "Systematic distribution of bioluminescence in living organisms." *J. Bioluminescence and Chemiluminescence* 1 : 147–63.

———, and J. G. Morin. 1978. "Bioluminescence in fishes." In *Bioluminescence in Action*. P. J. Herring, ed. New York: Academic Press, pp. 273–329.

Hidaka, T. 1972. "Biology of *Hyphantria cunea* Drury (Lepidoptera: Arctiidae) in Japan XIV Mating Behavior." *Appl. Ent. Zool.* 7 : 116–32.

Hildebrand, E. 1978. "Bacterial phototaxis." In *Taxis and Behavior*. G. L. Hazelbauer, ed. London: Chapman and Hall, pp. 35–73.

———, and N. Dencher 1975. "Two photosystems controlling behavioural responses of *Halobacterium halobium*." *Nature* 257 : 46–48.

———, and N. Schimz. 1985. "Behavioral pattern and its photosensory control in *Halobacterium halobium*." In *Sensing and Response in Microorganims*. M. Eisenbach and M. Balaban, eds. Amsterdam: Elsevier, pp. 129–42.

Hildebrand, J. G., and R. A. Montague. 1986. "Functional organization of olfactory pathways in the central nervous system of *Manduca sexta*." In *Mechanisms in Insect Olfaction*. T. L. Payne, M. C. Birch, and C. E. J. Kennedy, eds. Oxford: Clarendon pp. 279–85.

Himstedt, W., J. Kopp, and W. Schmidt. 1982. "Electroreception guides feeding behaviour in amphibians." *Naturwissenschaften* 69 : 552–53.

Hinde, R. A. 1966. *Animal Behavior*. New York: McGraw-Hill.

Hinton, H. E. 1973. "Natural deception." In *Illusion in Nature and Art*. R. L. Gregory and E. H. Gombrich, eds. London: Duckworth, pp. 97–159.

Hoffmann, G. 1983a. "The random elements in the search behavior of the desert isopod *Hemilepistus reaumuri*." *Behav. Ecol. and Sociobiol.* 13 : 81–92.

———. 1983b. "The search behavior of the desert isopod *Hemilepistus reaumuri* as compared with a systematic search." *Behav. Ecol. and Sociabiol.* 13 : 93–106.

Hölldobler, B. 1976. "Recruitment behavior, home range orientation and territoriality in harvester ants, *Pogonomyrmex*." *Behav. Ecol. and Sociobiol.* 1 : 3–44.

———. 1980. "Canopy orientation: a new kind of orientation in ants." *Science* 210 : 86–88.

———, and E. O. Wilson. 1990. *The Ants*. Cambridge, Mass.: Harvard University Press.

Højerslev, N. K. 1986. "Optical properties of sea water." In *Oceanography*. J. Sündermann, ed. New York: Springer-Verlag, pp. 383–462.

Holm, W. A. 1987. "Continuous wave radar." In *Principles of Modern Radar.* J. L. Eaves and E. K. Reedy, eds. New York: Van Nostrand Reinhold, pp. 397–421.

Holzmüller, W. 1984. *Information in Biological Systems: The Role of Macromolecules.* Cambridge: Cambridge University Press.

Hoober, J. K. 1984. *Chloroplasts.* New York: Plenum.

Hopkins, C. D. 1986. "Behavior of Mormyridae." In *Electroreception.* T. H. Bullock and W. Heilignenberg, eds. New York: Wiley, pp. .

———. 1988a. "Neuroethology of electric communication." *Ann. Rev. Neurosci.* 11 : 497–535.

———. 1988b. "Social communication in the aquatic environment." In *Sensory Biology of Aquatic Animals.* J. Atema, R. R. Fay, A. N. Popper, and W. N. Tavolga, eds. New York: Springer-Verlag, pp. 233–68.

Horton, J. W. 1957. *Fundamentals of Sonar.* Annapolis, Md.: U.S. Naval Institute.

Hsiao, T. H. 1985. "Feeding behavior." In Comprehensive Insect Physiology, Biochemistry, and Pharmacology. G. A. Kerkut and L. I. Gilbert, eds. Oxford: Pergamon, pp. 471–512.

Hudspeth, A. J., and D. P. Corey. 1977. "Sensitivity, polarity, and conductance change in the response of vertebrate hair cells to controlled mechanical stimuli." *Proc. Natl. Acad. Sci.* 74 : 2407–11.

Hughes, D. A. 1972. "On the endogeneous control of tide-associated displacements of pink shrimp, *Penaeus duorarum* Burkenroad." *Biol. Bull.* 142 : 271–80.

Hulse, S. H. 1989. "Comparative psychology and pitch pattern perception in songbirds." In *The Comparative Psychology of Audition: Perceiving Complex Sounds.* R. J. Dooling and S. H. Hulse, eds. Hillsdale, N. J.: Lawrence Erlbaum Associates, pp. 331–49.

Hutchings, M., and B. Lewis. 1983. "Insect sound and vibration receptors." In *Bioacoustics.* B. Lewis, ed. London: Academic Press, pp. 181–205.

Ingård, U. 1953. "A review of the influence of meteorological conditions on sound propagation." *J. Acoustical Soc. Am.* 25:405–11.

Ivanov, K. P. 1984. "Sensitivity and precision of thermoregulatory system functioning in the living organism." In *Thermal Physiology.* J. R. S. Hales, ed. New York: Raven, pp. 21–22.

Iversen, T. –H., and A. Rommelhoff. 1978. "The starch statolith hypothesis and the interaction of amyloplasts and endoplasmic reticulum in root geotropism." *J. Exp. Bot.* 29 : 1319–28.

Jackson, J., W. Ingram III, and H. W. Campbell. 1976. "The dorsal pigmentation pattern of snakes as an antipredator strategy: a multivariate approach." *Am. Naturalist* 110 : 1029–53.

Jander, R. 1975. "Ecological aspects of spatial orientation." *Ann. Rev. Ecol. and Syst.* 6 : 171–88.

Jander, R. 1977. "Orientation Ecology." In *Grzimek's Encyclopedia of Ethology.* B. Grzimek, ed. New York: Van Nostrand Reinhold, pp. 145–63.

Janzen, D. H. 1984. "Two ways to be a tropical big moth: Santa Rosa saturniids and sphingids." *Oxford Surveys in Evolutionary Biology* 1 : 85–140.

Jeans, J. 1941. *The Mathematical Theory of Electricity and Magnetism.* Cambridge: Cambridge University Press.

Jeffrey, A. 1979. *Mathematics for Engineers and Scientists.* Surrey, England: Thomas Nelson.

Jennings, H. S. 1904. Contributions to the study of the behavior of lower organisms. Washington, D.C.: Carnegie Institute.

Jerlov, N. G. 1968. *Optical Oceanography.* Amsterdam: Elsevier.

———. 1976. *Marine optics.* Amsterdam: Elsevier.

Jeske, H. 1988. "Meteorological optics and radiometeorology." In *Physical and Chemical Properties of the Air.* G. Fischer, ed. New York: Springer-Verlag, pp. 187–348.

Johnsen, P. B. 1984. "Establishing the physiological and behavioral determinants of chemosensory orientation." In *Mechanisms of Migration in Fishes.* J. D. McCleave, G. P. Arnold, J. J. Dodson, and W. H. Neill, eds. New York: Plenum, pp. 379–85.

Johnson, C. G. 1974. "Insect migration: aspects of its physiology." In *The Physiology of Insecta.* M. Rockstein, ed. New York: Academic Press, pp. 279–334.

Johnson, C. S. 1986. "Dolphin audition and echolocation capacities." In *Dolphin Cognition and Behavior: A Comparative Approach.* R. J. Schusterman, J. A. Thomas and F. G. Wood, eds. Hillsdale, N. J.: Lawrence Erlbaum Associates, pp. 115–36.

Johnson, G. D., and R. H. Rosenblatt. 1988. "Mechanisms of light organ occlusion in flashlight fishes, family Anomalopidae (Teleostei: Beryciformes), and the evolution of the group." *Zool. J. Linnean Soc.* 94 : 65–96.

Jolly, D. C. 1986. "Stunning whales." *Nature* 324 : 418.

Jones, C. D. 1983. "On the structure of instantaneous plumes in the atmosphere." *J. Hazardous Materials* 7 : 87–112.

Jones, F. R. H. 1981. "Fish migration: strategy and tactics." In *Animal Migration.* D. J. Aidley, ed. Cambridge: Cambridge University Press, pp. 139–65.

———. 1984a. "Could fish use inertial clues when on migration." In *Mechanisms of Migration in Fishes.* J. D. McCleave, G. P. Arnold, J. J. Dodson, and W. H. Neill, eds. New York: Plenum, pp. 67–78.

———. 1984b. "A view from the ocean." In *Mechanisms of Migration in Fishes.* J. D. McCleave, G. P. Arnold, J. J. Dodson, and W. H. Neill, eds. New York: Plenum,; pp. 1–26.

Jones, R. E., N. Gilbert, M. Guppy, and V. Nealis. 1980. "Long-distance movement of *Pieris rapae.* " *J. Anim. Ecol.* 49 : 629–42.

Judd, D. B., D. L. MacAdam, and G. Wyszecki. 1964. "Spectral distribution of typical daylight as a function of correlated color temperature." *J. Optical Soc. Am.* 54 : 1031–40.

———, and G. Wyszecki. 1963. *Color in Business, Science, and Industry.* New York: Wiley.

Judge, T. K. 1985. "A study of the homeward orientation of visually handicapped humans." In *Magnetite Biomineralization and Magnetoreception in Organisms.* J. L. Kirschvink, D. S. Jones, and B. J. MacFadden, eds. New York: Plenum, pp. 601–3.

Kaissling, K. -E. 1971. "Insect olfaction." In *Olfaction.* L. M. Beidler, ed. New York: Springer-Verlag, pp. 351–431.

Kalmijn, A. J. 1988a. "Detection of weak electric fields." In *Sensory Biology of Aquatic*

Animals. J. Atema, R. R. Fay, A. N. Popper, and W. N. Tavolga, eds. New York: Springer-Verlag, pp. 152–186.

———. 1988b. "Hydrodynamic and acoustic field detection." In *Sensory Biology of Aquatic Animals.* J. Atema, R. R. Fay, A. N. Popper, and W. N. Tavolga, eds. Springer-Verlag, New York: pp. 83–130.

Kalmring, K. 1985. "Vibrational communication in insects (reception and integration of vibratory information)." In *Acoustic and Vibrational Communication in Insects.* K. Kalmring and N. Elsner, eds. Berlin: Verlag Paul Parey, pp. 127–34.

Kamil, A. C. 1988. "Behavioral ecology and sensory biology." In *Sensory Biology of Aquatic Animals.* J. Atema, R. R. Fay, A. N. Popper, and W. N. Tavolga, eds. New York: Springer-Verlag, pp. 189–201.

Kamminga, C. 1988. "Echolocation of signal types of odontocetes." In *Animal Sonar.* P. E. Nachtigall and P. W. B. Moore, eds. New York: Plenum, pp. 9–22.

Kanz, J. E. 1977. "The orientation of migrant and non-migrant monarch butterflies, *Danaus plexippus* (L.)." *Psyche* 84 : 120–41.

Kasperbauer, M. J. 1987. "Far-red light reflection from green leaves and effects on phytochrome-mediated assimilate partitioning under field conditions." *Plant Physiol.* 85 : 350–54.

Katona, S. K. 1973. "Evidence for sex pheromones in planktonic copepods." *Limnology and Oceanography* 18 : 574–83.

Kauzmann, W. 1966. *Kinetic Theory of Gases.* W. A. New York: Benjamin.

Keeton, W. T. 1974. "Pigeon homing: no influence of outward-journey detours on initial orientation." *Monitore Zoologico Italiano* 8 : 227–34.

———. 1981. "The orientation and navigation of birds." In *Animal Migration.* D. J. Aidley, ed. Cambridge: Cambridge University Press, pp. 81–104.

Kelly, D. H. 1972. "Adaptation effects on spatio-temporal sine-wave thresholds." *Vision Research* 12 : 89–101.

Kennedy, J. S. 1983. "Zigzagging and casting as a programmed response to wind-borne odour: a review." *Physiol. Ent.* 8 : 109–20.

———. 1986. "Some current issues in orientation to odour sources." In *Mechanisms in Insect Olfaction.* T. L. Payne, M. C. Birch, and C. E. J. Kennedy, eds. Oxford: Clarendon, pp. 11–25.

Kennedy, J. S., C. O. Booth, and W. J. S. Kershaw. 1961. "Host finding by aphids in the field. III. Visual attraction." *Ann. Appl. Biol.* 49 : 1–21.

Kennedy, L. M., L. R. Saul, R. Sefecka, and D. A. Stevens. 1988. "Hodulcin: selective sweetness-reducing principle from *Hovenia dulcis* leaves." *Chem. Senses* 13 : 529–43.

Kenyon, K. W. and D. W. Rice. 1958. "Homing of Laysan albatrosses." *Condor* 60 : 3–6.

Kevan, P. G. 1972. "Floral colors in the high arctic with reference to insect-flower relations and pollination." *Can. J. Botany* 50 : 2289–2316.

———. 1978. "Floral coloration, its colorimetric analysis and significance in anthecology." In *The Pollination of Flowers by Insects.* A. J. Richards, ed. London: Academic Press, pp. 51–59.

Kiepenheuer, J. 1984. "The magnetic compass mechanism of birds and its possible association with the shifting course directions of migrants." *Behav. Ecol. Sociobiol.* 14 : 81–99.

Kirschfeld, K. 1976. "The resolution of lens and compound eyes." In *Neural Principles in Vision*. F. Zettler, and R. Weiler, eds. Berlin: Springer-Verlag, pp. 354–70.

Kirschvink, J. L., A. E. Dizon, and J. A. Westphal. 1986. "Evidence from strandings for geomagnetic sensitivity in cetaceans." *J. Exp. Biol.* 120 : 1–24.

———, D. S. Jones, and B. J. MacFadden, Eds. 1985. *Magnetite Biomineralization and Magnetoreception in Organisms*. New York: Plenum.

———, K. A. Peterson, M. Chwe, P. Filmer, and B. Roder. 1985. "An attempt to replicate the spinning chair experiment." In *Magnetite Biomineralization and Magnetoreception in Organisms*. J. L. Kirschvink, D. S. Jones, and B. J. MacFadden, eds. New York: Plenum, 1985, pp. 605–8.

Klinger, J. 1963. "Die Orientierung von *Ditylenchus dipsaci* in gemessenen Kunstlichen und biologischen CO_2–Gradienten." *Nematologica* 9 : 185–99.

———. 1965. "On the orientation of plant nematodes and of some other soil animals." *Nematologica* 11 : 4–18.

Klinke, R. 1977. "Physiology of hearing." In *Fundamentals of Sensory Physiology*. R. F. Schmidt, ed. New York: Springer-Verlag, pp. 180–203.

Knudsen, E. I. 1975. "Spatial aspects of the electric fields generated by weakly electric fish." *J. Comp. Physiology* 99 : 103–18.

———. 1983. "Space coding in the vertebrate auditory system." In *Bioacoustics*. B. Lewis, ed. New York: Academic Press, pp. 311–44.

Kober, R. 1988. "Echoes of fluttering insects." In *Animal Sonar*. P. E. Nachtigall and P. W. B. Moore, eds. New York: Plenum, pp. 477–81.

Köhler, K. -L. 1978. "Do pigeons use their eyes for navigation? A new technique!" In *Animal Migration, Navigation, and Homing*. K. Schmidt-Koenig and W. T. Keeton, eds. Berlin: Springer-Verlag, pp. 57–64 .

Konishi, M. 1983. "Neuroethology of acoustic prey localization in the barn owl." In *Neuroethology and Behavioral Physiology*. F. Huber and H. Markl, eds. Berlin: Springer-Verlag, pp. 303–17.

Koopman, B. O. 1980. *Search and Searching*. New York: Pergamon.

Kosugi, T., and K. Inouye. 1989. "Negative chemotaxis to ammonia and other weak bases by migrating slugs of the cellular slime moulds." *J. General Microbiology* 135 : 1589–98.

Krebs, J. R., and N. B. Davies. 1987. *An Introduction to Behavioral Ecology*. Oxford: Blackwell.

Krebs, W., and H. Kühn. 1977. "Structure of isolated bovine rod outer segment membranes." *Exp. Eye. Res.* 25 : 511–26.

Kreithen, M. L. 1978. "Sensory mechanisms for animal orientation—can any new ones be discovered?" In *Animal Migration, Navigation, and Homing*. K. Schmidt-Koenig and W. T. Keeton, eds. Berlin: Springer-Verlag, pp. 25–34.

———, and T. Keeton. 1974. "Detection of polarized light by the homing pigeon, *Columbia livia*." *J. Comp. Physiol.* 89 : 83–92.

Kuhn, G. F. 1979. "The pressure transformation from a diffuse sound field to the external ear and to the body and head surface." *J. Acoust. Soc. Am.* 65 : 991–1000.

Lachenbruch, A. H., and B. V. Marshall. 1986. "Changing climate: geothermal evidence from permafrost in the Alaskan Arctic." *Science* 234 : 689–96.

Lamb, B., A. Guenther, D. Gay, and H. Westberg. 1987. "A national inventory of biogenic hydrocarbon emissions." *Atmospheric Environment* 21 : 1695–1705.

Land, E. H. 1974. "The retinex theory of colour vision." *Proc. Roy. Inst. Great Britain* 47 : 23–58.

———. 1986. "An alternative technique for the computation of the designator in the retinex theory of color vision." *Proc. Natl. Acad. Sci.* 83 : 3078–80.

Land, M. F. 1981. "Optics and vision in invertebrates." In *Handbook of Sensory Physiology.* H. Autrum, ed. Berlin: Springer-Verlag, pp. 471–593.

———. 1983. "Sensory stimuli and behaviour." In *Animal Behavior.* Volume 1: *Causes and Effects.* T. R. Halliday and P. J. B. Slater, eds. New York: Freeman, pp. 11–39.

———. 1984. "Crustacea." In *Photoreception and Vision in Invertebrates.* M. A. Ali, ed. New York: Plenum, pp. 401–38.

———. 1984. "Molluscs." In *Photoreception and Vision in Invertebrates.* M. A. Ali, ed. New York: Plenum, pp. 699–725.

Land, M. F., and T. S. Collett. 1974. "Chasing behaviour of houseflies (*Fannia cani-cularis*)." *J. comp. Physiol.* 89 : 331–57.

Landau, B., H. Gleitman, and E. Spelke. 1981. "Spatial knowledge and geometric representation in a child blind from birth." *Science* 213 : 1275–78.

Lang, H. H. 1980. "Surface wave discrimination between prey and nonprey by the back swimmer *Notonecta glauca* L. (Hemiptera, Heteroptera). " *Behav. Ecol. Sociobiol.* 6 : 233–46.

Larkin, R. P., D. R. Griffin, J. R. Torre-Bueno, and J. Teal. 1979. "Radar observations of bird migration over the western North Atlantic Ocean." *Behav. Ecol. Sociobiol.* 4 : 225–64.

Laverack, M. S. 1988. "The diversity of chemoreceptors." In *Sensory Biology of Aquatic Animals* J. Atema, R. R. Fay, A. N. Popper, and W. N. Tavolga, eds. New York: Springer-Verlag, pp. 287–312.

Lednor, A. J. 1982. "Magnetic navigation in pigeons: possibilities and problems." In *Avian Navigation.* F. Papi and H. G. Wallraff, eds. Berlin: Springer-Verlag, pp. 109–19.

Lee, J. 1989. "Bioluminescence." In *The Science of Photobiology,* 2d ed. K. C. Smith, ed. New York: Plenum, pp. 391–417.

Leggett, W. C. 1984. "Fish migrations in coastal and estuarine environments: a call for new approaches to the study of an old problem." In *Mechanisms of Migration in Fishes.* J. D. McCleave, G. P. Arnold, J. J. Dodson, and W. H. Neill, eds. New York: Plenum, pp. 159–78.

Le Grand, Y. 1968. *Light, Colour and Vision.* London: Chapman and Hall.

Lenoir, A. 1982. "An informational analysis of antennal communication during tro-phallaxis in the ant *Myrmica rubra* L." *Behavioural Processes* 7 : 27–35.

Levi, L. 1968. *Applied Optics.* New York: Wiley.

———. 1980. *Applied Optics.* New York: Wiley.

Lewis, B. 1983. "Directional cues for auditory localization." In *Bioacoustics.* B. Lewis, ed. New York: Academic, pp. 223–57.

Lewis, D. 1966. "Stars of the sea road." *J. of Polynesian Society* 75 : 85–94.

———. 1970. "Polynesian and Micronesian navigation techniques." *J. of Institute of Navigation* 23 : 432–47.

———. 1971. "A return voyage between Puluwat and Saipan using Micronesian navigational techniques." *J. of Polynesian Society* 80 : 437–48.

———. 1972. *We, the Navigators.* Honolulu: University Press of Hawaii.

Lewis, D. B., and D. M. Gower. 1980. *Biology of Communication.* New York: Wiley.

Lighton, J. R. B. 1987. "Cost of tokking: the energetics of substrate communication in the tok-tok beetle, *Psammodes striatus.*" *J. Comp. Physiology* B 157 : 11–20.

Lindauer, M. 1976. "Foraging and homing flight of the honey-bee: some general problems of orientation." In *Insect Flight.* R. C. Rainey, ed. Oxford: Blackwell, pp. 199–216.

Linn, C. E., Jr., L. B. Bjostad, J. W. Du, and W. L. Roelofs. 1984. "Redundancy in a chemical signal: behavioral responses of male *Trichoplusiani* to a 6–component sex pheromone blend." *J. Chem. Ecol.* 10 : 1635–58.

———, M. G. Campbell, and W. L. Roelofs. 1986. "Male moth sensitivity to multi-component pheromones: critical role of female-released blend in determining the functional role of components and active space of the pheromone." *J. Chem. Ecol.* 12 : 659–68.

———. 1987. "Pheromone components and active spaces: what do moths smell and where do they smell it?" *Science* 237 : 650–52.

Lloyd, J. E. 1983. "Bioluminescence and communication in insects." *Ann. Rev. of Entom.* 28 : 131–60.

Locket, N. A. 1975. "Some problems of deep-sea fish eyes." In *Vision in Fishes.* M. A. Ali, ed. New York: Plenum, pp. 645–55.

Lockyer, C. H., and S. G. Brown. 1981. "The migration of whales." In *Animal Migration.* D. J. Aidley, ed. Cambridge: Cambridge University Press, pp. 105–37.

Löfstedt, C. J., Löfqvist, B. S. Lanne, J. N. C. Van Der Pers, and B. S. Hansson. 1986. "Pheromone dialects in European turnip moths *Agrotis segetum.*" *OIKOS* 46 : 250–57.

Lohmann, K. J. 1992. "How sea turtles navigate." *Scientific American* 266 (January) : 100–106.

Losey, G. S. 1978. "Information theory and communication." In *Quantitative Ethology.* P. W. Colgan, ed. New York: Wiley. pp. 43–78.

Lyon, E. P. 1904. "On rheotropism. I. - Rheotropism in fishes." *American Journal of Physiology* 12 : 149–61.

Lythgoe, J. N. 1979. *The Ecology of Vision.* Oxford: Clarendon.

———. 1980. "Vision in fishes: ecological adaptations." In *Environmental Physiology of Fishes.* M. A. Ali, ed. New York: Plenum, pp. 431–45.

———. 1984. "Visual pigments and environmental light." *Vision Res.* 24 : 1539–50.

———. 1988. "Light and vision in the aquatic environment." In *Sensory Biology of Aquatic Animals.* J. Atema, R. R. Fay, A. N. Popper, and W. N. Tavolga, eds. New York: Springer-Verlag, pp. 57–82.

Macagno, E. R., V. Lopresti, and C. Levinthal. 1973. "Structure and development of neuronal connections in isogenic organisms: variations and similarities in the optic system of *Daphnia magna.*" *Proc. Natl. Acad. Sci.* 70 : 57–61.

McCormick, C. A., and M. R. Braford, Jr. 1988. "Central connections of the oc-tavolateralis system: evolutionary considerations." In *Sensory Biology of Aquatic Animals.* J. Atema, R. R. Fay, A. N. Popper, and W. N. Tavolga, eds. New York: Springer-Verlag, pp. 733–56.

MacFadden, B. J., and D. S. Jones. 1985. "Magnetic butterflies, a case study of the monarch (Lepidoptera, Danaidae)." In *Magnetite Biomineralization and Magneto-*

reception in Organisms. J. L. Kirschvink, D. S. Jones, and B. J. MacFadden, eds. New York: Plenum, pp. 407-37.

Mac Leod, P. 1971. "Structure and function of higher olfactory centers." In *Olfaction.* L. M. Beidler, ed. New York: Springer-Verlag, pp. 182–204 .

McNeil, R., and J. Burton. 1977. "Southbound migration of shorebirds from the Gulf of St. Lawrence." *Wilson Bull.* 89 : 167–71.

Madge, D. S. 1961. "'Preferred temperature' of land arthropods." *Nature* 190 : 106–107.

Maeda, K., Y. Imae, J. -I. Shioi, and F. Oosawa. 1976. "Effect of temperature on motility and chemotaxis of *Escherichia coli.*" *J. Bact.* 127 : 1039–46.

Maier, I., and D. G. Müller. 1986. "Sexual pheromones in algae." *Biol. Bull.* 170 : 145–75.

Mangel, M., and C. W. Clark. 1988. *Dynamic Modeling in Behavioral Ecology.* Princeton, N. J.: Princeton University Press.

Mankin, R. W., K. W. Vick, M. S. Mayer, and J. A. Coffelt. 1980. "Anemotactic response threshold of the Indian meal moth, *Plodia interpunctella* (Hübner) (Lepidoptera: Pyralidae), to its sex pheromone." *J. Chem. Ecol.* 6 : 919–28.

Markl, L. 1977. "Adaptive radiation of mechanoreception." In *Sensory Ecology.* M. A. Ali, ed. New York: Plenum, pp. 319–44.

———. 1983. "Vibrational communication." In *Neuroethology and Behavioral Physiology.* F. Huber and H. Markl, eds. Berlin: Springer-Verlag, pp. 332–53.

———, and B. Hölldobler. 1978. "Recruitment and food-retrieving behavior in *Novomessor* (Formicidae, Hymenoptera). II. Vibration Signals." *Behav. Ecol. Sociobiol.* 4 : 183–216.

Marler, P. 1955. "Characteristics of some animal calls." *Nature* 176 : 6–8.

———. 1957. "Specific distinctiveness in the communication signals of birds." *Behaviour* 11 : 13–39.

———. 1959. "Developments in the study of animal communication." In *Darwin's Biological Work.* P. R. Bell, ed. Cambridge: Cambridge University Press, pp. 150–206.

———, and W. J. Hamilton III. 1966. *Mechanisms of Animal Behavior.* New York: Wiley.

Marr, D. 1982. *Vision.* New York: Freeman.

———, and E. Hildreth. 1980. "Theory of edge detection." *Proc. R. Soc. Lond.* B 207 : 187–217.

Marshall, D. A., L. Blumer, and D. G. Moulton. 1981. "Odor detection curves for n-pentanoic acid in dogs and humans." *Chem. Senses* 6 : 445–53.

———, and D. G. Moulton. 1981. "Olfactory sensitivity to α-ionone in humans and dogs." *Chem. Senses* 6 : 53–61.

Martin, A., and C. Simon. 1990. "Temporal variation in insect life cycles." *BioScience* 40 : 359–67.

Martin, R. L., C. Wood, W. Baehr, and M. L. Applebury. 1986. "Visual pigment homologies revealed by DNA hybridization." *Science* 232 : 1266–69.

Masters, W. M., and H. Markl. 1981. "Vibration signal transmission in spider orb webs." *Science* 213 : 363–65.

———, and A. J. M. Moffat. 1986. "Transmission of vibration in a spider's web." In *Spiders.* W. A. Shear, ed. Stanford, Calif.: Stanford University Press, pp. 49–69

Mather, J. G. 1985. "Magnetoreception and the search for magnetic material in rodents." In *Magnetite Biomineralization and Magnetoreception in Organisms.* J. L.

Kirschvink, D. S. Jones, and B. J. MacFadden, eds. New York: Plenum, pp. 509–33.

——, and R. R. Baker. 1981. "Magnetic sense of direction in woodmice for route–based navigation." *Nature* 291 : 152–55.

Matthews, G. V. T. 1953. "Navigation in manx shearwaters." *J. Exp. Zool.* 30 : 370–96.

——. 1968. *Bird Navigation.* Cambridge: Cambridge University Press.

Mautz, D. 1971. "Der Kommunikationseffekt der Schwänzeltänze bie *Apis mellifica carnica* (Pollm.). " *Zeitschrift fur vergleichende Physiologie* 72 : 197–220.

Mayer, M. S., and R. W. Mankin. 1985. "Neurobiology of pheromone perception." In *Comprehensive Insect Physiology, Biochemistry, and Pharmacology.* G. A. Kerkut and L. I. Gilbert, eds. Oxford: Pergamon, pp. 95–144.

Mead, C. 1983. *Bird Migration.* New York: Facts On File.

Medway, L. 1967. "The function of echonavigation among swiftlets." *Anim. Behaviour* 15 : 416–20.

Meier, A. H., and A. J. Fivizzani. 1980. "Physiology of migration." In *Animal Migration, Orientation, and Navigation.* J. S. A. Gauthreaux, ed. New York: Academic Press, pp. 225–82.

Menaker, M. 1989. "Extraretinal photoreception." In *The Science of Photobiology,* 2d ed. K. C. Smith, ed. New York: Plenum, pp. 215–30.

Mershon, D. H., and L. E. King. 1975. "Intensity and reverberation as factors in the auditory perception of egocentric distance." *Perception & Psychophysics* 18 : 409–15.

Mesibov, R., G. W. Ordal, and J. Adler. 1973. "The range of attractant concentrations for bacterial chemotaxis and the threshold and size of response over this range." *J. Gen. Physiol.* 62 : 203–23.

Meyer, E., and E. -G. Neumann. 1972. *Physical and Applied Acoustics.* New York: Academic Press.

Meyer, J. H. 1982. "Behavioral responses of weakly electric fish to complex impedances." *J. Comp. Physiol.* 145 : 459–70.

Meyer, P. W., I. J. Matus, and H. C. Berg. 1987. "Avoidance of Phycomyces in a controlled environment." *Biophys. J.* 51 : 425–37.

Michelsen, A. 1983. "Biophysical basis of sound communication." In *Bioacoustics.* B. Lewis, ed. New York: Academic Press, pp. 3–38.

——, B. B. Andersen, W. H. Kirchner, and M. Lindauer. 1989. "Honeybees can be recruited by a mechanical model of a dancing bee." *Naturwissenschaften* 76 : 277–80.

——, and O. N. Larsen. 1983. "Strategies for acoustic communication in complex environments." In *Neuroethology and Behavioral Physiology.* F. Huber and H. Markl, eds. Berlin: Springer-Verlag, pp. 321–31.

——, W. F. Towne, W. H. Kirchner, and P. Kryger. 1987. "The acoustic near field of a dancing honeybee." *J. Comp. Physiol.* A161 : 633–43.

Michener, M. C., and C. Walcott. 1967. "Homing of single pigeons—analysis of tracks." *J. Exp. Biol.* 47 : 99–131.

Mikkola, K. 1986. "Direction of insect migrations in relation to the wind." In *Insect Flight.* W. Danthanarayana, ed. Berlin: Springer-Verlag, pp. 152–71.

Miller, J. R., and M. O. Harris. 1985. "Viewing behavior-modifying chemicals in the context of behavior: lessons from the onion fly." In *Semiochemistry: Flavors and Pheromones.* New York: Walter de Gruyter, pp. 3–31.

———, and K. L. Strickler. 1984. "Finding and accepting host plants." In *Chemical Ecology of Insects*. W. J. Bell and R. T. Cardé, eds. London: Chapmann and Hall, pp. 128–157.

Miller, L. A. 1983. "How insects detect and avoid bats." In *Neuroethology and Behavioral Physiology*. F. Huber and H. Markl, eds. Berlin: Springer-Verlag, pp. 251–66.

Miller, S. E., and M. G. Hadfield. 1990. "Developmental arrest during larval life and life-span extension in a marine mollusc." *Science* 248 : 356–58.

Minnaert, M. 1954. *The Nature of Light and Colour in the Open Air*. New York: Dover.

Mitchell, E. R. 1979. "Migration by *Spodoptera exigua* and *S. frugiperda*, North American Style." In *Movement of Highly Mobile Insects: Concepts and Methodology in Research*. R. L. Rabb and G. G. Kennedy, eds. Raleigh, N.C.: N. C. State University Dept. of Entomology, pp. 386–93.

Mitchell, J. G., M. Martinez–Alonso, J. Lalucat, I. Esteve, and S. Brown. 1991. "Velocity changes, long runs, and reversals in *Chromatium minus* swimming response." *J. Bacteriol*. 173(3) : 997–1003.

Mitchell, J. M., Jr. 1976. "An overview of climatic variability and its causal mechanisms." *Quaternary Res*. 6 : 481–93.

Mittelstaedt, H. and M. -L. Mittelstaedt. 1982. "Homing by path integration." In *Avian Navigation*. F. Papi and H. G. Wallraff, eds. Berlin: Springer-Verlag, pp. 290–297.

Möhl, B. 1989. "Sense organs and the control of flight." In *Insect Flight*. G. J. Goldsworthy and C. H. Wheeler, eds. Boca Raton, Fla.: CRC Press, pp. 75–97.

Molitz, H. 1966. "Perturbation effects on rockets." In *The Fluid Dynamic Aspects of Ballistics*. ed. London: North Atlantic Treaty Organization, Advisory Group for Aerospace Research and Development, pp. 171–180.

Montgomery, J. C. and J. A. Macdonald. 1987. "Sensory tuning of lateral line receptors in Antarctic fish to the movements of planktonic prey." *Science* 235 : 195–96.

Moon, P. 1940. "Proposed standard solar-radiation curves for engineering use." *J. Franklin Institute* 230 : 583–617.

Moore, B. C. J. 1988. *An Introduction to the Psychology of Hearing*. New York: Academic Press.

Moore, B. R. 1988. "Magnetic fields and orientation in homing pigeons: Experiments of the late W. T. Keeton." *Proc. Natl. Acad. Sciences* 85 : 4907–9.

Moore, F. R. 1985. "Integration of environmental stimuli in the migratory orientation of the savannah sparrow (*Passerculus sandwichensis*)." *Anim. Behav*. 33 : 657–63.

Moore, R. and M. Evans. 1986. "How roots perceive and respond to gravity." *Am. J. Botany* 73 : 574–87.

Moulton, D. G. 1978. "Olfaction." In *Handbook of Behavioral Neurobiology*. R. B. Masterton, ed. New York: Plenum, pp. 91–117.

———, E. H. Ashton, and J. T. Eayrs. 1960. "Studies in olfactory acuity. 4. Relative detectability of n-aliphatic acids by the dog." *Anim. Behav*. 8 : 117–28.

———, and D. A. Marshall. 1976. "The performance of dogs in detecting α-ionone in the vapor phase." *J. Comp. Physiol*. 110 : 287–306.

Moulton, J. M. 1960. "Swimming sounds and the schooling of fishes." Woods Hole, Mass.: *Biol. Bull. Mar. Biol. Lab*. 119 : 210–23.

Munn, R. E. 1966. *Descriptive Micrometeorology*. New York: Academic Press.

Murlis, J. 1986. "The structure of odour plumes." In *Mechanisms in Insect Olfaction*. T. L. Payne, M. C. Birch, and C. E. J. Kennedy, eds. Oxford: Clarendon, pp. 27–38.

————, and C. D. Jones. 1981. "Fine-scale structure of odour plumes in relation to insect orientation to distant pheromone and other attractant sources." *Physiol. Entomol.* 6 : 71–86.

Myrberg, A. A., Jr. 1981. "Sound communication and interception in fishes." In *Hearing and Sound Communication in Fishes.* W. N. Tavolga, A. N. Popper, and R. R. Fay, eds. New York: Springer-Verlag, pp. 395–425.

————, C. R. Gordon, and A. P. Klimley. 1976. "Attraction of free ranging sharks by low frequency sound, with comments on its biological significance." In *Sound Reception in Fish.* S. Dijkgraaf, ed. Amsterdam: Elsevier, pp. 205–28.

————, S. J. Ha, and J. C. Banbury. 1972. "Effectiveness of acoustic signals in attracting epipelagic sharks to an underwater sound source." *Bull. Marine Science* 22 : 926–49.

————, and J. Y. Spires. 1980. "Hearing in damselfishes: an analysis of signal detection among closely related species." *J. Comp. Physiol.* 140 : 135–44.

Nakaoka, Y., and F. Oosawa. 1977. "Temperature-sensitive behavior of *Paramecium caudatum.*" *J. Protozool.* 24 : 575–80.

Nathans, J., T. P. Piantanida, R. L. Eddy, T. B. Shows, and D. Hogness. 1986. "Molecular genetics of inherited variation in human color vision." *Science* 232 : 203–10.

Neitz, M., J. Neitz, and G. H. Jacobs. 1991. "Spectral tuning of pigments underlying red–green color vision." *Science* 252 : 971–74.

Nelson, D. R., and R. H. Johnson. 1976. "Some recent observations on acoustic attraction of Pacific reef sharks." In *Sound Reception in Fish.* S. Dijkgraaf, ed Amsterdam: Elsevier, pp. 229–39.

Neuweiler, G. 1983. "Echolocation and adaptivity to ecological constraints." In *Neuroethology and Behavioral Physiology.* F. Huber and H. Markl, eds. Berlin: Springer-Verlag, pp. 280–302.

————, and M. B. Fenton. 1988. "Behaviour and foraging ecology of echolocating bats." In *Animal Sonar.* P. E. Nachtigall and P. W. B. Moore, eds. New York: Plenum, pp. 535–549.

Newman, E. A., and P. H. Hartline. 1982. "The infrared "vision" of snakes." *Scientific American* 246(3) : 116–27.

Newton, I. 1730. *Opticks.* London: Innys.

Nicol, J. A. C. 1978. "Bioluminescence and vision." In *Bioluminescence in Action.* P. J. Herring, ed. New York: Academic Press, pp. 367–417.

Nielsen, E. T. 1961. "On the habits of the migratory butterfly *Ascia monuste* L." *Biologiske Meddelelser Det Kongelige Danske Videnskabernes Selskab* 23(11) : 1–81.

Nisbet, I. C. T. 1970. "Autumn migration of the blackpoll warbler: evidence for long flight provided by regional survey." *Bird-Banding* 41(3) : 207–40.

Nobel, P. S. 1974. *Biophysical Plant Physiology.* San Francisco: Freeman.

Nordlund, D. A. 1981. "Semiochemicals: a review of the terminology." In *Semiochemicals: Their Role in Pest Control.* D. A. Nordlund, R. L. Jones, and W. J. Lewis, eds. New York: Wiley pp. 13–28.

Northcote, T. G. 1984. "Mechanisms of fish migration in rivers." In *Mechanisms of Migration in Fishes.* J. D. McCleave, G. P. Arnold, J. J. Dodson, and W. H. Neill, eds. New York: Plenum, pp. 317–55.

Northmore, D. P. M. 1977. "Spatial summation and light adaptation in the goldfish visual system." *Nature* 268 : 450–51.

Nultsch, W., and D. -P. Häder. 1988. "Photomovement in motile microorganisms—II." *Photochem. Photobiol.* 47 : 837–69.

O'Brien, W. J. 1979. "The predator-prey integration of planktivorous fish and zooplankton." 67 : 572–81.

Oertel, G. W. and L. Treiber. 1969. "Metabolism and excretion of C19- and C18-steroids by human skin." *European J. Biochemistry* 7 : 234–38.

Ogden, J. C., and T. P. Quinn. 1984. "Migration in coral reef fishes: ecological significance and orientation mechanisms." In *Mechanisms of Migration in Fishes.* J. D. McCleave, G. P. Arnold, J. J. Dodson, and W. H. Neill, eds. New York: Plenum, pp. 293–308.

Osuji, F. N. C. 1975. "The distribution of the larvae of *Dermestes maculatus* (Coleoptera: dermestidae) in a radial temperature gradient." *Ent. Exp. and Appl.* 18 : 313–20.

Otte, D. 1974. "Effects and functions in the evolution of signaling systems." *Ann. Rev. Ecol. and Systematics* 5 : 385–417.

Owen, D. 1980. *Camouflage and Mimicry.* Chicago: University of Chicago Press.

Palmer, J. D. 1990. "The rhythmic lives of crabs." *BioScience* 40 : 352–58.

Papi, F. 1982. "Olfaction and homing in pigeons: ten years of experiments. " In *Avian Navigation.* F. Papi and H. G. Wallraff, eds. Berlin: Springer-Verlag, pp. 149–159.

———, P. Ioalé, V. Fiaschi, S. Benvenuti, and N. E. Baldaccini. 1978. "Pigeon homing: cues detected during the outward journey influence initial orientation." In *Animal Migration, Navigation, and Homing.* K. Schmidt-Koenig and W. T. Keeton, eds. Berlin: Springer-Verlag, pp. 65–77.

Partridge, B. L., and T. J. Pitcher. 1980. "The sensory basis of fish schools: relative roles of lateral line and vision." *J. Comp. Physiol.* 135 : 315–25.

Pasquill, F. 1962. *Atmospheric Diffusion.* London: D. van Nostrand.

Patlak, C. S. 1953. "A mathematical contribution to the study of orientation of organisms." *Bull. Math. Biophys.* 15 : 311–38.

Pawlik, J. R., C. A. Butman, and V. R. Starczak. 1991. "Hydrodynamic facilitation of gregarious settlement of a reef-building tube worm." *Science* 251 : 421–24.

Payne, K. 1990. "Elephant talk." *National Geographic* 176 : 264–77.

Pearson, R. G. 1972. *The Avian Brain.* London: Academic Press.

Pelosi, P., and R. Viti. 1978. "Specific anosmia to 1-carvone: the minty primary odour." *Chemical Senses and Flavour* 3 : 331–337.

Pelosi, P., and A. M. Pissanelli. 1981. "Specific anosmia to 1,8-cineole: the camphor primary odour." *Chem. Senses* 6 : 87–93.

Perrins, C. M., M. P. Harris, and C. K. Britton. 1973. "Survival of manx shearwaters *Puffinus puffinus.*" *The Ibis* 115 : 535–48.

Perry, J. N., and C. Wall. 1986. "The effect of habitat on the flight of moths orienting to pheromone sources." In *Mechanisms in Insect Olfaction.* T. L. Payne, M. C. Birch, and C. E. J. Kennedy, eds. Oxford: Clarendon Press, pp. 91–96.

Perutz, M. F. 1987. "Physics and the riddle of life." *Nature* 326 : 555–58.

Peters, B. G. 1952. "Toxicity tests with vinegar eelworm. I. Counting and culturing." *J. Helminthol.* 26 : 97–110.

Petranka, J., L. B. Kats, and A. Sih. 1987. "Predator-prey interactions among fish and larval amphibians: use of chemical cues to detect predatory fish." *Anim. Behav.* 35 : 420–25.

Pettigrew, J. D. 1978. "A role for the avian pecten oculi in orientation to the sun?" In *Animal Migration, Navigation, and Homing.* K. Schmidt-Koenig and W. T. Keeton, eds. Berlin: Springer-Verlag, pp. 42–54.

Phillips, J. B. 1986. "Two magnetoreception pathways in a migratory salamander." *Science* 233 : 765–67.

Piddington, R. W. 1972. "Auditory discrimination between compressions and rarefactions by goldfish." *J. Exp. Biol.* 56 : 403–19.

Pierce, J. R. 1961. *Symbols, Signals and Noise: The Nature and Processes of Communication.* New York: Harper & Brothers.

Platt, J. R. 1964. "Strong inference." *Science* 146 : 347–53.

Pline, M., J. A. Diez, and D. B. Dusenbery. 1988. "Extremely sensitive thermotaxis of the nematode *Meloidogyne incognita*." *J. Nematol.* 20 : 605–608.

———, and D. B. Dusenbery. 1987. "Responses of the plant-parasitic nematode *Meloidogyne incognita* to carbon dioxide determined by video camera-computer tracking." *J. Chem. Ecol.* 13 : 873–88.

Poff, K. L., and M. Skokut. 1977. "Thermotaxis by pseudoplasmodia of *Dictyostelium discoideum*." *Proc. Natl. Acad. Sciences* 74 : 2007–10.

Poinar, G. O., Jr. 1983. *The Natural History of Nematodes.* Englewood Cliffs, N. J.: Prentice-Hall.

Poor, H. V. 1988. *An Introduction to Signal Detection and Estimation.* New York: Springer-Verlag.

Popper, A. N., P. H. Rogers, W. M. Saidel, and M. Cox. 1988. "Role of the fish ear in sound processing." In *Sensory Biology of Aquatic Animals.* J. Atema, R. R. Fay, A. N. Popper, and W. N. Tavolga, eds. New York: Springer-Verlag, pp. 688–710.

Popper, K. R. 1959. *The Logic of Scientific Discovery.* London: Hutchinson.

Pratt, L. H., and M. -M. Cordonnier. 1989. "Photomorphogenesis." In *The Science of Photobiology.* K. C. Smith, ed. New York: Plenum, pp. 273–304.

Presti, D. E. 1985. "Avian navigation, geomagnetic field sensitivity, and biogenic magnetite." In *Magnetite Biomineralization and Magnetoreception in Organisms.* J. L. Kirschvink, D. S. Jones, and B. J. MacFadden, eds. New York: Plenum, pp. 455–82.

———, and P. Galland. 1987. "Photoreceptor biology of Phycomyces." In *Phycomyces.* E. Cerdá-Olmedo and E. D. Lipson, eds. Cold Spring Harbor, N.Y.: Cold Spring Harbor Laboratory, pp. 93–126.

Prestwich, K. N., and T. J. Walker. 1981. "Energetics of singing in crickets: effect of temperature in three trilling species (Orthoptera: Gryllidae)." *J. Comp. Physiol.* B 143 : 199–212.

Purves, P. E., and G. E. Pilleri. 1983. *Echolocation in whales and dolphins.* London: Academic Press.

Purves, W. K., and G. H. Orians. 1987. *Life, the Science of Biology.* Sunderland, Mass.: Sinouer Associates.

Pye, J. D. 1983. "Echolocation and countermeasures." In *Bioacoustics*. B. Lewis, ed. New York: Academic Press, pp. 407–29.

Quine, D. B., and M. L. Kreithen. 1981. "Frequency shift discrimination: can homing pigeons locate infrasounds by doppler shifts?" *J. Comp. Physiol.* 141 : 153–55.

Quinn, T. P. 1984. "An experimental approach to fish compass and map orientation." In *Mechanisms of Migration in Fishes*. J. D. McCleave, G. P. Arnold, J. J. Dodson, and W. H. Neill, eds. New York: Plenum, pp. 113–23 .

———, 1984. "Homing and straying in Pacific salmon." In *Mechanisms of Migration in Fishes*. J. D. McCleave, G. P. Arnold, J. J. Dodson, and W. H. Neill, eds. New York: Plenum, pp. 357–62.

Rainey, R. C. 1976. "Flight behavior and features of the atmospheric environment." *Symp. Royal Entom. Soc. of London* 7 : 75–112.

———. 1979. "Interactions between weather systems and populations of locusts and noctuids in Africa." In *Movement of Highly Mobile Insects: Concepts and Methodology in Research*. R. L. Rabb and G. G. Kennedy, eds. Raleigh, N. C.: N. C. State University Dept. of Entomology, pp. 109–19.

Raisbeck, G. 1964. *Information Theory*. Cambridge, Mass.: M. I. T.

Ralston, J. V., and L. M. Herman. 1989. "Dolphin auditory perception." In *The Comparative Psychology of Audition: Perceiving Complex Sounds*. R. J. Dooling and S. H. Hulse, eds. Hillsdale, N. J.: Lawrence Erlbaum Associates, pp. 295–328.

Raper, J. R. 1970. "Chemical ecology among lower plants." In *Chemical Ecology*. E. Sondheimer and J. B. Simeone, eds. New York: Academic Press, pp. 21–42.

Raper, K. 1940. "Pseudoplasmodium formation and organization in *Dictyostelium discoideum*." *J. Elisha Mitchell Sci. Soc.* 56 : 241–82.

Ratliff, F. 1978. "A discourse on edges." In *Visual Psychophysics and Physiology*. J. C. Armington, J. Krauskopf, and B. R. Wooten, eds. New York: Academic Press, pp. 299–314.

Rayleigh, Baron (John William Strutt). 1945. *The Theory of Sound*. New York: Dover.

Rhoades, D. F. 1985. "Pheromonal communication between plants." In *Chemically Mediated Interactions Between Plants and Other Organisms*. G. A. Cooper-Driver and T. Swain, eds. New York: Plenum, pp. 195–218.

Richardson, E. G. 1953. *Technical Aspects of Sound*. Amsterdam: Elsevier.

Richardson, W. J. 1976. "Autumn migration over Puerto Rico and the western Atlantic: a radar study." *The Ibis* 118 : 309–32.

———. 1978. "Southeastward shorebird migration over Nova Scotia and New Brunswick in autumn: a radar study." *Can. J. Zool.* 57 : 107–24.

———. 1978. "Timing and amount of bird migration in relation to weather: a review." *Oikos* 30 : 224–72.

Richerson, J. V. and J. H. Borden. 1972. "Host finding by heat perception in *Coeloides brunneri* (Hymenoptera: Braconidae)." *Can. Ent.* 104 : 1877–81.

Ricklefs, R. E. 1990. *Ecology*. New York: Freeman.

Riechert, S. E., and R. G. Gillespie. 1986. "Habitat choice and utilization in web–building spiders." In Spiders; W. A. Shear, ed. Stanford, Calif.: Stanford University Press, pp. 23–48.

Riley, J. R., and D. R. Reynolds. 1986. "Orientation at night by high-flying insects." In *Insect Flight*. W. Danthanarayana, ed. Berlin: Springer-Verlag, pp. 71–87 .

Robacker, D. C., B. J. D. Meeuse, and E. H. Erickson. 1988. "Floral aroma." *BioScience* 38 : 390–98.

Robberecht, R. 1989. "Environmental photobiology." In *The Science of Photobiology*, 2d ed. K. C. Smith, ed. New York: Plenum, pp. 135–54.

Roberts, O. F. T. 1923. "The theoretical scattering of smoke in a turbulent atmosphere." *Proc. Royal Soc. London* A 104 : 640–54.

Roeder, K. D. 1967. *Nerve Cells and Insect Behavior.* Cambridge, Mass.: Harvard University Press.

Rogers, P. H., and M. Cox. 1988. "Underwater sound as a biological stimulus." In *Sensory Biology of Aquatic Animals.* J. Atema, R. R. Fay, A. N. Popper, and W. N. Tavolga, eds. New York: Springer-Verlag, pp. 131–49.

Roitblat, H. L., R. H. Penner, and P. E. Nachtigall. 1990. "Matching-to-sample by an echolocating dolphin (*Tursiops truncatus*)." *J. Exp. Psychol.* 16 : 85–95.

Rose, A. 1973. *Vision: Human and Electronic.* New York: Plenum.

Rossel, S., and R. Wehner. 1982. "The bee's map of the e-vector pattern in the sky." *Proc. Natl. Acad. Sci.* 79 : 4451–55.

———. 1986. "Polarization vision in bees." *Nature* 323 : 128–31.

Rossotti, H. 1983. *Colour.* Princeton, N.J.: Princeton University Press.

Ryan, M. J., and W. Wilczynski. 1988. "Coevolution of sender and receiver: effect of local mate preference in cricket frogs." *Science* 240 : 1786–88.

Sabelis, M. W., and P. Schippers. 1984. "Variable wind directions and anemotactic strategies of searching for an odour plume." *Oecologia* 63 : 225–28.

Saidel, W. M. 1988. "How to be unseen: an essay in obscurity." In *Sensory Biology of Aquatic Animals.* J. Atema, R. R. Fay, A. N. Popper, and W. N. Tavolga, eds. New York: Springer-Verlag, pp. 487–513.

Salisbury, F. B., and C. W. Ross. 1978. *Plant Physiology.* Belmont, Calif.: Wadsworth.

Salomonsen, F. 1967. "Migratory movements of the arctic tern." *Biologiske Meddelelser Det Kongelige Danske Videnskabernes Selskab* 24 : 3–42.

Sandberg, R., J. Pettersson, and T. Alerstam. 1988. "Shifted magnetic fields lead to deflected and axial orientation of migrating robins, *Erithacus rubecula*, at sunset." *Anim. Behav.* 36 : 877–87.

Saunders, J. C., and W. J. Henry. 1989. "The peripheral auditory system in birds: structural and functional contributions to auditory perception." In *The Comparative Psychology of Audition: Perceiving Complex Sounds.* R. J. Dooling and S. H. Hulse, eds. Hillsdale, N. J.: Lawrence Erlbaum Associates, pp. 35–64.

Savina, V. P., N. L. Sokolov, and E. A. Ivanov. 1975. "Composition of the volatile compounds of sweat and urine in man." *Kosm. Biol. Aviakosmicheskaya Med.* 9 : 76–78.

Schaeffer, A. A. 1928. "Spiral movement in man." *J. Morphol.* 45 : 293–398.

Schimz, A., W. Sperling, E. Hildebrand, and D. Köhler-Hahn. 1982. "Bacteriorhodopsin and the sensory pigment of the photosystem 565 in *Halobacterium halobium.*" *Photochem. Photobiol.* 36 : 193–96.

Schluger, J. H., and C. D. Hopkins. 1987. "Electric fish approach stationary signal sources by following electric current lines." *J. Exp. Biol.* 130 : 359–67.

Schlunegger, U. P. 1972. "Distribution patterns of n-alkanes in human liver, urine, and sweat." *Biochim. Biophys. Acta* 260 : 339–44.

Schmid–Hempel, P. 1987. "Foraging characteristics of the desert ant Cataglyphis." *Experientia Supplementum* 54 : 43–61.

Schmidt, R. F. 1978. *Fundamentals of Sensory Physiology*. New York: Springer-Verlag.

Schmidt–Koenig, K. 1985. "Migration strategies of monarch butterflies." *Contributions in Marine Science* 27 : 786–98.

———, and C. Walcott. 1978. "Tracks of pigeons homing with frosted lenses." *Anim. Behav.* 26 : 480–86.

Schmitt, B. C. and B. W. Ache. 1979. "Olfaction: responses of a decapod crustacean are enhanced by flicking." *Science* 205 : 204–6.

Schnitzler, H. –U., E. Kalko, L. Miller, and A. Surlykke. 1988. "How the bat, *Pipistrellus kuhli*, hunts for insects." In *Animal Sonar*. P. E. Nachtigall and P. W. B. Moore, eds. New York: Plenum, pp. 619–23.

———, D. Menne, R. Kober, and K. Heblich. 1983. "The acoustical image of fluttering insects in echolocating bats." In *Neuroethology and Behavioral Physiology*. F. Huber and H. Markl, eds. Berlin: Springer-Verlag, pp. 235–50.

Schnitzler, M. J., S. M. Block, H. C. Berg, and E. M. Purcell. 1990. "Strategies for chemotaxis." In *Biology of the Chemotactic Response*. J. P. Armitage and J. M. Lackie, eds. Cambridge: Cambridge University Press, pp. 15–34.

Schöne, H. 1984. *Spatial Orientation*. Princeton, N. J.: Princeton University Press.

Schuijf, A. 1976. "The phase model of directional hearing in fish." In *Sound Reception in Fish*. S. Dijkgraaf, ed. Amsterdam: Elsevier, pp. 63–86.

———. 1981. "Models of acoustic localization." In *Hearing and Sound Communication in Fishes*. W. N. Tavolga, A. N. Popper, and R. R. Fay, eds. New York: Springer-Verlag, pp. 267–310.

———, and A. D. Hawkins. 1983. "Acoustic distance discrimination by the cod." *Nature* 302 : 143–44.

Schwartz, L. S. 1963. *Principles of Coding, Filtering, and Information Theory*. Baltimore: Spartan Books.

Scientific American. [Anon.]. 1991. "50 and 100 years ago." 264(2) : 12.

Searle, C. L. 1982. "A model of auditory localization: peripheral constraints." In *Localization of Sound: Theory and Applications*. R. W. Gatehouse, ed. Groton, Conn.: Amphora Press, pp. 42–50.

Seeley, T. 1977. "Measurement of nest cavity volume by the honey bee (*Apis mellifera*)." *Behav. Ecol. Sociobiol.* 2 : 201–27.

Seeley, T. D. 1985. *Honeybee Ecology*. Princeton, N. J.: Princeton University Press.

———. 1988. "The effectiveness of information collection about food sources by honey bee colonies." *Anim. Behav.* 35 : 1572–74.

———. 1989. "The honey bee colony as a superorganism." *Am. Scientist* 77 : 546–53.

Segall, J. E., S. M. Block, and H. C. Berg. 1986. "Temporal comparisons in bacterial chemotaxis." *Proc. Natl. Acad. Sci.* 83 : 8987–91.

Selin, I. 1965. *Detection Theory*. Princeton, N. J.: Princeton University Press.

Semm, P., and R. C. Beason. 1987. "Magnetic responses of the trigeminal nerve system of the bobolink (*Dolichonyx oryzivorus*)." *Neuroscience Letters* 80 : 229–34.

Shapas, T. J., and W. E. Burkholder. 1978. "Patterns of sex pheromone release from adult females, and effects of air velocity and pheromone release rates on theoretical communication distances in *Trogoderma glabrum*." *J. Chem. Ecol.* 4 : 395–408.

Shaw, E. A. G. 1982a. "External ear response and sound localization." In *Localization of Sound: Theory and Applications*. R. W. Gatehouse, ed. Groton, Conn.: Amphora Press, pp. 30–41.

———. 1982b. "1979 Rayleigh medal lecture: the ellusive connection." In *Localization of Sound: Theory and Applications*. R. W. Gatehouse, ed. Groton, Conn.: Amphora Press, pp. 13–29.

Shepherd, G. M. 1988. *Neurobiology*. Oxford: Oxford University Press.

Sherry, D. F. 1985. "Food storage by birds and mammals." 15 : 153–88.

Shlaer, R. 1972. "An eagle's eye: quality of the retinal image." *Science* 176 : 920–22.

Shrödinger, E. 1920. "Theorie der Pigmente von grösster Leuchtkraft." *Ann. Physik* 62 : 603.

Shropshire, W., Jr., and J. -F. Lafay. 1987. "Sporangiophore and mycelial responses to stimuli other than light." In *Phycomyces*; E. Cerdé-Olmedo and E. D. Lipson, eds. Cold Spring Harbor, N. Y.: Cold Spring Harbor Laboratory, pp. 127–54.

Siedler, G., and H. Peters. 1986. "Physical properties general. of sea water." In *Oceanography*. J. Sündermann, ed. New York: Springer-Verlag, pp. 233–64.

Simmons, J. A. 1989. "A view of the world through the bat's ear: the formation of acoustic images in echolocation." *Cognition* 33 : 155–99.

———, and L. Chen. 1989. "The acoustic basis for target discrimination by FM echolocating bats." *J. Acoust. Soc. Am.* 86 : 1333–50.

———, and A. D. Grinnell. 1988. "The performance of echolocation: acoustic images perceived by echolocating bats " In *Animal Sonar*. P. E. Nachtigall and P. W. B. Moore, eds. New York: Plenum, pp. 353–85.

———, and S. A. Kick. 1983. "Interception of flying insects by bats." In *Neuroethology and Behavioral Physiology*. F. Huber and H. Markl, eds. Berlin: Springer-Verlag, pp. 267–79.

———, and C. F. Moss. 1990. "Convergence of temporal and spectral information into acoustic images of complex sonar targets perceived by the echolocating bat, *Eptesicus fuscus*." *J. Comp. Physiol.* A166 : 449–70.

Sivian, L. J., and S. D. White. 1933. "On minimum audible sound fields." *J. Acoust. Soc.* April : 288–321.

Skiles, D. D. 1985. "The geomagnetic field." In *Magnetite Biomineralization and Magnetoreception in Organisms*. J. L. Kirschvink, D. S. Jones, and B. J. MacFadden, eds. New York: Plenum, pp. 43–102.

Slagsvold, T. 1976. "Arrival of birds from spring migr. ion in relation to vegetational development." *Norw. J. Zool.* 24 : 161–73.

Sliney, D., R. T. Wangemann, and J. K. Franks. 1976. "Visual sensitivity of the eye to infrared laser radiation." *J. Opt. Soc. Am.* 66 : 339–41.

Snyder, A. W., S. B. Laughlin, and D. G. Stavenga. 1977. "Information capacity of eyes." *Vision Res.* 17 : 1163–75.

———, L. G. Stavenga, and S. B. Laughlin. 1977. "Spatial information capacity of compound eyes." *J. Comp. Physiol.* 116 : 183–207.

Soll, D. R. 1990. "Behavioral studies into the mechanism of eukaryotic chemotaxis." *J. Chem. Ecol.* 16 : 133–50.

Song, P. -S. 1983. "Protozoan and related photoreceptors: molecular aspects." *Ann. Rev. Biophys. Bioeng.* 12 : 35–68.

Sotthibandhu, S., and R. R. Baker. 1979. "Celestial orientation by the large yellow underwing moth, *Noctua pronuba* L." *Anim. Behav.* 27 : 786–800.

Southwood, T. R. E. 1962. "Migration of terrestrial arthropods in relation to habitat." *Biol. Rev.* 37 : 171–214.

———. 1981. "Ecological aspects of insect migration." In *Animal Migration.* D. J. Aidley, ed. Cambridge: Cambridge University Press, pp. 197–208.

Starr, M., J. H. Himmelman, and J. -C. Therriault. 1990. "Direct coupling of marine invertebrate spawning with phytoplankton blooms." *Science* 247 : 1071–74.

Stavenga, D. G., and J. Schwemer. 1984. "Visual pigments of invertebrates." In *Photoreception and Vision in Invertebrates.* M. A. Ali, ed. New York: Plenum, pp. 11–61.

Steck, W. F., E. W. Underhill, B. K. Bailey, and M. D. Chisholm. 1982. "A 4-component sex attractant for male moths of the armyworm, *Pseudaletia unipuncta.*" *Entomol. Exp. Appl.* 32 : 302–4.

Steen, W. J. v. d. 1984. "Methodological aspects of migration and orientation in fishes." In *Mechanisms of Migration in Fishes.* J. D. McCleave, G. P. Arnold, J. J. Dodson, and W. H. Neill, eds. New York: Plenum, pp. 421–44.

Stephens, K. 1986. "Pheromones among the procaryotes." *Critical Revs. Microbiol.* 13 : 309–34.

Stevens, S. S., and E. B. Newman. 1936. "The localization of actual sources of sound." *Am. J. Psychol.* 48 : 297–306.

Stone, L. D. 1975. *Theory of Optimal Search.* New York: Academic Press.

Stoneman, M. G., and M. B. Fenton. 1988. "Disrupting foraging bats: the clicks of arctiid moths." In *Animal Sonar.* P. E. Nachtigall and P. W. B. Moore, eds. New York: Plenum, pp. 635–638.

Stonier, T. 1990. *Information and the Internal Structure of the Universe.* New York: Springer-Verlag.

Stowe, M. K. 1986. "Prey specialization in the Araneidae." In *Spiders.* W. A. Shear, ed. Stanford, Calif.: Stanford University Press, pp. 101–31.

Stroup, E. D. 1985. "Navigating without instruments: the voyages of Hokule'a." *Oceanus* 28 : 69–75.

Strutt, John William. *See* Baron Rayleigh.

Sukuki, A., M. Mori, Y. Sakagami, A. Isogai, M. Fujino, C. Kitada, R. A. Craig, and D. B. Clewell. 1984. "Isolation and structure of bacterial sex pheromone, cPD1." *Science* 226 : 849–850.

Sündermann, J. 1986. *Oceanography.* New York: Springer-Verlag.

Supa, M., M. Cotzin, and K. M. Dallenbach. 1944. "Facial vision: the perception of obstacles by the blind." *Am. J. Psychol.* 57 : 133–83.

Surlykke, A. 1988. "Interaction between echolocating bats and their prey." In *Animal Sonar.* P. E. Nachtigall and P. W. B. Moore, eds. New York: Plenum, pp. 551–66.

Sutcliffe, J. F. 1986. "Black fly host location: a review." *Can. J. Zool.* 64 : 1041–53.

Sutton, O. G. 1953. *Micrometeorology.* New York: McGraw-Hill.

Suzuki, A., M. Mori, Y. Sakagami, and A. Isogai. 1984. "Isolation and structure of bacterial sex pheromone, cPD1." *Science* 226 : 849–50.

Swihart, S. L. 1970. "The neural basis of colour vision in the butterfly, *Papilio troilus*." *J. Insect Physiol.* 16 : 1623–36.

Taliaferro, W. H. 1920. "Reactions to light in *Planaria maculata*, with special reference to the function and structure of the eyes." *J. Exp. Zool.* 31 : 59–116.

Tamaki, Y. 1985. "Sex pheromones." In *Comprehensive Insect Physiology Biochemistry and Pharmacology.* G. A. Kerkut, and L. I. Gilbert, eds. Oxford: Pergamon, pp. 145–91.

Tavolga, W. N. 1976. "Acoustic obstacle detection in the sea catfish (*Arius felis*)." In *Sound Reception in Fish.* S. Dijkgraaf, ed. Amsterdam: Elsevier, pp. 185–203.

Taylor, L. R. 1986. "The four kinds of migration." In *Insect Flight.* W. Danthanarayana, ed. Berlin: Springer-Verlag, pp. 265–80.

Tennekes, H., and J. L. Lumley. 1972. *A First Course in Turbulence.* Cambridge, Mass.: MIT.

Thomas, S. D. 1987. *The Last Navigator.* New York: Henry Holt.

Thompson, J., and A. Taylor. 1980. *Polynesian Canoes and Navigation.* Laie, Hawaii: Institute for Polynesian Studies, Brigham Young University-Hawaii Campus.

Thomson, J. A. 1980. "How do we use visual information to control locomotion?" *Trends in Neurosciences* 3 : 247–50.

Thorpe, W. H. 1950. "A note on detour experiments with *Ammophila pubescens* Curt (Hymenoptera; Sphecidae)." *Behaviour* 13 : 257–63.

———. 1963. *Learning and Instinct in Animals.* Cambridge, Mass.: Harvard University Press.

Tinbergen, N. 1967. *The Herring Gull's World.* Garden City, N. Y.: Anchor Books.

———. 1972. *The Animal in its World.* Cambridge, Mass.: Harvard University Press.

Tinbergen, N., M. Impekoven, and D. Franck. 1967. "An experiment on spacing-out as a defence against predation." *Behaviour* 28 : 307–21.

Tinbergen, N., and A. C. Perdeck. 1951. "On the stimulus situation releasing the begging response in the newly hatched herring gull chick (*Larus argentatus argentatus* Pont.)." *Behaviour* 3 : 1–39.

Toerring, M. -J., and P. Moller. 1984. "Locomotor and electric displays associated with electrolocation during exploratory behavior in Mormyrid fish." *Behavioural Brain Research* 12 : 291–306.

Tolmazin, D. 1985. *Elements of Dynamic Oceanography.* Boston: Allen and Unwin.

Tomchik, K. J., and P. N. Devreotes. 1981. "Adenosine 3',5'-monophospate waves in *Dictyostelium discoideum*: a demonstration by isotope dilution-fluorography." *Science* 212 : 443–46.

Towne, W. F., and J. L. Gould. 1985. "Magnetic field sensitivity in honeybees." In *Magnetite Biomineralization and Magnetoreception in Organisms.* J. L. Kirschvink, D. S. Jones, and B. J. MacFadden, eds. New York: Plenum, pp. 385–406.

———. 1988. "The spatial precision of the honey bees' dance communication." *J. Insect Behavior* 1 : 129–55.

Towne, W. F., and W. H. Kirchner. 1989. "Hearing in honey bees: detection of air-particle oscillations." *Science* 244 : 686–88.

Truax, B. 1984. *Acoustic Communication.* Norwood, N. J.: Ablex.

Tucker, V. A. 1969. "Wave-making by whirligig beetles (gyrinidae)." *Science* 166 : 897–99.

Tyndall, J. 1874. "On the atmosphere in relation to fog-signalling." *Contemporary Rev.* 24 : 819–41.

Ullyott, P. 1936. "The behaviour of *Dendrocoelum lacteum*. II. Responses in non-directional gradients." *J. Exp. Biol.* 13 : 265–78.

Urick, R. J. 1983. *Principles of Underwater Sound.* New York: McGraw-Hill.

Urquhart, F. A. 1960. *The Monarch Butterfly.* Toronto: University of Toronto Press.

———. 1966. "A study of the migrations of the Gulf Coast population of the monarch butterfly (*Danaus plexippus* L.) in North America." *Ann. Zool. Fenn.* 3 : 82–87.

Urquhart, F. A., and N. R. Urquhart. 1976a. "Ecological studies of the monarch butterfly (*Danaus p. plexippus*)." *National Geographic Soc. Research Reports*, 437–43.

———. 1976b. "Migration of butterflies along the Gulf Coast of northern Florida." *J. of Lepidopterists' Soc.* 30 : 59–61.

———. 1976c. "A study of the peninsular Florida populations of the monarch butterfly (*Danaus p. Plexippus*; Danaidae)." *J. of Lepidopterists' Soc.* 30 : 73–87.

———. 1978. "Autumnal migration routes of the eastern population of the monarch butterfly (*Danaus p. plexippus* L.; Danaidae; Lepidoptera) in North America to the overwintering site in the Neovolcanic Plateau of Mexico." *Can. J. Zool.* 56 : 1759–64.

———. 1979. "Vernal migration of the monarch butterfly (*Danaus p. plexippus*, Lepidoptera: Danaidae) in North America from the overwintering site in the Neovolcanic plateau of Mexico." *Can. Ent.* 111 : 15–18.

Vale, G. A., and D. R. Hall. 1985. "The role of 1-octen-3-ol, acetone and carbon dioxide in the attraction of tsetse flies, Glossina spp. (Diptera: Glossinidae), to ox odour." *Bull. Ent. Res.* 75 : 209–17.

van der Steen, W. J., and A. ter Maat. 1979. "Theoretical studies on animal orientation. I. Methodological appraisal of classifications." *J. Theor. Biol.* 79 : 223–34.

Varley, G. C., G. R. Gradwell, and M. P. Hassell. 1974. *Insect Population Ecology; An Analytical Approach.* Berkeley: University of California Press.

Viaud, G. 1940. "Phototropisme des rotifères." *Bull. Biol. France et Belgique* 74 : 249–308.

Vicker, M. G., W. Schill, and K. Drescher. 1984. "Chemoattraction and chemotaxis in *Dictostelium discoideum*: myxamoeba cannot read spatial gradients of cyclic adenosine monophosphate." *J. Cell Biol.* 98 : 2204–14.

Visser, J. H. 1986. "Host odor perception in phytophagous insects." *Ann. Revs. Entom.* 31 : 121–44.

Vogel, S. 1981. *Life In Moving Fluids.* Princeton, N. J.: Princeton University Press.

Volkenstein, M. V. 1982. *Physics and Biology.* New York: Academic Press.

von Campenhausen, C., I. Riess, and R. Weissert. 1981. "Detection of stationary objects by the blind cave fish *Anoptichthys jordani* (Characidae)." *J. Comp. Physiol.* A143 : 369–74.

von Frisch, K. 1938. "The sense of hearing in fish." *Nature* 141 : 8–11.

———. 1953. *The Dancing Bees.* New York: Harcourt, Brace and World.

———. 1967. *The Dance Language and Orientation of Bees.* Cambridge, Mass.: Harvard University Press.

———. 1974. "Decoding the language of the bee." *Science* 185 : 663–68.

von Saint Paul, U. 1982. "Do geese use path integration for walking home?" In *Avian Navigation*. F. Papi and H. G. Wallraff, eds. Berlin: Springer-Verlag, pp. 298–307.

Voous, K. H., and J. Wattel. 1963. "Distribution and migration of the greater shearwater." *Ardea* 51 : 143–57.

Waddington, K. D. 1983. "Foraging behavior of pollinators." In *Pollination Biology*. L. Real, ed. Orlando, Fla.: Academic Press, pp. 213–39.

Wagner, G. 1972. "Topography and pigeon orientation." In *Animal Orientation and Navigation*. S. R. Galler, K. Schmidt-Koenig, G. J. Jacobs, and R. E. Belleville, eds. Washington, D. C.: National Aeronautics and Space Administration, pp. 259–73.

———. 1978. "Homing pigeons' flight over and under low stratus." In *Animal Migration, Navigation, and Homing*. K. Schmidt-Koenig and W. T. Keeton, eds. Berlin: Springer-Verlag, pp. 163–70.

Wagner, H. 1982. "Flow-field variables trigger landing in flies." *Nature* 297 : 147–48.

Walcott, C. 1982. "Is there evidence for a magnetic map in homing pigeons?" In *Avian Navigation*. F. Papi and H. G. Wallraff, eds. Berlin: Springer-Verlag, pp. 99–108.

Walcott, C., and R. P. Green. 1974. "Orientation of homing pigeons altered by a change in the direction of an applied magnetic field." *Science* 184 : 180–82.

Wald, G. 1945. "Human vision and the spectrum." *Science* 101 : 653–58.

———, P. K. Brown, and I. R. Gibbons. 1963. "The problem of visual excitation." *J. Opt. Soc. Am.* 53 : 20–35.

Wales, W. 1975. "Extraretinal photosensitivity in fish larvae." In *Vision in Fishes*. M. A. Ali, ed. New York: Plenum, pp. 445–57.

Walker, M. G., F. R. H. Jones, and G. P. Arnold. 1978. "The movements of plaice (*Pleuronectes platessa* L.) tracked in the open sea." *J. du Conseil—Conseil International pour l'exploration de la Mer* 38 : 58–86.

Walker, M. M., J. L. Kirschvink, and A. E. Dizon. 1985. "Magnetoreception and biomineralization of magnetite fish." In Magnetite Biomineralization and Magnetoreception in Organisms; J. L. Kirschvink, D. S. Jones, and B. J. MacFadden, eds. New York: Plenum, pp. 417–37.

Walker, T. J. 1979. "Migrating Lepidoptera: Are butterflies better than moths?" *Florida Entomologist* 63(1) : 79–98.

Wall, C., and J. N. Perry. 1987. "Range of action of moth sex-attractant sources." *Entomologica Experimentalis et Applicata* 44 : 5–14.

Wallach, H. 1939. "On sound localization." *J. Acoust. Soc. Am.* 10 : 270–74.

Wallraff, H. G. 1959. "Örtlich und zeitlich bedingte Variabilität des Heimkehrverhaltens von Brieftauben." *Zeitschrift für Tierpsychologie* 16 : 513–44.

———. 1984. "Migration and navigation in birds: a present-state survey with some digressions to related fish behaviour." In *Mechanisms of Migration in Fishes*. J. D. McCleave, G. P. Arnold, J. J. Dodson, and W. H. Neill, eds. New York: Plenum, pp. 509–44.

Walters, K. F. A., and A. F. G. Dixon. 1982. "The effect of host quality and crowding on the settling and take-off of cereal aphids." *Ann. Appl. Biol.* 101 : 211–18.

Ward, S. 1973. "Chemotaxis by the nematode *Caenorhabditis elegans*: Identification of attractants and analysis of the response by use of mutants." *Proc. Natl. Acad. Sci.* 70 : 817–21.

Waser, N. M. 1983. "The adaptive nature of floral traits: ideas and evidence." In *Pollination Biology*. L. Real, ed. Orlando, Fla.: Academic Press, pp. 241–85.

Waser, P. M., and C. H. Brown 1984. "Is there a "sound window" for primate communications?" *Behav. Ecol. and Sociobiol* 15 : 73–76.

———. 1986. "Habitat acoustics and primate communication." *Am. J. Primatol.* 10 : 135–54.

Waterman, T. H. 1984. "Natural polarized light and vision." In *Photoreception and Vision in Invertebrates*. M. A. Ali, ed. New York: Plenum, pp. 63–114.

———. 1989. *Animal Navigation*. New York: Scientific American Library.

Watkins, W. A., K. E. Moore, C. W. Clark, and M. E. Dahlheim. 1988. "The sounds of sperm whale calves." In *Animal Sonar*. P. E. Nachtigall and P. W. B. Moore, eds. New York: Plenum, pp. 99–107.

Watkins, W. A., P. Tyack, K. E. Moore, and J. E. Bird. 1987. "The 20-Hz signals of finback whales (*Balaenoptera physalus*)." *J. Acoust. Soc. Am.* 82 : 1901–12.

Weast, C. 1985. *CRC Handbook of Chemistry and Physics*. Boca Raton, Fla.: CRC Press.

Wehner, R. 1981. "Spatial vision in arthropods." In *Comparative Physiology and Evolution of Vision in Invertebrates*. Berlin: Springer-Verlag, pp. 287–616.

———. 1983. "Celestial and terrestrial navigation: human strategies–insect strategies." In *Neuroethology and Behavioral Physiology*. F. Huber and H. Markl, eds. Berlin: Springer-Verlag, pp. 366–81.

———. 1984. "Astronavigation in insects." *Ann. Rev. Entomol.* 29 : 277–98.

———. 1987. "'Matched filters'—neural models of the external world." *J. Comp. Physiology* A 161 : 511–31.

———, and P. Duelli. 1971. "The spatial orientation of desert ants, *Cataglyphis bicolor*, before sunrise and after sunset." *Experientia* 27 : 1364–66.

———, and I. Flatt. 1972. "The visual orientation of desert ants, *Cataglyphis bicolor*, by means of terrestrial cues." In *Information Processing in the Visual Systems of Arthropods*. R. Wehner, ed. Berlin: Springer-Verlag, pp. 295–302.

———, R. D. Harkness, and P. Schmid–Hempel. 1983. "Foraging strategies in individually searching ants *Cataglyphis bicolor* (Hymenoptera: Formicidae)." In *Information Processing in Animals*. M. Lindauer, ed. Stuttgart: Gustav Fischer Verlag, pp. 1–79.

———, and S. Rossell. 1985. "The bee's celestial compass—A case study in behavioural neurobiology." *Fortschritte der Zoologie* 31 : 11–53.

———, and M. V. Srinivasan. 1981. "Searching behavior of desert ants, genus Cataglyphis (Formicidae, Hymenoptera)." *J. Comp. Physiol.* 142 : 315–38.

Weihs, D. 1984. "Bioenergetic considerations in fish migration." In *Mechanisms of Migration in Fishes*. J. D. McCleave, G. P. Arnold, J. J. Dodson, and W. H. Neill, eds. New York: Plenum, pp. 487–508.

Wellington, W. G. 1974. "Changes in mosquito flight associated with natural changes in polarized light." *Can. Ent.* 106 : 941–48.

Wells, M. J., and S. K. L. Buckley. 1972. "Snails and trails." *Anim. Behav.* 20 : 345–55.

Wenner, A. M., and P. H. Wells. 1990. *Anatomy of a Controversy*. New York: Columbia University Press.

Wesseling, J. 1962. "Some solutions of the steady state diffusion of carbon dioxide through soils." *Neth. J. Agric. Sci.* 10 : 109–17.

Westby, G. W. M. 1984. "Electroreception and communication in electric fish." *Science Progress, Oxford* 69 : 291–313.

Westerberg, H. 1984. "The orientation of fish and the vertical stratification at fine- and micro-structure scales." In *Mechanisms of Migration in Fishes.* NATO Advanced Research Institute on Mechanisms of Migration in Fishes, ed. New York: Plenum, pp. 179–203.

Weydemeyer, W. 1973. "The spring migration pattern at Fortine, Montana." *Condor* 75 : 400–13.

Wheeler, J. W. 1976. "Insect and mammalian pheromones." *Lloydia* 39 : 53–59.

Whitaker, B. D., and K. L. Poff. 1980. "Thermal adaptation of thermosensing and negative thermotaxis in Dictyostelium." *Exp. Cell. Res.* 128 : 87–93.

White, F. A. 1975. *Our Acoustic Environment.* New York: Wiley.

Whittaker, R. H., and P. P. Feeny. 1971. "Allelochemics: chemical interactions between species." *Science* 171 : 757–70.

Wiersma, H. 1988. "The short-time-duration narrow-bandwidth character of odontocete echolocation signals." In *Animal Sonar.* P. E. Nachtigall and P. W. B. Moore, eds. New York: Plenum, pp. 129–45.

Wilcox, R. S. 1988. "Surface wave reception in invertebrates and vertebrates." In *Sensory Biology of Aquatic Animals.* J. Atema, R. R. Fay, A. N. Popper, and W. N. Tavolga, eds. New York: Springer-Verlag, pp. 644–63.

Wiley, R. H., and D. G. Richards. 1978. "Physical constraints on acoustic communication in the atmosphere: implications for the evolution of animal vocalizations." *Behav. Ecol. Sociobiol.* 3 : 69–94.

———. 1982. "Adaptations for acoustic communication in birds: sound transmission and signal detection." In *Acoustic Communication in Birds.* D. E. Kroodsma, E. H. Miller and H. Ouellet, eds. New York: Academic Press, pp. 131–81.

Wille, P. 1986. "Acoustical properties of the ocean." In *Oceanography.* J. Sündermann, ed. New York: Springer-Verlag, pp. 265–382.

Williams, T. C. and J. M. Williams. 1978. "Orientation of transatlantic migrants." In *Animal Migration, Navigation, and Homing.* K. Schmidt-Koenig and W. T. Keeton, eds. Berlin: Springer-Verlag, pp. 239–51.

———, J. M. Williams, L. C. Ireland, and J. M. Teal. 1977. "Autumnal bird migration over the western North Atlantic Ocean." *American Birds* 31 : 251–67.

Willows, A. O. D. 1967. "Behavioral acts elicited by stimulation of single, identifiable brain cells." *Science* 157 : 570–74.

Wilson, E. O. 1962. "Chemical communication among workers of the fire ant *Solenopsis saevissima* (Fr. Smith). 1. The organization of mass-foraging." *Anim. Behav.* 10 : 134–47.

———. 1970. "Chemical communication within animal species." In *Chemical Ecology.* E. Sondheimer and J. B. Simeone, eds. New York: Academic Press, pp. 133–55.

———, and W. H. Bossert 1963. "Chemical communication among animals." In *Recent Progress in Hormone Research.* G. Pincus, ed. New York: Academic Press, pp. 673–716.

Wiltschko, W., and R. Wiltschko. 1972. "Magnetic compass of European robins." *Science* 176 : 62–64.

———. 1982. "The role of outward journey information in the orientation of homing pigeons." In *Avian Navigation*. F. Papi and H. G. Wallraff, eds. Berlin: Springer-Verlag, pp. 239–52.

———. 1988. "Magnetic orientation in birds." In *Current Ornithology*. R. F. Johnston, ed. New York: Plenum, pp. 67–121.

Winn, H. E., T. J. Thompson, W. C. Cummings, J. Hain, J. Hudnall, H. Hays, and W. W. Steiner. 1981. "Song of the humpback whale—population comparisons." *Behav. Ecol. and Sociobiol.* 8 : 41–46.

Wittmers, L. E., Jr., R. S. Pozos, G. Fall, and L. Beck. 1987. "Cardiovascular responses to face immersion (the diving reflex) in human beings after alcohol consumption." *Ann. Emergency Medicine* 16(9) : 1031–36.

Wolfheim, J. H. 1983. *Primates of the World: Distribution, Abundance, and Conservation.* Seattle: University of Washington Press.

Wright, R. H. 1958. "The olfactory guidance of flying insects." *Can. Entom.* 90 : 81–89.

Würsig, B. 1989. "Cetaceans." *Science* 244 : 1550–57.

Yocum, C. S., L. H. Allen, and E. R. Lemon. 1964. "Photosynthesis under field conditions. VI. Solar radiation balance and photosynthetic efficiency." *Agronomy J.* 56 : 249–59.

Yorke, E. D. 1985. "Energetics and sensitivity considerations of ferromagnetic magnetoreceptors." In *Magnetite Biomineralization and Magnetoreception in Organisms.* J. L. Kirschvink, D. S. Jones, and B. J. MacFadden, eds. New York: Plenum, pp. 233–42.

Yoshida, K., T. Hisatomi, and N. Yanagishima. 1989. "Sexual behavior and its pheromonal regulation in ascosporogenous yeasts." *J. Basic Microbiol.* 29(2) : 99–128.

Young, J. F. 1971. *Information Theory.* New York: Wiley Interscience.

Zakon, H. H. 1988. "The electroreceptors: diversity in structure and function." In *Sensory Biology of Aquatic Animals.* J. Atema, R. R. Fay, A. N. Popper, and W. N. Tavolga, eds. New York: Springer-Verlag, pp. 814–50.

Zigmond, S. H. 1974. "Mechanisms of sensing chemical gradients by polymorphonuclear leukocytes." *Nature* 249 : 450–52.

———, and S. J. Sullivan. 1979. "Sensory adaptation of leukocytes to chemotactic peptides." *J. Cell. Biol.* 82 : 517–27.

Zimmermann, M. 1978. "Neurophysiology of sensory systems." In *Fundamentals of Sensory Physiology.* R. F. Schmidt, ed. New York: Springer-Verlag, pp. 31–79.

Index

··